The Teaching of Astronomy

Allegory of Astronomy, by Francois Denis Née. Collection of Jay M. Pasachoff. Further information on page 431.

International Astronomical Union
Union Astronomique International

The Teaching of Astronomy

Proceedings of the 105th colloquium of the
International Astronomical Union,
Williamstown, Massachusetts, 26-30 July 1988

Edited by

Jay M. Pasachoff
Hopkins Observatory, Williams College, Massachusetts, USA

John R. Percy
Erindale College and Department of Astronomy
University of Toronto, Ontario, Canada

*The right of the
University of Cambridge
to print and sell
all manner of books
was granted by
Henry VIII in 1534.
The University has printed
and published continuously
since 1584.*

CAMBRIDGE UNIVERSITY PRESS

Cambridge

New York   Port Chester   Melbourne   Sydney

ASTRONOMY   0386698 1

Published by the Press Syndicate of the University of Cambridge
The Pitt Building, Trumpington Street, Cambridge CB2 1RP
40 West 20th Street, New York, NY 10011, USA
10 Stamford Road, Oakleigh, Melbourne 3166, Australia

First published 1990

Printed in Great Britain at the University Press, Cambridge

British Library cataloguing in publication data available

Library of Congress cataloguing in publication data available

ISBN 0 521 35331 9

Cover: *Aristotle, Ptolemy and Copernicus* by Stefano della Bella; Frontispiece of *Dialogo*, by Galileo Galilei. Collection of Jay M. Pasachoff. Further information on page 434.

International Astronomical Union
Union Astronomique International

The following Colloquia of the International Astronomical Union are published for the Union by Cambridge University Press.

82.   Cepheids. *Edited by Barry F. Madore.* 0 521 30091 6. 1985

91.   History of Oriental Astronomy. *Edited by G. Swarup, A.K. Bag and K.S.Shukla.* 0 521 34659 2. 1987

92.   Physics of Be Stars. *Edited by A. Slettebak and T.P. Snow.* 0 521 33078 5. 1987

101. Supernova Remnants and the Interstellar Medium. *Edited by R.S. Roger and T.L. Landecker.* 0 521 35062 X 1988

105. The Teaching of Astronomy. *Edited by Jay M. Pasachoff and John R. Percy.* 0 521 35331 9. 1990

106. Evolution of Peculiar Red Giant Stars. *Edited by Hollis Johnson and Ben Zuckerman* 0 521 36617 8. 1989

111. The Use of Pulsating Stars in Fundamental Problems of Astronomy. *Edited by Edward G. Schmidt* 0 521 37023 X. 1989

CONTENTS

3. The Teaching Process

4. Student Projects

8. Conceptions/Misconceptions

9. High-School Courses

10. Teacher Training

12. Planetariums

13. Developing Countries

Preface

Astronomy may well be the most appealing science to students and the general public. It therefore plays an important role in promoting public interest, appreciation, and understanding of science in general. There is widespread interest in such exciting topics as cosmology, black holes, and the exploration of the solar system. There is also a deep appreciation of the practical, aesthetic and philosophical aspects of astronomy, including the ever-changing beauty of the night sky. Astronomy is taught to some extent in schools, colleges, and universities. Interested members of the public also learn about astronomy from newspaper and magazine articles, from very occasional television programs, and from visits to planetariums. Amateur astronomers are an especially receptive audience for astronomical information, and a valuable ally in passing this information on to the public.

From July 26 to 30, 1988, 162 astronomers from 31 countries gathered at Williams College in Williamstown, Massachusetts, to discuss The Teaching of Astronomy. Although they came from many levels of very diverse educational systems in vastly different countries, they found much in common, and many ideas to share. They compared problems in teaching, and solutions that have been attempted. They became more aware of the diversity of astronomy education, and made new contacts and friendships with "kindred spirits" from many parts of the world. This book is not only a record of the meeting but also an independent and lasting work in its field. We hope that it will be found interesting and useful by teachers everywhere, and will stimulate the further development of all aspects of astronomy education.

The program of the meeting consisted of invited papers and many contributed oral and poster papers, almost all of which are included in these proceedings. We have also included, as best we could, the comments and questions from the audience after each oral paper. We thank Bill Luzader, Deborah Pasachoff, and Eloise Pasachoff for their help in recording these. We also thank Andrew Fraknoi for organizing an extensive display of books and other teaching material.

This was the first international conference on the specific topic of astronomy education. The idea of the meeting was planted in 1984 by Robert Dukes and Joseph Meyer of the American Association of Physics Teachers, and developed in 1985 through informal discussions between Robert Dukes and John Percy. It was enthusiastically endorsed by the International Astronomical Union (IAU) Commission 46: The Teaching of Astronomy, at the 1985 General Assembly of the IAU in New Delhi, and was approved as IAU Colloquium 105 by the IAU Executive Committee in 1986. As the U.S. and Canadian National Representatives to IAU Commission 46, it was natural that we would play major roles in the meeting. The meeting was also endorsed by the American Association of Physics Teachers, the American Astronomical Society, the Astronomical Society of the Pacific, the Royal Astronomical Society (UK) and the Royal Astronomical Society of Canada. The support of these organizations is much appreciated.

The Scientific Organizing Committee was chaired by one of us (JRP), who is deeply grateful to the SOC for advice and assistance with the planning of the meeting: Lucienne Gouguenheim (France), Syuzo Isobe (Japan), Cecylia Iwaniszewska (Poland), Josip Kleczek (Czechoslovakia), Derek McNally (UK), Mazlan Othman (Malaysia), Jay Pasachoff (USA), Aage Sandqvist (Sweden), and Silvia Torres-Peimbert (Mexico). JRP also thanks Erindale College and the Department of Astronomy, University of Toronto (especially Esther Oostdyk, Joan Tryggve, and Maria Wong) for their assistance and support.

The Local Organizing Committee was chaired by the other of us (JMP), who thanks, for local support, the Conference Office of Williams College, especially Judith Grinnell and Lynn Chick; the Office of Public Information of Williams College, especially Ellen Berek, James Kolesar, Tom Bleezarde, and Ann-Rita Congello; the staff who worked on the colloquium, especially Darrel Hoff and Ardith Hoff, Diane Gordon, Bradford Behr, Amy Steele, Alex Steele, and Susan Kaufman; Williams College staff members Barbara Madden and Alice Seeley; and Mike Martys and Brian Quinn of the Williams College Computer Center. Darrel Hoff, in particular, devoted many weeks to the planning of the meeting, and to seeing that it ran smoothly. We also thank George R. Clarkson and Nancy P. Kutner of the Rensselaer Polytechnic Institute, and Kevin Reardon of Williams College for digitizing and labelling the group photograph for use in this book. JMP thanks Jean Audouze and the Institut d'Astrophysique, Paris, for their hospitality during the editing of these Proceedings.

The meeting was financially supported by the Perkin-Elmer Corporation, the International Astronomical Union, the Royal Astronomical Society (U.K.), the Royal Astronomical Society of Canada, the American Association of Physics Teachers, the National Science Foundation (U.S.A.), and the international currency program administered by the Smithsonian Institution. We especially thank Professor William A. Fowler of the California Institute of Technology, whose enthusiasm and support for the meeting was instrumental in attracting some of this support.

We thank Simon Mitton of Cambridge University Press for his encouragement and support. It has been a pleasure to work with him and his staff. The manuscripts submitted by the authors have been edited, and word-processed in LaTeX. We especially thank Marie Glendenning for her excellent typing and LaTeXing of a difficult manuscript.

Jay M. Pasachoff
Williams College

John R. Percy
University of Toronto

July 1989

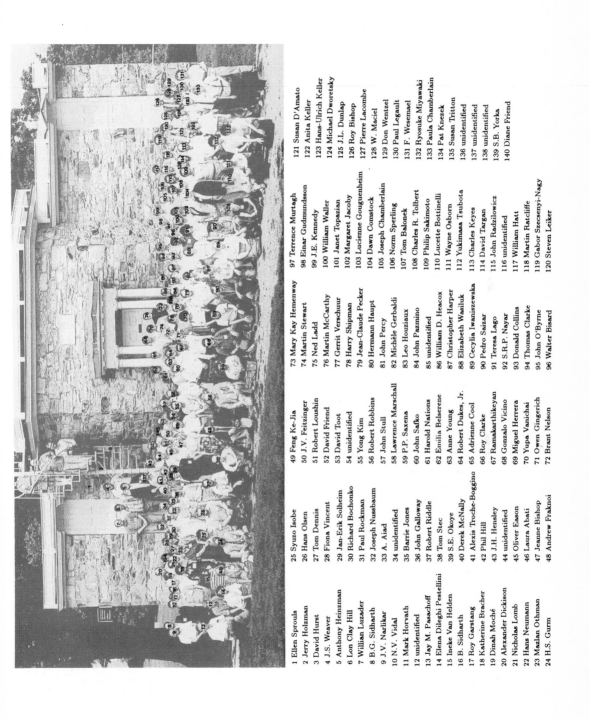

1 Ellen Sprouls
2 Jerry Holzman
3 David Hurst
4 J.S. Weaver
5 Anthony Heinzman
6 Lon Clay Hill
7 William Luzader
8 B.G. Sidharth
9 J.V. Narlikar
10 N.V. Vidal
11 Mark Horvath
12 unidentified
13 Jay M. Pasachoff
14 Elena Dileghi Pestellini
15 Ineke Van Helden
16 B. Sidharth
17 Roy Garstang
18 Katherine Bracher
19 Dinah Moché
20 Alexander Dickison
21 Nicholas Lomb
22 Hans Neumann
23 Mazlan Othman
24 H.S. Gurm

25 Syuzo Isobe
26 Hans Olsen
27 Tom Dennis
28 Fiona Vincent
29 Jan-Erik Solheim
30 Richard Bochonko
31 Paul Rockman
32 Joseph Nussbaum
33 A. Aiad
34 unidentified
35 Barrie Jones
36 John Galloway
37 Robert Riddle
38 Tom Stec
39 S.E. Okoye
40 Derek McNally
41 Alexis Troche-Boggino
42 Phil Hill
43 J.H. Hensley
44 unidentified
45 Oliver Eason
46 Laura Abati
47 Jeanne Bishop
48 Andrew Fraknoi

49 Feng Ke-Jia
50 J.V. Feitzinger
51 Robert Loushin
52 David Friend
53 David Toot
54 unidentified
55 Yong Kim
56 Robert Robbins
57 John Stull
58 Lawrence Marschall
59 P.P. Saxena
60 John Safko
61 Harold Nations
62 Emilia Belserene
63 Anne Young
64 Robert Dukes, Jr.
65 Adrienne Cool
66 Roy Clarke
67 Ramakarthikeyan
68 Gonzalo Vicino
69 Miguel Herrera
70 Yupa Vanichai
71 Owen Gingerich
72 Brant Nelson

73 Mary Kay Hemenway
74 Martin Stewart
75 Ned Ladd
76 Martin McCarthy
77 Gerrit Verschuur
78 Harry Shipman
79 Jean-Claude Pecker
80 Hermann Haupt
81 John Percy
82 Michèle Gerbaldi
83 Leo Houziaux
84 John Pazmino
85 unidentified
86 William D. Heacox
87 Christopher Harper
88 Elizabeth Wasiluk
89 Cecylia Iwaniszewska
90 Pedro Saizar
91 Teresa Lago
92 S.R.P. Nayar
93 Donald Collins
94 Thomas Clarke
95 John O'Byrne
96 Walter Bisard

97 Terrence Murtagh
98 Einar Gudmundsson
99 J.E. Kennedy
100 William Waller
101 Janet Topazian
102 Margaret Jacoby
103 Lucienne Gouguenheim
104 Dawn Comstock
105 Joseph Chamberlain
106 Norm Sperling
107 Tom Balonek
108 Charles R. Tolbert
109 Philip Sakimoto
110 Lucette Bottinelli
111 Wayne Osborn
112 Yukimasa Tsubota
113 Charles Keyes
114 David Targan
115 John Radzilowicz
116 unidentified
117 William Hatt
118 Martin Ratcliffe
119 Gabor Szecsenyi-Nagy
120 Steven Leiker

121 Susan D'Amato
122 Anita Keller
123 Hans-Ulrich Keller
124 Michael Dworetsky
125 J.L. Dunlap
126 Roy Bishop
127 Pierre Lacombe
128 W. Maciel
129 Don Wentzel
130 Paul Legault
131 F. Wesemael
132 Ryosuke Miyawaki
133 Paula Chamberlain
134 Pat Knezek
135 Susan Triton
136 unidentified
137 unidentified
138 unidentified
139 S.B. Yorka
140 Diane Friend

IOB. Cap. IX. v. 9.
Plejades, Orion, Ursa minor.

Buch Hiob Cap. IX. v. 9.
Siebengestirn, Orion, kleine Bär.

Plejades, Orion, Ursa Minor, by Johann Georg Pintz; from the *Kupfer-Bibel*. Chapin Library of Rare Books, Williams College. Further information on page 434.

Prologue

DIVERSE STRUCTURES, SAME SCIENCE

Donat G. Wentzel
University of Maryland, College Park, Maryland 20742, U.S.A.

This is truly an international gathering, as befits the IAU. We have a wide range of educational backgrounds. Most of us hope to adapt each others' ideas to suit our own needs. That adaption will be much easier if we appreciate each others' individual circumstances, and especially if we know something about each others' different educational systems. To ease the exchange of information, I shall outline two very different educational schemes. I shall call them the "traditional" and the "U.S." system. I suspect that about half of us will identify more closely with the traditional system, and half with the U.S. system.

The table outlines the main features of the two systems according to age, because children and students think alike at a given age no matter where they live. The differences between the two educational systems arise mainly from different social goals.

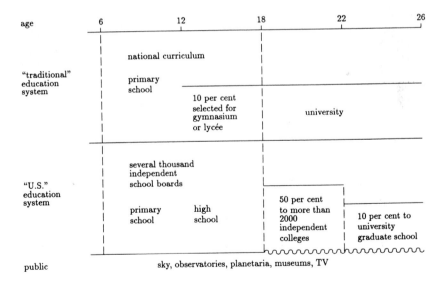

The two systems are fairly similar in the primary grades, until age 12. Then the traditional system identifies several pathways for further education. The children are sent to separate schools according to their general ability in academic subjects.

A rather small fraction is selected for university studies. This fraction has grown in the last two decades. Nevertheless, university study still implies a privilege. It also implies a duty for rigorous study, starting already at age 12. If children and older students study astronomy, they consider astronomy as a rigorous physical science.

The U.S. system has a very much broader goal. For social reasons, the U.S. system seeks to provide the same education for almost all through high school, usually age 18. The brightest children may take more advanced subjects, but they usually remain in a school that must accommodate the needs of all children.

At age 18, about half the U.S. population continues into college. Therefore, a college education serves a very different group of people than does a university education in the traditional system. Studies in most U.S. colleges are much less rigorous than in traditional universities. In particular, college astronomy courses for non-astronomy, non-physics majors are rather descriptive and contain little mathematics.

The U.S. college system has emphasized those aspects of the physical sciences that can be made popularly attractive. The system attempts to treat science as a part of culture, much as is literature, history, and painting. Astronomy is perhaps the physical science most suitable for such a cultural treatment. There is a substantial overlap with the goals of public education in astronomy. I have shown that overlap by the wiggly line in the table between college education and public education.

The "popularized" college astronomy courses have a very broad impact: probably one out of eight young adults in the United States has had an astronomy course, and they tend to remember that course with some pleasure. Many politicians have had a smattering of astronomy even if they always dreaded math and physics. Many of these politicians have learned to appreciate the goals of the physical sciences, and they can influence policy all the way from local governments to the federal government. They may willingly vote funds to support science in principle because of their astronomy experience. Unfortunately, they may not appreciate the rigor needed to carry out that science. They do not appreciate that science requires support over many years. Consequently, the support of science in the U.S. changes from year to year, depending on the political winds. This is very obvious in the U.S. space program, where some major projects have been turned on, off, and then on again. But other parts of U.S. science are also insecure. For instance, the budgets of many state universities tend to rise in an election year, and then languish the next year. The support of science at those universities changes similarly.

The U.S. educational system has probably become too popularized. There is now a widespread complaint that U.S. college graduates are scientifically illiterate. A video program has been made about students who are just finishing college at the prestigious Harvard University. They are asked to explain the seasons. They explain with much conviction that summer occurs because the Earth is then closer to the sun. (see paper by P. Sadler.) More rigorous study is again becoming fashionable. But the system of providing nearly the same education for everyone will remain. That goal is probably the ultimate limitation of the U.S. educational system.

There is another important difference between the two systems: the traditional educational systems tend to be structured on a national basis. The important examinations are nationwide and certificates are nationwide. Some countries represented here have very advanced standards in astronomy; some countries are just beginning with astronomy. But in nearly all countries there is an effort to establish a national program.

In the U.S., a national program is avoided for historical reasons. Education is local. The class content for primary and high schools is decided by several thousand independent school boards. The members of most of these school boards are elected. They must respond to the wishes of all the local voters. If the local voters see no need for the physical sciences, then the schools offer hardly any science at all. If the local voters are well educated, they may vote for a superb school system. Some school districts have established schools that specialize in the teaching of science. There are a few high schools in the United States (such as the Bronx High School of Science in New York City) that are as rigorous and difficult as any German Gymnasium or a Lycée in France, *etc.*

The course content in U.S. colleges is decided by several thousand independent colleges and several hundred universities. Most of them are publicly supported, typically by the county or by the state government. These colleges must respond to the needs of the local community, just as the high schools do. Some colleges and universities are private and obtain much of their income from investments. They can set more lofty educational goals, but they may have to limit those goals when the value of their investments declines.

The independence of all these schools and colleges makes the U.S. system rather complex. One must learn to rank the colleges and even some high schools. Where the German student simply says that he made the Abitur, the U.S. student says that he graduated from Williams College, or he received a "bachelors degree" from the University of Chicago, or he went to the Bronx High School of Science, and so on.

My goal in this outline is to stress the tremendous diversity of our needs at this colloquium because we come from many countries and many kinds of institutions. When we exchange experiences, we should always identify our background and the kind of students we teach.

Fortunately, we do have a common science. We have the same media to reach the public. The same questions on astronomy education are probably asked in every country represented at this colloquium.

One common question is: Just how much astronomy can school children really understand? For example, at what age can they understand the phases of the moon? That explanation requires children to imagine the world simultaneously from the different perspectives of the Earth and the sun. Apparently, some minimum age is needed before children can imagine even one perspective that is beyond their immediate experience. In teaching lunar phases, we ask children to imagine two perspectives! We astronomers tend to ask more of children than we should expect. We must ask more often when our subject can be introduced in a useful fashion. We should also consider whether tools like a planetarium can make the learning easier.

Problems with logical ability are not limited to the primary grades. In the U.S., at least, we also run into limited logical ability at the university level. I shall cite one obvious example. It is easy for students to understand that the parallax of a star is smaller the further the star. But what would happen if we could also double the baseline? That extra choice already confuses many college students. And what could we measure if we also improved the angular resolution of the telescope? That question involves three variables. So many variables tend to boggle the student's mind. The student cannot sort out the logical steps. Yet we astronomers use such complex arguments all the time. We discuss the H-R diagram in that way, and the age of the universe. When we exceed the students' logical ability, we stop educating and merely cause frustration. I hope we learn from each other at this colloquium how to teach astronomy at a level that is effective for the student.

Another common question is: What aspect of astronomy should we teach if only part of a course is available for astronomy? Traditionally, the answer has often been: celestial mechanics. There is a practical reason for this answer. Celestial mechanics is easily integrated into mathematics and physics courses. But it is a rather abstract subject, and it is logically quite difficult. For other topics, during the last two decades, hands-on experiments have been devised that many students can carry out themselves. Even computers can now be used for informative experiments. Such experiments provide a new avenue for introducing astronomy into many different courses of instruction. They permit students to discover the world by themselves. Such learning is vastly more effective than rote learning from books. I hope we shall hear of many interesting hands-on and computer experiments during this colloquium.

Perhaps the greatest challenge to astronomy teaching lies not in the detailed choices of topics or exercises, but in presenting astronomy as an intellectual stimulation to the students. Astronomy can lead students to create for themselves a view of the world, using the perspectives that have been forged by astronomers over the last few centuries. I think students must be stimulated to form some world view. I hope we learn at this colloquium how to stimulate the students more effectively.

Finally, at the university level, we must teach future astronomers and astronomy teachers. In the United States, education beyond college is called "graduate school". At this level, the traditional and U.S. systems are quite similar. Unfortunately, rather few U.S. students choose rigorous studies in physics during college. Therefore, only a few can later study physics or astronomy in graduate school. About half the U.S. graduate students in physics and astronomy are now from outside the U.S., mostly from Asia. Both American and European universities have many students from developing countries. One subject for discussion at this colloquium should be the job opportunities for astronomers returning to their own countries, and the training that is really appropriate for astronomers returning to astronomically developing countries.

The IAU has created the Commission 46 on the Teaching of Astronomy. It has the goal of enhancing opportunities for astronomy education internationally. The major Commission projects have been:

- the International Schools for Young Astronomers (ISYA), of which there have been sixteen,
- the Visiting Lecturers Program (VLP), so far operating in Peru and Paraguay, and awaiting a start in Nigeria,
- the Commission Newsletter, of which there have been two dozen,
- the list of Astronomy Educational Materials and the National Reports on Astronomy Teaching, prepared for seven IAU General Assemblies,
- meetings with local teachers on the day before IAU General Assemblies,
- and this colloquium.

The ISYA and VLP support the professional training of astronomers. The Newsletter has carried several articles on the astronomy training of teachers, but so far the IAU has no formal arrangement supporting teacher training.

I hope that this colloquium leads to additional concrete ways in which the IAU and Commission 46 can aid astronomy education. We have good reason to teach astronomy: it is an exciting subject that spurs the imagination. We should capitalize on that opportunity.

Discussion

J.-C. Pecker: *Your fine diagram puts a vertical boundary line at the age of 26. But we should, I believe, consider also the importance of adult-oriented teaching of astronomy. This concerns the astronomy teachers — who have to learn all their life — as well as the "general" public (including politicians!); TV broadcasts, for example, may be a positive asset (although, in general, they often tend to destroy the good effects of education given to the people when they were younger!).*

J.C. LoPresto: *Public education and professional science education are difficult to accomplish simultaneously. Colleges are not the only format for public education. What is your opinions?*

D. Wentzel: Definitely museums, TV, *etc.*, are a major branch of science education for astronomy. That is why they have their own column in my table. The wiggly line between U.S. colleges and public education merely indicates one of the few connections between the systems.

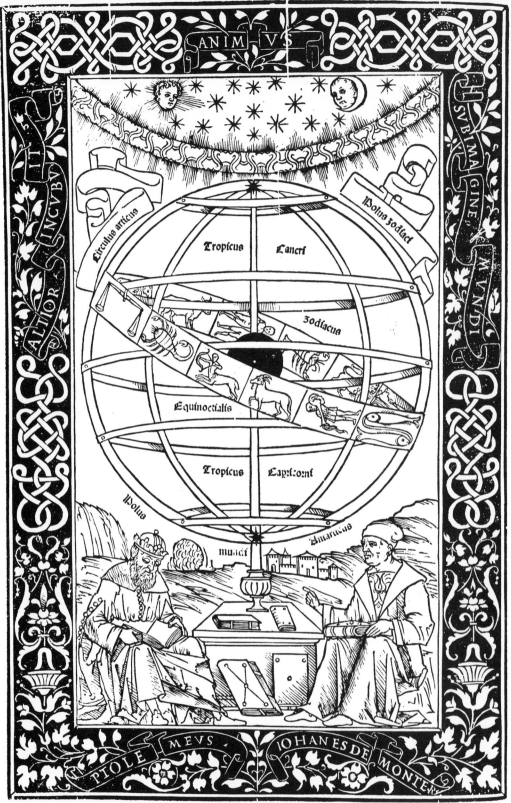

Ptolemy and Regiomontanus with Armillary Sphere, artist unknown; Frontispiece of
Epytoma in Almagestum Ptolomei, by Johannes Regiomontanus. Chapin Library of
Rare Books, Williams College. Further information on page 435.

Chapter 1

Curriculum

Teaching astronomy depends on deciding what to teach. In addition to a general discussion of desiderata, the papers in this section discuss particular programs in the U.K., China, Portugal, Hungary, and Brazil — countries that are broadly representative of world astronomy. Some related papers appear in the chapter on the developing countries.

CURRICULUM FOR THE TRAINING OF ASTRONOMERS

Jayant V. Narlikar
Tata Institute of Fundamental Research, Bombay 400005 and
Inter-University Centre for Astronomy and Astrophysics, Pune 411007, India

1. Introduction

In a discussion on the training of astronomers, the question naturally arises: "Training...but at what level?" After all, the subject of astronomy happens to be unique amongst all sciences in that it interests children and educated general public at one end and professionals at the other. Before coming to these specifics, certain general remarks will be in order.

Astronomy is a branch of science. This fact is not readily appreciated at the "lay" end of the above spectrum. Unlike other sciences, which have Earth-based laboratories in which controlled experiments are performed, astronomy has its labs located in the heavens far beyond the reach, let alone "control," of the terrestrial observer. So how can one be sure that what one sees out there is subject to the same rules and disciplines of science that govern the local laboratory experiments of physics and chemistry? Yet that happens to be so!

The basic thrill of learning and teaching astronomy lies in appreciating this fact. With suitable ingenuity, scientists have tried to overcome the disadvantages of a remote and uncontrollable laboratory: we have harnessed the latest available technology to augment our capability for collecting information. And with equal facility scientists have looked for explanations of very unusual and often mystifying cosmic events in terms of known laws of physics. The measure of success we have achieved has been such as to prompt Albert Einstein to wonder: "The most incomprehensible thing about the universe is that it is comprehensible."

Recently, the Government of India has set up the Inter-University Centre for Astronomy and Astrophysics (IUCAA). The Centre will serve as a centralized national facility for research workers in astronomy and astrophysics from Indian universities. Among its other programs, IUCAA will conduct pedagogical activities in these subjects at various levels. The ideas presented here form part of the on-going discussions about how astronomy teaching could be integrated with other courses in schools and universities.

2. Observational Astronomy

From ancient to modern times, light remains the main (if not the only) agent bringing information from remote corners of the universe. This monopoly of light has been broken today (but only marginally) by cosmic rays, neutrinos, and gravitational radiation. It is natural therefore that the part of the curriculum dealing with observational astronomy will contain mostly the role played by light; and so the following topics should find a logical place in "what every astronomer ought to know."

a) The nature of light

i) A description of the electromagnetic theory of Maxwell that led to the appreciation of light as an electromagnetic wave.

ii) The manifestations of electromagnetic waves of different wavelengths. Although human eyes respond only to a limited optical window, astronomers today cannot afford to ignore the other wavelengths.

iii) The role of the "spectrum," from the more familiar "rainbow" to the invisible radio or x-ray spectra. Continuum radiation and line radiation.

iv) The quantum nature of light as exhibited by the line spectrum of an astronomical object. It is important to emphasize that one of the crucial sources of evidence for quantum theory came from the Fraunhofer lines in the spectrum of the sun.

b) The role of the atmosphere

How and why does the atmosphere admit only selectively the cosmic radiation that bombards our planet? The role of space astronomy *vis-a-vis* that of ground-based astronomy.

c) The types of radiation

Although stars are the most familiar radiating sources, the astronomer encounters other sources that radiate quite differently. Thus it is desirable to know the different kinds of radiation one is looking for, *e.g.*, black-body radiation, synchrotron radiation, bremsstrahlung, inverse-Compton effect, *etc.* Again, it is worth emphasizing that our information of these radiation types is based on laboratory studies.

d) Detector systems

From the human eye to the CCD system lies a wide span of technology of detectors of astronomical radiation. The curriculum should ideally emphasize the difference of detector systems for radiation of different frequencies. Why are radio telescopes so enormous in size while x-ray detectors are relatively tiny? What are the differences in techniques for measuring line radiation and continuum radiation? Why is it useful to know if the radiation is polarized?

e) Star gazing

None of the above items in the curriculum can be really appreciated without some background of the traditional method of introducing astronomy, *viz* star gazing. But modern inputs into such a program should include the invisible universe...the world of galaxies and quasars invisible to the human eye, of giant molecular clouds, infrared sources, X-ray emitters, and huge radio galaxies. Appropriately introduced, a star-gazing program not only brings the thrills of observing but it will also convey to the student the rich storehouse of information and puzzles that lie beyond the range of naked-eye observing.

3. Theoretical Astronomy

Without its theoretical component, astronomy would not be a science. The interaction between physics and astronomy has been two-fold. Contrary to the belief generally held by laboratory physicists, astronomy has contributed to the growth of our understanding of physics. Thus not only has physics helped in the elucidation of astronomical phenomena but it has also in turn been helped by the inputs coming from astronomy. Examples are the discovery of the law of gravitation (where the credit should go to the information provided by the motion of the moon and the planets instead of to the falling apple), the viability of nuclear fusion (demonstrated in the sun and the stars and yet to be reproduced in controlled form on the Earth), and the more recent vigorous brainstorming going on between particle physicists and cosmologists.

In a typical text, astronomy is described in terms of things encountered as we go progressively away from the Earth in a sequence: the planets, the sun, the stars, the Milky Way, the galaxies, clusters, superclusters, quasars, and finally, the expanding universe. This may be the correct approach in a very elementary text, but it fails to depict the physics-astronomy interaction described above.

Fred Hoyle and I wrote an elementary text (Hoyle and Narlikar, 1980) that attempts to describe this interaction. Drawing on the basic interactions of physics, we grouped the astronomical phenomena according to the interactions playing crucial roles therein. Typically such a grouping would be along the following lines:

a) The electromagnetic interaction

The physics of radiation mechanisms and their examples, stellar radiation (black body), radio sources (synchrotron), X-ray sources (inverse-Compton), *etc.*, acceleration of charged particles (pulsars, cosmic rays), absorption and scattering of electromagnetic waves (interstellar dust), and so on.

b) The strong and weak interactions

Nucleosynthesis in stars, cosmic rays, supernovae, the primeval universe, *etc.*

c) Gravity

Motion of planets, satellites and binary stars, stellar structure and evolution, N-body motions in clusters of stars and galaxies, black holes, the expanding universe.

For integrating astronomy with physics an approach of this kind is called for. Only then will the student really appreciate how "scientific" astronomy is.

4. The Level of Presentation

It could be argued that the above approach may be difficult to incorporate in an elementary curriculum. I do not share this point of view. Admittedly the extent of mathematics and the depth of physics that can be used in front of school children or lay audiences are very limited. But nevertheless attempts should be made to inject some science into the purely descriptive syllabus that is often found in such situations. The "how" and "why" are just as important as the "what" in astronomy.

The task is not easy! I myself made an attempt to describe stellar evolution at a purely descriptive level (Narlikar, 1984). I had to introduce a few simple formulae...but perhaps one can do better with more analogies instead.

At the high-school and undergraduate levels, one can introduce simple algebra and perhaps some calculus also, to make the presenter's task easier! The level of mathematics can be higher still at the graduate level courses.

5. Conclusion

To do justice to astronomy and to would-be astronomers, the curriculum should be such as to bring out the aim of astronomy to understand the physics behind the cosmic events and to thereby enrich our understanding of not only the cosmos but of the science as well.

References

Hoyle, F. and Narlikar, J.V. *The Physics-Astronomy Frontier.* W.H. Freeman, New York. 1980.

Narlikar, J.V. *From Black Clouds to Black Holes.* World Scientific, Singapore. 1984.

Discussion

J.V. Feitzinger: *Can you comment on TV teaching of basic astronomy in India and what plans has your institution to do this?*

J.V. Narlikar: Our Institution will be located next to the Education Media Research Centre, which has well equipped TV studios. We plan to use EMRC facilities extensively for making astronomy programs for students and the general public.

AN ASTRONOMY DEGREE COURSE IN THE U.K.: SYLLABUS AND PRACTICAL WORK

M.M. Dworetsky
Department of Physics and Astronomy, University College London, Gower Street, London WC1E 6BT, U.K.

1. Introduction

What is the purpose of an astronomy degree? Why should students wish to take such a course? What will they do after graduation? In what way would such a course uniquely differ from a physics degree with a little astronomy tossed in? And given that we are called upon to provide such a course, what syllabus might we teach? These are some of the questions that occurred to me as I was preparing this paper.

One obstacle to giving clear answers is that the higher education systems of various countries differ greatly in structure. As one who was trained in one system (U.S.A.) and who teaches in another (U.K.), I am perhaps in a better position than most to appreciate the differences in approach, and to weigh the advantages and shortcomings of each system. But, as Shakespeare's Dogberry said, "Comparisons are odorous," and I do not propose to do this! What I describe refers to current practice in the university system of England and Wales, and I will use my own institution's long-standing astronomy degree as an example.

2. The Astronomy Degree

The course is a three-year honours degree Bachelor of Science. There is no "breadth" requirement for students to take courses in other subjects, in contrast to the usual practice in North American colleges and universities. Normally, students arrive at our front gates at age 18, having recently obtained good results in national "A-level" examinations in physics and mathematics, roughly equivalent to successful completion of the first year in typical North American universities. Students generally apply for their degree courses during the year preceding their university entrance. They are informed that a binding commitment is not required because, as a joint Department of Physics and Astronomy, we are able to permit unhindered internal transfers between the physics, the astronomy, and the joint astronomy/physics streams at various times in the first year with a minimum of paperwork. There is also a mathematics/astronomy degree course in which the teaching is shared equally by the two separate departments.

In the United Kingdom, a good B.Sc. honours degree in astronomy is sufficient qualification for applying to undertake a postgraduate Ph.D. research studentship in the subject without any further qualifying examinations. Effectively, we now have the answer to our first question; the degree we teach must provide sufficient training in physics, mathematics, and astronomy to permit an outstanding graduate to undertake an original research project, lasting three years (or perhaps four), that leads to an acceptable Ph.D. thesis. The three-year undergraduate course is, therefore, of necessity a rigorous and intensive one, in which the final year specialized options are taught at a level typically found in postgraduate courses in North America. However, one of our graduates could not be expected to have the extensive broad exposure provided by the usual extra year or two of study required in North American universities before advancement to Ph.D. candidacy.

But, we admit many more students to study astronomy than could (or should!) ever go on to do Ph.D. research or — far more difficult — eventually obtain employment as professional astronomers. So, why do so many students apply for the course? In the overwhelming majority of cases, their motivation is intellectual curiosity and a fascination with the subject. On their application forms, roughly one-third indicate an interest in an astronomy-related career, another third indicate an interest in occupations open to any graduate with a "numerate" science or engineering degree, while the other third indicate "undecided," perhaps the wisest answer a 17-year-old might give. Thus, most are realistic enough to understand that "professional astronomer" may be a career ideal, but that the road is long and the obstacles formidable. For these reasons, we must ensure that our astronomy syllabuses always stimulate and motivate the intellectual curiosity and best efforts of our students, while providing a sound basis in physics, mathematics, and practical skills useful in a wide variety of careers.

For about a third of our graduates, some form of further training follows the award of their honours degree. This may be a Ph.D. studentship, a taught M.Sc. course in a related subject like remote sensing or atmospheric physics, or (far too rarely!) training as a secondary school physics teacher. The great majority of the remainder successfully obtain positions open to physics or engineering graduates generally, because industrial employers and the Civil Service usually are aware of the strong physics and mathematics content of the astronomy degree, and normally treat it on essentially equal terms. A surprising number of our graduates enter careers in the financial institutions of the City of London; several have trained to be accountants.

3. The Syllabus

The astronomy degree course must necessarily involve a good deal of physics and mathematics, including nearly all the basic material involved in the introductory and intermediate years of a physics degree. The uniquely astronomical content of each of the first two years at University College London is only about 30 percent of the total lecture and practical material met by the student. In our astronomy/physics joint degree, this is reduced to about 20 per cent. It is the final year's specialized courses that slake most completely the students' thirst for astronomical material, because they may enroll entirely for courses in astronomy, although they are encouraged to take one or two physics lecture courses if they wish.

A very brief summary of the astronomy degree syllabus is given in Tables 1 and 2 for each of the first two years. These courses involve 32 lectures and 4 problem sessions. In addition, practical and laboratory courses (which are discussed in more detail in the next section of this paper) constitute one quarter of the workload in each of the first two years.

In their third (*i.e.*, final) year, astronomy degree students undertake an independent and substantial research project dealing with some aspect of astronomy or space science (for details, see McNally 1989). Formally, the project counts as one-quarter of the year's work. In addition, all astronomy degree students undertake an advanced observational and practical course at our University of London Observatory in the north London suburb of Mill Hill. The remaining five lecture courses (of 30 lectures each) may be selected from among the following courses: Planetary Geology, Planetary Atmospheres, Solar Physics, Interstellar Physics, Stellar Atmospheres, Stellar Structure and Evolution, High Energy Astrophysics, Extragalactic Astronomy and Cosmology, Observational Astronomy (a practical), and Mathematics for General Relativity. Most of these courses are taught at the same level as graduate-degree lecture courses in North American universities: for example, "Stellar Atmospheres" covers most of the material in the first half of Mihalas (1978) although the order, emphasis, and some of the treatments are different, while the observational discussions are brought up to date each year. Astronomy students may opt to enroll in one or two physics courses with their Tutor's permission: Physics of the Earth, Atomic and Molecular Physics, and Plasma Physics are the most popular of these.

One of the keys to the appeal and excitement of the astronomy degree course is that, with few exceptions, each of these courses is taught by an acknowledged expert who is actively doing research in the particular subject being taught. The students, who often find their second year to be hard going and rather too dry and mathematical, frequently tell us that they find their final year to be the most intellectually stirring and stimulating of all, largely because of the depth and immediacy of these courses.

Our astronomy and physics joint degree is generally similar to the astronomy degree in the first two years, except that only one astronomy lecture course is taken instead of two each year. Waves, optics, and electromagnetism receive more thorough treatment, as does nuclear physics. In the third year all options within the Department are available to these joint honours students, subject only to a minimum of two astronomy and two physics lecture courses. These students are free to do the Observatory practical course as an option and must also choose between the astronomy and physics projects. About half of them choose one or the other option; several choose both. Thus they have the freedom almost to emulate the astronomers, or very nearly to follow the straight physics course, according to their inclination and possible choice of career or postgraduate course.

4. Practical and Laboratory Work

The degrees taught in the Physics and Astronomy Department at University College London include components of practical and laboratory work of several types, and together with the project in the final year add up to 25 per cent or more of the required course units; for the astronomy degree, the total is nearly 30 percent because of the mandatory Observational Astronomy course.

Why this strong emphasis on practical and experimental work? Firstly, the College has a long tradition of experimental work in physics, and was the first to introduce systematic laboratory instruction in England (1866). We can inculcate formally subjects that we consider to be essential to a good general scientific education, such as statistical analysis of data, computer programming, electronics, and numerical methods. We feel that these subjects are best comprehended in a laboratory or observatory environment, where repeated applications of the principles learned are immediately possible. At the same time, we find that these courses allow us to explore those dimensions of a student's performance not measured by examinations, yet which may be even more relevant than examination grades when predicting future success in research, industry, or commerce.

The units in the first two years are not confined to doing set experiments. About 40 per cent of the workload in each year involves "minicourses." In the first year, all physics and astronomy students attend a series of lectures on elementary statistics and data analysis, and undertake a self-paced tutorial Fortran course, as part of their astronomy practical and physics laboratory training. They also spend one week of afternoons working in the student machine shops learning the rudiments of metal-working, and they attend a short engineering drawing course following their examinations. Second year students attend minicourses on electronics and numerical methods, and do associated experiments and computations.

5. First-Year Course

First year astronomy and astronomy/physics students attend at the University of London Observatory for twenty sessions of three hours, from October through March. On clear evenings, they have the use of an excellent 6 inch (15 cm) refractor and two 8 inch (20 cm) Celestron reflectors. The 6 inch telescope is a temporary substitute for an 8 inch refractor that is undergoing refurbishment in our shops. These telescopes can be used for a variety of visual observing sessions or for photography. Theodolites are used for experimental determination of latitude and longitude, and a coelostat and small solar spectrograph are available for solar spectroscopy experiments during clear afternoons. On cloudy evenings, they undertake more formal set experiments that draw on a wide variety of sources, such as the *Sky & Telescope* series of laboratory exercises, Culver's (1984) book of exercises, Kleczek's (1987) re-publication and extension of Minnaert's (1969) *Exercises in Astronomy*, and the Edinburgh Teaching Package (Tritton 1989). We also use several of our own "home-grown" experiments, often based on material obtained at the Observatory (see Dworetsky 1989 for an example).

Our approach to practical work is governed by the benefits we feel our students ought to derive from it. In recent years the teaching of physics in British state schools has been under severe pressure due to a shortage of qualified teachers. Although they usually come to us well-qualified theoretically, we find that an increasing proportion of our intake have had no experience of laboratory work as part of their school courses. This lack of experience is generally quite apparent in about 25 per cent of our students in the first few sessions.

All students are required to keep a laboratory notebook in which all experimental measurements, sketches, calculations, and notes are kept. This notebook is not ordinarily to be marked, although it is subject to inspection at any time in order to verify that the student has actually done the work presented, or if we encounter problems when marking a written-up experiment. It need not be neat — but it had better exis! Students jotting down readings or calculations on scraps of paper quickly learn that these are subject to confiscation. Lesson one: a scientist (or science student) always keeps a laboratory notebook.

The average astronomy degree student will complete, and present for detailed marking, about eight to twelve experiments during the course. These must be written up in detail in other notebooks, which are then marked and returned the following week. Students are given constructively critical comments on each write-up; these are intended to give guidance in order to improve subsequent presentations. A numerical mark is also given.

Aside from the formal minicourse work, there are several specific skills we attempt to foster. The most important of these skills is undoubtedly the writing of clear complete sentences and paragraphs in good formal English. The set experiments contain several "thought" questions which usually require short written answers or (occasionally) mathematical derivations of formulae (usually quite straightforward). From time to time some remarkably twisted syntax emerges from students; the proportion of students who commit frequent spelling errors is depressingly large. During the academic session, improvement is usually noted in the majority of cases, presumably as a direct result of the markers' detailed comments.

Accuracy of calculation is also a virtue to be encouraged, as is clear presentation of results by means of tables and graphs. An appreciation of significant figures and the need to quote an error (whether formal or estimated) is illustrated by the results in Table 3, which are students' calculations of the distance (about 10^4 pc) to the globular cluster M15, based on RR Lyr mean photographic magnitudes. Two students quoted this result accurate to 0.01 pc; three rounded to the nearest parsec; five more correctly rounded to the nearest 100 pc; but of the ten who attempted the experiment in 1987-88, *only one* quoted an estimated error! The absence of an error estimate for a measured quantity is invariably noted, and commented on, by the markers.

Table 3. Examples of Students' Expression of Significant Figures

Student Code	Distance (parsecs) to M15 (Estimated from RR Lyrs)
A	10500
B	10924
C	1.03×10^4
D	10597.42
E	11263.33
F	10600
G	10600 ± 500
H[a]	9100
J[b]	10666
K	10375

[a] Low value due to arithmetic blunder.

[b] Also quoted as 3.5×10^4 ly.

A graph is often an excellent way to communicate a considerable amount of complicated information clearly and quickly. The ability to impart information in this way is something we try to encourage at every opportunity. As a reader, how might you have reacted to six *superimposed* RR Lyr light curves, without individual data points, coded only by color? One student submitted a graph drawn exactly this way! Contrast such a confusing approach with another, in which six light curves are shown separately in six small plots, with observed data points clearly shown by circled dots and error bars.

The ability to draw a clear sketch can also be useful, especially in note-taking and in describing a complicated object or pointing out particular details or features. From time to time, students are asked to make drawings, either of objects viewed telescopically (*e.g.*, Jupiter or the moon) or images on photographs. In all cases, we try to impress upon students the need to note basic details such as the actual size of the original photographic image, or the angular size of a telescopic object (bifilar micrometers may be used for the latter purpose).

Students are trained in the use of a variety of expensive astronomical instruments and measuring machines. Considerable care must be taken to supervise their work both from the point of view of personal safety when working in darkness, and to protect the equipment from damage. We employ postgraduate student demonstrators for this purpose, who also act as markers for specific experiments. These marks are reviewed by the academic staff member in charge of the course.

The syllabus of set experiments is intended to complement and illustrate lecture material. For example, classification of galaxies is only very briefly discussed in lectures, but is covered in some depth by one of our set experiments. Familiarization with the more important astronomical charts, atlases, and catalogues is another task specifically assigned to the practical course, while positional astronomy is amplified by three experiments of graded difficulty: *Apparent Motions of Stars* (easy), *Determination of Photographic Stellar Coordinates* (a bit more difficult), and *Determination of Latitude and Longitude by Theodolite* (an observational and computational challenge). One of our locally-produced experiments, *Star Counts*, provides an opportunity to study an open cluster statistically, while a suitable extension of Minnaert's (1969) *Motion of the Hyades* experiment permits the student to construct a realistic color-magnitude diagram. (It is astounding how many students reverse the conventional absolute magnitude axis!) Another London experiment (Dworetsky 1989) helps to illustrate some of the underlying physics behind the classification of stellar spectra. There are also set experiments that involve the study of Schmidt telescope photographs from Palomar and the U.K. Schmidt in Australia (Tritton 1989); our students especially enjoy the study of Palomar prints of the Cygnus region, and the study of asteroid trails on a U.K. Schmidt photograph of the Virgo Cluster. In addition, every student in the course is introduced to photography and darkroom skills.

Several experiments require some calculations, which the students may wish to perform on a computer. We encourage this, provided that the effort involved in programming is not greater than that required if a pocket calculator is used instead!

6. Second-Year Course

The second-year course is completely based in the laboratories at University College. The set experiments for Practical Astrophysics are intended to complement the course work on atomic physics, and involve training in laboratory spectroscopy, photoelectric instruments, interferometry, and other optical equipment. Each student also undertakes a six-week project, that (optionally) may be undertaken at the Observatory, although usually these are laboratory-based. The educational goals are, otherwise, the same as those in the first year; *e.g.*, to keep careful notes of scientific work; to write up the results of an experiment in an organized, complete, clear manner; to present the results of calculations or measurements in clear tables and graphs; and to provide an adequate analysis of experimental errors.

7. Third-Year Course

The third-year course is an advanced course in observational astronomy. Each student attends one evening per week for twenty weeks. There is a strong emphasis on computing work and numerical techniques for data analysis and theoretical modeling. Students are trained on our 24 inch (60 cm) reflector in the use of a modern astronomical spectrograph (Dworetsky 1980) and in astronomical photoelectric photometry. Spectrographic experiments include MK spectral classification, stellar mass loss, measurement of radial velocity (including solar reduction), micro-densitometry of stellar spectra, *e.g.*, Balmer line profiles, and nebular spectroscopy. Photometry of variable stars is sometimes undertaken, weather permitting, and pho-tography using our 24 inch (60 cm) refracting telescope may lead to a determination of proper motion (by measuring an archive plate as well as a new plate) or lunar libration. There are also set experiments (all locally produced) on spectrophotome-try of Seyfert galaxies, UV spectroscopy of the solar chromosphere, stellar rotation, abundance analysis of stellar spectra, Fourier transforms and convolutions, and the orbit of a spectroscopic binary, among others. One very ambitious student derived the orbit of an asteroid from his own observations and computer program.

The quality of the written reports is even more important in the final year course. At this level, the goal is to apply all the experience gained in the first two years, and to submit complete reports of the work done, based on the model of scientific papers.

8. Concluding Remarks

Our degree course is extremely challenging, and necessarily so. Examinations are important, and good results indicate high levels of ability, but our experience over many years has tended to indicate that a strong showing in practical and project work is generally a more reliable indicator of research potential than good examination results alone. Our ideal research student in astronomy has not only ob-tained good marks in theoretical subjects, but has consistently shown that his or her practical work has been carried out with originality, care, energy, and enthusiasm.

In my current capacity as Tutor to Astronomy Students, I am asked to provide a large number of references. It helps to be able to inform prospective employers of an individual's ability to write clear reports, to carry out a meticulous scientific investigation, or to write computer programs that work. We are also able to observe their ability to put in a strong, sustained effort under pressure, when they do their final year projects.

While there is no doubt about the value of, and need for, the well-presented lecture courses in mathematics, physics, and astronomy which make up the bulk of our degree, the strong practical element adds a vital extra dimension to the training and assessment of astronomy students which cannot be provided solely by lectures and examinations.

References

Boas, M.L., *Mathematical Methods in the Physical Sciences*. Wiley, New York. 1983.

Culver, R.B. *An Introduction to Experimental Astronomy*. W.H. Freeman, New York. 1984.

Dworetsky, M.M. *Quarterly Journal of the Royal Astronomical Society*, **21**, 50. 1980.

Dworetsky, M.M. IAU Colloquium No. 105, "The Teaching of Astronomy." 1989.

Green, R. *Spherical Astronomy.* Cambridge University Press, Cambridge. 1985.

Jones, B.W. *The Solar System.* Pergamon Press, Oxford. 1984.

Kleczek, J. (ed.) *Exercises in Astronomy.* D. Reidel, Dordrecht, Holland. 1987.

Kleppner, D. and Kolenkow, R.J. *An Introduction to Mechanics.* McGraw-Hill, London. 1978.

Leighton, R.B. *Principles of Modern Physics.* McGraw-Hill, New York. 1959.

McNally, D. IAU Colloquium No. 105, "The Teaching of Astronomy." 1989.

Mihalas, D. *Stellar Atmospheres.* W.H. Freeman, San Francisco. Second Edition 1978.

Minnaert, M.G.J. *Practical Work in Elementary Astronomy.* D. Reidel, Dordrecht, Holland. 1969.

Roy, A.E. and Clarke, D. *Astronomy: Structure of the Universe.* Adam Hilger, Bristol. 1982.

Roy, A.E. and Clarke, D. *Astronomy: Principles and Practices.* Adam Hilger, Bristol. 1988.

Tritton, S. IAU Colloquium No. 105, "The Teaching of Astronomy." 1989.

Zeilik, M. and Smith, E.v.P. *Introductory Astronomy and Astrophysics.* Saunders, Philadelphia. Second Edition 1987.

Table 1. First-Year Lecture Course Structure for the
Astronomy Degree.

First Semester

Mathematics for Physics & Astronomy

Vectors and vector differentiation. Velocity and acceleration. Revision of elementary functions. Partial differentiation. Complex numbers, Argand diagram. Review of integration. Line and surface integrals. First order separable differential equations. Second order linear D.E.'s. Vector calculus. Grad, div, curl, ∇ operator, Gauss's and Stokes's theorems.

Text: Boas (1983), Chapters 1-7.

Introductory Classical, Relativistic, and Quantum Mechanics

Review of Newton's laws and basic definitions. Equations of motion. Harmonic motion including damped, forced oscillators. Angular momentum, torques, equations of motion. Vectors in mechanics. Kepler's laws. Coriolis effect.

Special relativity: Einstein's postulates, Lorentz transformations, time dilation. Quantum theory. Historical background, Bohr model of H atom, Heisenberg, wave functions, quantum numbers, Pauli exclusion.

Text: Kleppner and Kolenkow (1978).

Table 1 (cont'd). First-Year Lecture Course Structure for the
Astronomy Degree.

Foundations of Modern Astronomy

Observational techniques: outline of optical, UV, x-ray, IR, Radio astronomy.
Stellar and Solar astronomy. Interstellar matter. The Galaxy. Extragalactic
astronomy and cosmology.

Texts: Zeilik and Smith (1987), Chapters 8-24; *or* Roy and Clarke (1982),
Chapters 2, 7-18.

Second Semester

Mathematics for Physics and Astronomy Students I

(Mathematics Department)

Algebra: finite groups, linear transformations, matrices, determinants. Analysis: sequences, series, functions of a real variable, further differential and
integral calculus. Taylor series.

Electromagnetism, Waves, and Optics

Electric charges and currents. Conductors. Capacitance. Dielectrics. Magnetism. Lorentz force. Inductance. AC theory and circuits. Waves and vibrations, superposition, boundary conditions. Reflection, impedance.

Fourier analysis. Doppler effect; Dispersion. Wave description of electromagnetic radiation. Interference phenomena. Diffraction (Fraunhofer and Fresnel).

Classical and Solar System Astronomy

Spherical astronomy: trigonometry, coordinate systems, navigation. Time and
calendars. Star positions. Orbits, Kepler's laws, Kepler's equation and solution.
Tidal effects.

Planetary interiors. The Earth-moon system. Ages of rocks and meteorites.
Planetary atmospheres. Planets: description, spacecraft results.

Texts: Green (1985), *or* Roy and Clarke (1988) part 2. Jones (1984); Zeilik
and Smith (1987) part 1.

Table 2. Second-Year Lecture Course Structure for the
Astronomy Degree.

First Semester

Mathematical Methods in Physics

Vector algebra, curvilinear coordinates. Partial differential equations. Ordinary differential equations. Laplace's equation. Wave equation. Spherical harmonics. Bessel functions. Matrices and applications.

Quantum Physics for Astronomy

One-dimensional Schoredinger equation. Wave functions. Solution of infinite square well. Step potential: examples of tunnelling. Finite square well. More formal quantum mechanics: eigenvalues, eigenfunctions, expectation values. Angular momentum. Central field problem. Solution of H atom. Perturbation theory, Zeeman effect. Atomic and molecular spectroscopy.

Level and scope similar to Leighton (1959), Chapters 2-9.

Electromagnetism and Thermodynamics for Astronomy

Electromagnetism: review. Maxwell's equations, EM waves, solutions. Statistical and radiation thermodynamics. Classical thermodynamics. Kinetic theory of gases.

Second Semester

Mathematics for Physics and Astronomy Students II
(Mathematics Department)

Functions of a complex variable. Linear vector spaces. Analytical dynamics. Hamilton's equations.

OPTIONAL COURSES (in lieu of Mathematics II)

(1) Earth Resources.
(2) Digital Circuits.
(3) Computer and Microprocessor Systems.

Stellar Astrophysics

Stellar interiors: Basic equations, approximate solutions, energy generation, stellar evolution. Astrophysical processes: absorption, scattering, emission, line broadening, opacity sources. Stellar atmospheres: radiative transfer, grey atmosphere, "simple" LTE line formation.

Techniques in Modern Astronomy

Information acquisition and analysis of data. Optical astronomy. UV, x-ray, gamma-ray, infra-red, radio techniques and detectors. Gravitational waves. Cosmic rays.

Discussion

J.-C. Pecker: *Do you introduce in the curriculum any practical teaching with respect to bibliographical retrieval — how to find one's way in a library, how to read textbooks and review papers towards more specialized information, etc.? I have often had a bad experience with this problem; and, consequently, I introduced in Nice, in 1965-69 (graduate school), a specialized practical training in using documentation tools of all kinds. This has been, unfortunately, discontinued.*

M.M. Dworetsky: All our students write an essay in the first year, for which the proper use of the library and the preparation of a formal bibliography are important. Bibliography and library research are also emphasized in the project.

L. Gouguenheim: *What percentage of your students become professional astronomers? And what happens to the others? Are these 3 years of education in astronomy useful for them?*

M.M. Dworetsky: About 20 per cent undertake Ph.D. research. Of these about two thirds complete a thesis, and usually find a postdoctoral position. Permanent positions are harder to obtain; many of our Ph.D.'s obtain posts abroad. The other graduates generally compete successfully for jobs in industry and commerce. Graduate unemployment is very low.

M.K. Hemenway: *How large are your classes?*

M.M. Dworetsky: The astronomy entry group is usually about fifteen students each year; the astronomy/physics group is usually larger; about twenty enter each year.

C.R. Chambliss: *Your comments on the inability of many students to deal properly with significant figures are most interesting. I have found this to be one of the most difficult subjects to get across to many students in my basic course.*

M.M. Dworetsky: One thing to note is that where students are criticized for using too many digits, they tend to over-react and use too few significant figures in subsequent calculations.

P.P. Saxena: *Do your students get teaching in spherical trigonometry and spherical astronomy when they pursue their studies for B.Sc. degrees in astronomy?*

M.M. Dworetsky: All students attend a short series of five lectures on spherical trigonometry and basic spherical astronomy as part of the practical course. There are further lectures on positional astronomy given in the Classical and Solar System Astronomy course (Table 1).

THE TEACHING OF ASTRONOMY IN CHINA

Feng Ke-Jia
Department of Astronomy, Beijing Normal University,
Beijing 100875, China

1. General Information

Astronomical education has been developing at an increasing rate in China since 1977. Many Chinese astronomers think that the development and the popularization of astronomical education are future human needs. For this reason, we use radio and TV broadcasts as well as planetariums to popularize astronomical education in our country. The teaching of astronomy is enhanced in schools step by step. For elementary schools many astronomical topics are included in a course under the general title of *Nature*. Some activities such as astronomical observation and courses of astronomical lectures are organized in secondary schools. In universities, elective courses of astronomy are arranged not only in some departments of natural science but also in some departments of liberal arts. Some students in other scientific departments are encouraged to take astronomical courses, so that universities can supply frontier science with researchers.

In recent years, education in astronomy has made remarkable progress in China. Two new planetariums have been built, one in Nanjing, Jiangsu Province, and the other in Taiyuan, Shanxi Province. Two kinds of ordinary astronomical telescopes, an 80-mm refractor and a 120-mm reflector with equatorial mounting, both made in China, are supplied on the domestic market, which has played an important role in astronomical education. From 1984 to 1987, about 70 different books on astronomy were published. The astronomical center for children in Suzhou, Jiangsu Province, is supplied with an equatorially mounted reflector of 400-mm aperture. The observatory in Beijing No. 4 High School houses an equatorially mounted reflector of 350-mm aperture. These facts show the improvement in spreading astronomical knowledge.

2. Elementary Schools

Astronomical education at the elementary level is included in a course under the general title of *Nature*. From grade four to grade six, some general knowledge of astronomy is introduced. In grade four, for instance, pupils are required to recognize constellations, to use sundials, and to determine the solar altitude. In grade five, pupils are taught movement of constellations in the four seasons. In grade six, pupils are introduced to the moon, solar eclipses, lunar eclipses, the solar system, the Milky Way, galaxies, and the universe in general. Sometimes pupils may do some observations, or visit astronomical clubs. It is expected that some elementary schools will set up their own observatories with small telescopes.

3. Secondary Schools

Astronomical education at the secondary level is included in courses under the general titles of *Geography* and *Physics*. At the senior high level, chapters on the Earth and the universe are incorporated in *Geography*. In the chapter on the celestial bodies, students are required to learn about the celestial sphere, stars, and nebulae. In the chapter on the sun, students are required to learn some knowledge of the structure of the solar atmosphere, solar activity, solar-terrestrial relationships, and the region of solar nuclear reactions. In the chapter on the solar system, students are initiated into the movement and structural characteristics of planets, the phases of the moon, and the Earth-moon system. In the chapter on movement of the Earth, students are introduced to the rotation of the Earth, the axis of rotation, the obliquity of the ecliptic, and seasonal variation.

There is a chapter on gravitation in the course of *Physics* at the senior-high level. As a matter of course, planetary motion, the law of gravitation, and artificial satellites are introduced in this chapter. It is expected that astronomy will be given a greater share at the senior-high level in future.

4. University Education

At present **there are** four major facilities for education in astronomy: the Department of Astronomy at Nanjing University, the Department of Astronomy at Beijing Normal University, the Speciality of Astronomy at Beijing University, and the Center for Astrophysics at the University of Science and Technology of China.

Since 1984 about 135–140 undergraduates and 55–60 graduates have graduated from the four major facilities.

In China, other universities offer astrophysics and/or general astronomy as elective courses. Usually, astrophysics is taught in departments of physics, and general astronomy is taught in departments of geography, departments of philosophy, and other departments. It is anticipated that more and more universities will introduce these programs.

The length of the degree course is four years for undergraduates in the astronomy departments and there are many resemblances among the courses at other universities.

There are three principal kinds of courses in an astronomy department. They are the obligatory course, the restricted elective course, and the elective course. The undergraduates in the astronomy department must learn many kinds of obligatory courses, such as foreign languages, mathematical analysis, differential equations, linear algebra, general physics, experimental physics, methods of mathematics and physics, theoretical mechanics, fundamental astronomy, *etc.* Those courses are arranged in the freshman and sophomore years (first and second years). There are three kinds of restricted elective courses; astrophysics, astrometry, and celestial mechanics are offered in the junior and senior years (third and fourth years) in the astronomy department. The following restricted elective courses in astrophysics can be selected by undergraduates specializing in astrophysics: fluid mechanics, electrodynamics, practical astrophysics, methods of radio astronomy, thermodynamics and statistical physics, *etc.* The following restricted elective courses of astrometry can be selected by undergraduates specializing in astronomy or astrometry: methods of data processing, theory of probability and data processing, measurement of the rotation of the Earth, theory of the rotation of the Earth, star catalogues and astronomical constants, fundamentals of celestial mechanics, *etc.* The following restricted elective courses of celestial mechanics can be selected by undergraduates specializing in celestial mechanics: functions of real variables, elementary topology, advanced differential equations, celestial mechanics, methods of celestial mechanics, theory of the orbits of artificial satellites, quantitative theory of celestial mechanics, *etc.* By the way, the undergraduate can also select three or more of the following topics of the special elective courses: plasma physics, theory of stellar atmospheres, solar physics, radio astrophysics, solar-terrestrial physics, stellar physics, astrophysics, cosmical physics, *etc.*

5. Public Education

There are four planetariums in China. A number of astronomical clubs can be found all over the country. These clubs usually are connected with the local Association of Science and Technology.

The Chinese Astronomical Society has always paid great attention to bringing up young amateur astronomers. These young astronomical enthusiasts pay regular visits to observatories and planetariums and are directly trained by astronomers there. For instance they observed Halley's Comet in 1986 and the annular solar eclipse on September 23, 1987.

The researchers of planetariums and clubs help to encourage public interest in astronomy and to provide the newspapers, radio, and TV with necessary astronomical information.

References

(1) *Amateur Astronomer*, Published Monthly by Amateur Astronomer Publishing House, Beijing Planetarium, Beijing, China, Nos. 122–158.
(2) "The Present State of Astronomical Education in China" by Feng Ke-Jia, *Supplement to Proceedings of the Third Asian — Pacific Regional Meeting of the International Astronomical Union*, Sept. 30–Oct. 6, 1984, Kyoto, Japan, pp 18–22.

THE TEACHING OF ASTRONOMY AT UNIVERSITY LEVEL IN PORTUGAL

M.T.V.T. Lago

Grupo de Matemática Aplicada, University of Porto, Portugal

1. Introduction

In 1984, the University of Porto started a program for the first degree in astronomy in a Portuguese university. Jointly offered by the Physics and Applied Mathematics Departments at the School of Sciences, the degree is strongly marked by the interdisciplinary character essential for the teaching of modern astronomy. It includes 37 per cent in physics, 32 per cent in mathematics, 25 per cent in astronomy, and 6 per cent in chemistry, geology, mathematics, or physics. Its structure in course units comprises:

	45 to 50 %	Mathematics
1st year	28 to 30 %	Physics
	11 to 12 %	Applied Mathematics
	11 to 12 %	Astronomy
	49 to 50 %	Physics
2nd year	24 to 26 %	Mathematics
	26 %	Applied Mathematics
	63 %	Physics
3rd year	25 %	Astronomy
	12 %	Applied Mathematics
	75 %	Astronomy
4th year	25 %	Physics, Applied Mathematics, Chemistry, or Geology.

Most of the initial three years provides basic training in mathematics and physics, except for an introductory course (first year) intended as an overview of modern astronomy and aiming at keeping alive the student's enthusiasm. In the 3rd year, Astronomy I and II appear as basic courses. Finally, the 4th year includes 6 courses in various astronomical topics, such as Formation and Evolution of Stars, Stellar Structure, Cosmology, Astrometry, Extragalactic Astronomy, *etc*. The list of these topics changes every year and is naturally strongly dependent on the availability of lecturers (local and invited). The inclusion in some of these courses of units of 10 to 15 hours delivered by visiting professors or researchers has proven very stimulating. These units have the great advantage of exposing the students to different people and styles of presentation and help to compensate for the lack of "people around," considering that the number of astronomers in Portugal is presently so small. We are at least one order of magnitude below the European average of 1 to 2 astronomers per 100,000 inhabitants.

2. The European Astrophysical Doctoral Network

Further to the astronomy degree, complementary education is also starting to be provided through the European Astrophysics Summer Schools organized by the European Astrophysical Doctoral Network. This consortium of 14 European countries represented by one or two universities from each nation is aimed at:

– the exchange of doctoral students, on a short time scale or for a full thesis work;

– post-doctoral interchange and collaboration; and

– through the organization of regular European Astrophysics Summer Schools to offer the European students beginning doctorates in astrophysics a common level in the most fundamental fields as well as an opportunity to meet with colleagues and experts in those fields. It has both national and European community support.

3. The Center for Astrophysics at the University of Porto

Yet another component relevant for the teaching of astronomy at the University of Porto has to do with recent developments in the University itself. As a result of both National Research Council and University support, a Center for Astrophysics has been funded and is starting its activities at the University of Porto. Although its main objective is to provide infrastructure and support for research, it also specifically includes a component relating to the teaching of astronomy, both at

i) University level — providing the conditions for the visitors at the Center to collaborate in teaching through the previously mentioned short course units or longer ones; involving the terminal-year students in the Center's research projects;

ii) lower education level — from primary schools to higher education; there is a set of well-defined projects based on the interaction between the Center, the astronomy students, and the schools; however, the space allocated to this paper does not allow further details.

Of no less importance is the scheme for grants associated with the Center for Astrophysics. The first nationwide competition has awarded to selected candidates grants for a Ph.D. in astronomy at well known institutions, on the condition of their future association with the Center, where they must return once they have finished the doctorate. This condition, we hope, will allow us to gain the critical mass so necessary not only for research but also for good quality education as well.

The University of Porto's Center for Astrophysics is also involved in other very interesting and important actions in education in astronomy. However, they are still in a negotiation stage.

4. Conclusions

For those of you who read the report *"1987 — The Year of Astronomy in Portugal?"* published in the January issue of the *IAU Commission 46 Newsletter*, I would like to conclude that 1988 seems to be as important for astronomy in Portugal as 1987. Most of our proposals have been approved and already partially financed; others seem well on their way. And I will finish with the same sentence as in that report:

> *"Years of low profile and little activity will take time and effort to be replaced. Mentalities probably will take even longer to change. But the times are of optimism and strong hopes in a brighter future."*

5. Acknowledgments

The partial support of travel grants from the IAU and of JNICT through Projecto 87/27 is gratefully acknowledged.

ASTRONOMY EDUCATION IN HUNGARY

Gábor Szécsényi-Nagy
ELTE Csillagászati Tanszék, Department of Astronomy,
Eötvös University, H-1083 Budapest, Kun Béla tér 2, Hungary

1. Introduction: The Hungarian Educational System

Our educational system, like that of any other country, has grown up to meet the needs of the environment in which it developed. Perhaps its most distinctive feature is its emphasis on education of the masses rather than on education of the intellectuals.

The philosophy of the Hungarian educational system is that a democracy depends upon a well-informed electorate, and that therefore each citizen should receive the best education possible. As a result, in our country most children in the same community attend school together from kindergarten through secondary school practically regardless of differences in intellectual ability or in family background.

Educational policies and curricula are set up by the state organizations established by the government, and consequently the general plan varies only slightly from school to school. The school year is nine months in length, beginning in early September and continuing until about the tenth of June, with a vacation of a week or two at Christmas and a shorter one in the spring.

2. Astronomy in Elementary Education

There are eight years of elementary schooling not including kindergarten, which is an optional part of the public-school system. Only the last grade of kindergarten is now compulsory. Schoolchildren enter first grade at the age of six (or seven) and attendance is compulsory until the age of sixteen or until the student has finished the eighth grade. All elementary schools in Hungary are divided into two sections. The first, second, third, and fourth grades form the junior section while the other four grades represent the senior section.

Pupils of the junior section learn about the shape of the Earth, the Moon, the planets, the Sun, and their apparent motions. In the higher grades, the planetary orbits, inclination of the axes, relative positions, eclipses and phases, geometry of the seasons, tides, celestial and geographical coordinates, solar and sidereal time, and calendars are discussed mainly during lessons of geography and physics.

Many pupils attend special performances at the Budapest Planetarium (the best in Hungary), which can be visited during one-day excursions from about forty to fifty per cent of the country. Others may attend presentations at some smaller planetariums in the provinces or visit observatories (local and "Urania" observing stations).

Astronomical subjects are often demonstrated and explained making use of the smallest personal (so called school-) computers.

For the time being, approximately eighty per cent of Hungarian citizens complete their elementary school studies. The others can continue and finish at evening adult schools, which are very popular in Hungary.

3. Astronomy in Secondary Education

Elementary school is followed by four years of optional secondary school or high school. Admission to the Hungarian grammar and other secondary schools is automatic on completion of elementary school. The two kinds of secondary schools providing academic courses in Hungary are the so-called *gimnázium* (or grammar-school) and the special secondary school that provides technical courses as well. Their students are between fourteen and eighteen. Astronomy is taught as a section of physics mainly during the second (level 10) and fourth (level 12) years. The first part contains a bit of history of astronomy from Eratosthenes to Newton, Kepler's laws, the law of gravitation, angular momentum, proofs of the rotation of the Earth, kinematics, and dynamics of the solar system. The second part contains some

Table 1. Astronomy in Compulsory Education in Hungary

Age groups (years)	Level	Educational Institutions	Astronomy Taught as Part of …
3 –	—	Kindergarten	—
4 –	—	Kindergarten	—
5 –	0	Infant-school	Our World
6 –	1	Primary-school	Elements of Natural History
7 –	2	(Junior	Elements of Natural History
8 –	3		Introduction to Natural
9 –	4	Section)	Philosophy
10 –	5	Primary-school	
11 –	6	(Senior	Geography
12 –	7		Geography
13 – 14	8	Section)	Physics
. .			
16		(Repeaters)	

astrophysics (stellar colors and temperatures), spectroscopy (classification of stellar spectra, Doppler-shift *etc.*), theories of planetary and stellar evolution, radioastronomy, and cosmology. Unfortunately, all of these subjects are scheduled for the last semester of the secondary school's curriculum and they only seldom get the necessary emphasis because in that period both schoolchildren and teachers are concentrating upon the final examinations.

Recent changes in the curriculum of our secondary schools made it possible to increase a bit the number of lessons dealing with astronomical subjects. Teachers are allowed to choose from different units to be taught (which can be about atomic physics, biophysics, *etc.*, but about astronomy as well) although they may omit these extra units and use the surplus lessons to prepare schoolchildren for the final examinations and university entrance exams — both of which are indispensable to admission to the Hungarian universities.

In order to support secondary-school teachers in their efforts and to give them up-to-date information, a national meeting was organized in 1987. Professors of Eötvös University and research fellows of other institutes and observatories contributed to the success of the conference held under the title: "The Teaching of Astronomy and Space Research in Secondary Schools." Secondary-school teachers and Urania observatory staff led laboratory practices and workshops, which were very popular.

Both elementary and secondary schools are completely free.

Table 2. Astronomy in Secondary Education in Hungary

Age groups (years)	Level	Educational Institutions	Astronomy Taught as Part of ...
14 –	I(9)	Comprehensive	Geography
15 –	II(10)	or Grammar	Physics
16 –	III(11)	School	Introduction to Philosophy
17 – 18	IV(12)	(Gimnázium)	Physics
18 –		Compulsory military service (12–18 months), but for males only	

4. The Teaching of Astronomy at the Universities

Some courses of astronomy (in the first place, introductory astronomy and astrophysics or general astronomy and astronomical geography) are offered in all Hungarian universities to students in mathematics/physics and geography/geophysics, but the only institution of higher education in our country with its own Department of Astronomy is the Eötvös University of Budapest. At this institution, a comprehensive set of astronomy courses is offered to science undergraduate and graduate students, while at any other Hungarian university only the above-mentioned introductory courses are offered. The introductory courses are given as a rule by a professor of geography or physics and never exceed one semester. Astronomical geography is taught to students of geography during the first cycle whereas introductory astronomy or astrophysics is given during the second cycle for students of math/physics. A textbook is available for learning astronomical geography and one is in press for general astronomy.

At some universities, these introductory courses are coupled with practical work in a lab or dome.

5. Astronomy at the Eötvös University

The Eötvös University is the oldest Hungarian university; it was founded in 1635 (more than 350 years ago!) in Nagyszombat (which is now in Czechoslovakia and is called Trnava) by the archbishop of Hungary, Cardinal Péter Pázmány — who taught Johannes Kepler at Graz University — and was moved to Buda in the eighteenth century. For a very long period it was the only scientific institution that employed astronomers. After World War II, according to Stalin's ideas higher education and research institutes were split from each other in Hungary too and now there are at least ten different facilities where astronomers also work. Hungarian universities, colleges, secondary schools, observatories, and planetariums employ about one hundred astronomers. Consequently, the country needs at least two or three recent astronomy graduates per year to replace those who resign or retire.

Table 3. Astronomy in Hungarian Universities other than the
Eötvös University (Courses available for science students)

Cycle/*Semester*	Courses	Lessons
First Cycle		
First	Intro. to Astronomical Geography	42
Second	Astronomical Geography	26
Third	—	
Fourth	—	
Second Cycle		
Fifth	—	
Sixth	—	
Seventh	Intro. to Astronomy and Astrophysics	28
Eighth	—	
Third Cycle		
Ninth	—	
Tenth	—	

Since some people change careers or leave the country, our quota was fixed at four
graduating astronomers a year. But quite recently a more substantial and important
need arose. As astronomy became so popular in Hungary that every thousandth
citizen learns astronomy at home, makes telescopes, or observes celestial objects in
his spare time, many communities need an educated person trained in these fields
to run local astronomy clubs or amateur societies. To satisfy these needs, Eötvös
University and the Society for the Dissemination of Knowledge (TIT) initiated a
new program two years ago.

Table 4. Teaching of Astronomy at the Eötvös University of
Budapest (Courses suggested to science students)

Cycle/*Semester*	Courses	Lessons
First Cycle		
First	Introduction to Astronomy	42
	Introductory Planetology	28
	Introduction to Astronomical Geography	42
Second	Introduction to Astrophysics	39
	The Solar System	26
	Astronomical Geography	39
Third	Introductory Astronomy	42
	Solar Physics	28
	Astronomy for Meteorologists	28
Fourth	Introductory Astrophysics	39

Accordingly, the university offers astronomy courses at three different levels. In the same way as other universities and colleges, the Department of Astronomy provides many introductory courses to freshmen and sophomores.

Those science students who wish to specialize in astronomy have to accomplish a complete set of introductory courses as well as six semesters of mathematics and eight semesters of physics. After finishing the first cycle, they apply for specialization, and during the second and third cycles they learn as would-be astronomers. They have to follow at least seven astronomy courses and pass the exams. After the seventh semester, everyone selects a topic and works it out. In order to get their degrees, they have to write and defend a thesis too. An Astronomy degree can be earned after five years of university studies or more.

Table 5. The Training of Astronomers at the Eötvös University of Budapest (Compulsory Courses)

Subject Semester:	5th	6th	7th	8th	9th	10th
Celestial Mechanics	+	+	+	+	−	+
General Astrophysics	+	+	+	+	+	+
General Astronomy	+	+	+	+	+	+
Astronomical Instruments	+	+	−	−	−	−
Astronomical Laboratory	+	+	−	−	−	−
Astronomical Techniques	−	−	+	+	+	+
Recent Results of Astronomy	+	+	+	+	+	+
Galactic Astronomy	−	−	+	+	+	+
Extragalactic Astronomy	−	−	−	+	−	−
Extraterrestrial Astronomy	−	−	−	−	+	−
History of Astronomy	+	+	+	+	−	−
Practical Research Work	−	−	−	+	+	−
Computational Astronomy	−	−	+	−	−	+

Eötvös is the only university in Hungary entitled to award Ph.D. degrees in astrophysics. It awards about two degrees a year in this discipline.

Our latest activity in the teaching of astronomy is the training of club and amateur society leaders. This is a triennial postgraduate program. Our freshmen are active school-teachers, engineers, and economists who study the set of compulsory subjects at a correspondence course. They have to attend tutorial classes regularly, pass the exams, and defend their theses. When all of these are done, their postgraduate degrees are awarded.

Table 6. The Training of Astronomy Club Leaders at the
Eötvös University of Budapest (Compulsory Courses)

Subject Semesters:	1st	2nd	3th	4th	5th	6th
General Astronomy	+	+	+	+	−	−
General Astrophysics	+	+	+	+	−	−
History of Astronomy	+	+	+	+	+	+
Celestial Mechanics	−	−	+	+	−	−
Mathematics	+	+	−	−	−	−
Physics	+	+	−	−	−	−
Computers in Astronomy	−	−	+	+	−	−
Astronomical Telescopes	−	−	+	+	−	−
Astronomical Photography	−	−	−	−	+	+
Astronomical Measurements	−	−	−	+	+	−
Methods of Popularization of Astronomy	−	−	−	−	+	+
The Use of Visual Aids	−	−	−	−	−	+

GRADUATE ASTRONOMY STUDIES AT THE UNIVERSITY OF SÃO PAULO, BRAZIL

Walter J. Maciel

Departamento de Astronomia, Instituto Astronômico e Geofisico da USP,
Caixa Postal 30.627, 01051 São Paulo SP, Brazil

1. Introduction

The University of São Paulo (USP), founded in 1934, is the largest university in Brazil, having about 50,000 students and 5,000 teachers/researchers distributed among 33 institutes and 184 departments.

Astronomical work was already developed at the older São Paulo Observatory (founded 1912), which was later attached to the university. The observatory has been renamed as Instituto Astronômico e Geofisico (IAG), and presently houses the departments of Astronomy, Geophysics, and Meteorology. Graduate astronomy courses, intended to provide Master (M.Sc.) and Doctor (D.Sc.) degrees, started in 1973.

2. Astronomical Research at the USP

The astronomical research at the USP is basically conducted by the IAG. The main fields of research include theoretical and observational projects in the following areas: Fundamental Astronomy, Mathematical and Dynamical Astronomy, and Astrophysics (solar system, stars, the interstellar medium, galaxies, and astrophysical plasmas). The main instruments include a meridian circle, a Danjon astrolabe, and a 60-cm Boller and Chivens reflector, located at the nearby "Abrahão de Moraes Observatory." Instruments of the National Astrophysical Laboratory (LNA), which include a 1.6-m Boller and Chivens reflector, are normally used by the staff, and several projects are conducted at the ESO and U.S. observatories in Chile. The present staff consists of 32 people, 17 of whom have a Ph.D. or equivalent degree; the remaining are masters still working on their doctoral theses.

3. Astronomy Courses

The university does not provide a college degree in astronomy, and candidates for graduate studies are strongly recommended to complete a 4-year Physics program. Several one-semester undergraduate courses in astronomy are offered to students majoring in Physics, thus forming an introductory basis to further graduate work.

The candidate for the M.Sc. degree must take a series of one-semester courses from a minimum of six. Additional short courses by visiting scientists are often scheduled. The average student takes about two years to conclude the courses, and then takes a qualifying examination. At the same time, he or she engages in a research project for the M.Sc. dissertation, which is finished on average after 4 years of graduate work. The D.Sc. candidate takes 3 more courses plus a new qualifying examination. After a minimum of 3 more years, he or she will have produced a thesis based on an original investigation, which will be published in paper form in international astronomical journals.

4. Discussion

Through the end of 1987, the total number of students enrolled in the M.Sc. and D.Sc. programs were 138 and 65, respectively; 58 of them have obtained the M.Sc. degree, and 14 obtained the D.Sc. degree. The present number of students (December 1987) is 26 (M.Sc. course) and 41 (D.Sc. course).

After an initial period when most students were enrolled in the M.Sc. program, in recent years this tendency has been inverted. There were 138 candidates for the M.Sc. degree in the 15-year period, out of which 54 (39%) left the program and 58 (42%) concluded it. The attrition rate is much lower among candidates for the D.Sc. degree, where out of 65 candidates, 10 (15%) have left the program, and 14 (22%) concluded it. However, several students who left the program have in fact enrolled in another one in the country, so that the trial period effectively occurs during the M.Sc. program. All candidates for the D.Sc. degree who concluded the program have obtained jobs in the country. Among the masters, 27 (59%) have also obtained jobs, which high-percentage reflects the lack of qualified astronomers in the country.

(Work partly supported by FAPESP and CNPq — Brazil.)

CURRICULUM FOR THE TRAINING OF ASTRONOMERS: COMMENTS

Daniel B. Caton
Department of Physics and Astronomy, Appalachian State University,
Boone, North Carolina 28608 U.S.A.

In this first international meeting on the teaching of astronomy, we should not only look at many specific techniques and approaches but also examine the overall process. In doing so, several general problems come to light and need to be commented upon:

1. Introductory astronomy course lab exercises are often lacking in rigor, compared to labs in other physical sciences. Students are often asked to do simple, qualitative exercises like drawing the moon or constellations – projects that bear more resemblance to 19th-century astronomy than to the work of modern science. Lab programs should be modernized, taking advantage of modern telescopes and ancillary instrumentation.

2. A survey taken of U.S. astronomy department chairs, in preparation for an American Astronomical Society roundtable discussion, revealed a wide spectrum of approaches to undergraduate astronomy instruction. The one single obvious result of the survey was the recognition of a need for an international survey, with the results distributed and discussed by the participants. The dispersion of programs may also suggest another need ....

3. The astronomy instructional community lacks a central journal for the publication of pedagogical articles. The physicists have the *American Journal of Physics* and the *Physics Teacher* for advanced and lower-level articles, respectively. While astronomical articles appear in these from time to time (as well as in other publications), there is no single publication that educators can depend upon to contain important articles. While there is probably too little material available to form a new journal or newsletter, perhaps educational sections could be started in the *Publications of the Astronomical Society of the Pacific, Mercury,* or *Sky & Telescope*[1].

4. Astronomy students (majors) are often told to "get a physics (undergraduate) degree" in preparation for becoming an astronomer, yet strongly desire to take *astronomy* courses. This dual-program requirement results in either larger course loads (to include the astronomy), or the possibility of losing them to other disciplines. Students can perhaps be kept interested by involving them in astronomy *research* while they are learning their basic math and physics.

These four points should be carefully examined by the international community

---

[1] Ed. note: In late 1989, *Astronomy* has begun a teachers' insert.

of astronomy educators, and solutions devised at the appropriate levels to solve the problems that they address.

Discussion

J.-C. Pecker: *I am surprised that no one has so far mentioned the need for introducing in all curricula some coherent teaching about the history of ideas, of instruments, of astronomy in general. In my view, this is essential.*

M. Zeilik: *How does one rally the support for the practical, project, and experimental aspects of the University College London curriculum, which I admire?*

D. McNally: It is a matter of tradition. The degree, when first established by C.W. Allen, had a considerable committment to practical work. The project came later, but at a time when funding could still be considered "liberal." The amount of practical work per student has been reduced because of increase in student numbers without concomitant increase in provision of facilities — however, the demand on resources is still about the same as in the original scheme. I seriously doubt we could get such support if starting from scratch in today's straitened times.

W. Bisard: *Is astronomy a required course of* all *elementary teachers in your teacher education colleges?*

D. McNally: No. Astronomy in the Education Colleges only exists where enthusiasts bring astronomy into their science curriculum — there is no formal requirement unfortunately.

J.V. Narlikar: Unfortunately, no, but we are trying to change that.

B.W. Jones: *The University College London astronomy degree is primarily aimed at producing professional astronomers, yet only ∼ 15 per cent of the graduates enter astronomy in some form. Is this viewed with disquiet, or is it thought that "yuppies[2] " who know some astronomy will be better "yuppies" than those who don't?*

Reply: The latter — where "yuppies" = all sorts of non-astronomical professionals.

W. Bisard: *The results of well-founded educational research firmly supports project-oriented or hands-on science and astronomy projects over lecture techniques. Unfortunately, university science educators do not realize this fundamental finding of science education research.*

[2]Ed. note: "Young urban professionals" led recently to the word "yuppie," connoting shallow lives with much available money.

Fertilization of Egypt, by William Blake; Plate in *The Botanic Garden*, by Erasmus Darwin. Chapin Library of Rare Books, Williams College. Further information on page 433.

Chapter 2

Astronomy and Culture

Astronomy can be linked with many other topics, often in ways that especially interest students. Examples from astronomy in the history of science are perhaps most straightforward. Links of astronomy with classical mythology in art and music also interest many nonscientists. Science fiction, music, and poetry are other examples of fields with astronomical contexts. Further papers discuss how certain students can be brought to study astronomy from considerations of the northern lights and, for business students, the commercial use of space technology. A final, inspirational talk discusses the thrill of discovery and how students would benefit from understanding it.

THE USE OF HISTORY IN THE TEACHING OF ASTRONOMY

Owen Gingerich
Harvard-Smithsonian Center for Astrophysics, Cambridge, Massachusetts 02138, U.S.A.

Three good reasons for using historical materials in teaching astronomy are:

- The simplest concepts are introduced first in a natural sequence;
- For non-science students, history can build onto other interests;
- The historical perspective shows the changing and iterative nature of scientific explanatory structures.

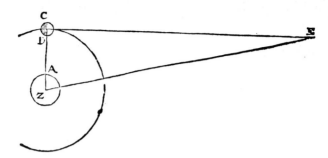

In which E is put for the Center of the Sun, Z for the Center of the Earth, D is the Moon. C Z is the common Interfection of the Plane of the Circle of the Moon's Illumination with the Plane of the Ecliptic : The Line CE, in the Plane of the Ecliptic, is the Diftance of the Center of the Moon from that of the Sun ; and the Line ZE in the fame Plane, is the Diftance of the Earth from the Center of the Sun. From

Fig. 1. Whiston's Astronomical Lectures *of 1828 shows the Aristarchan method for getting the ratio of the lunar and solar distances, which came out as 1:19 for the 3rd-century B.C. Ionian cosmographer.*

Introducing Concepts in Historical Sequence

Aristarchus' method of using the lunar day-night dichotomy to get the relative distance of the sun and moon has long been used as a pedagogic tool. Our Fig. 1 is from William Whiston's *Astronomical Lectures* (London, 1728), but clear diagrams and presentations are also found in several modern textbooks. Teaching this point requires that students learn how the phases of the moon work; they can see that astronomers begin from simple principles to make deductions about the scale of the universe, they can recognize how even the ancient Greek astronomers could begin to build an understanding of the universe, and they can readily grasp the crucial role of observational error. As Whiston remarked, "Notwithstanding that great Subtilty of Wit and Reason, yet many defects are seen in [this method], which forbids us to expect an accurate Investigation of this Parallax by means [of it]." Adventuresome teachers can ask the students if Aristarchus' result, that the sun is 19 times farther than the moon, is "wrong." (If so, can we claim that our current measured distance to the Andromeda Galaxy is "right"?)

Eratosthenes' method for estimating the size of the Earth is probably even more impressive for beginning students, partly for the spurious reason overemphasized by some authors concerning the accuracy of the ancient Alexandrian's result. Here is a good opportunity to introduce the meaning of significant figures!

Un missionnaire du moyen âge raconte qu'il avait trouvé le point
où le ciel et la Terre se touchent...

*Fig. 2. Purported medieval view of the cosmos, as conceived by Camille Flammarion
and published in his* L'Atmosphére: Météorologie Populaire *(Paris, 1988).*

The definitions involved with the celestial sphere (equator, ecliptic, solstice,
height of the pole star, *etc.*) are most readily taught with a geocentric armillary
sphere, which disconcerts some students. It is very useful to take a medieval view of
a fixed Earth for the first several days of an astronomy class, to challenge students
to produce a proof that the Earth moves.

Building Bridges with History

One of the most widely reproduced views of the early sixteenth-century cosmos
appears in Fig. 2. It shows a medieval traveler at the end of the Earth peering beyond
the dome of the sky to see the celestial machinery beyond. Unfortunately, it is pure
art nouveau, the creation of famed French popularizer Camille Flammarion, and first
published by him in 1888. The graphic is philosophically anachronistic: in that age
such curiosity would have been objectionable, especially since all the iconography
of the age put God and the heavenly hosts immediately outside the starry sphere,
much as Dante had portrayed it with his word pictures.

Far better depictions of the cosmos around 1500 are found, for example, in
the full-page illustration at the end of the creation sequence in the 1493 *Nuremburg
Chronicle*, or, more subtly, in Hans Holbein's "Ambassadors." Holbein's magnificent
oil hangs in the National Gallery in London, full of Renaissance symbolism. On

one level, it depicts the advanced college curriculum, the quadrivium comprising astronomy, geometry, arithmetic, and music. On another level it is a morality play testifying to the vanity and mortality of human learning in contrast to the half-hidden eternal truths. A sympathetic discussion of such an artistic treasure sets the stage for Copernicus and, for at least some students, bridges to a richer cultural context of the Scientific Revolution.

Even relatively recent history, such as the Shapley-Curtis debate on the scale of the universe, or the race to find the spiral arms of the Milky Way, can enrich an introductory course, making the role of astronomers more human and more exciting, and therefore more memorable and educational.

The Historical Perspective and the Nature of Science

Why teach Ptolemy? Some students get annoyed when they have to learn "wrong" theories. Ptolemy is often presented as being a wrong-headed geocentrist, whereas Kepler has the Right Stuff. But are the Keplerian ellipses "right"? Newton did not discover the universality of gravitation until fairly late in the game, in the fall of 1684. In one draft of his essay on gravity he refers to the planets as moving in ellipses, and in the next draft, a month later, this claim is excised. What Newton finally realized was that every body in the universe attracts every other body, and because of these perturbations, no planet actually moves in a perfect elliptical orbit.

The historical material can help make the point that science is a dynamic inter-action between observations and theory, or, in Einstein's words, "never completely final, always subject to question and doubt."

A year ago, *Science* magazine carried an interesting critique of introductory science courses for non-science students, under the provocative title, "Are Our Universities Rotten at the 'Core'?" (the 'Core' punning on Harvard's "Core Curriculum" of basic courses). My Harvard colleague, chemist Frank Westheimer, who is somewhat suspicious of historically oriented presentations, wrote that "Perhaps it serves a purpose to recite some of the intellectual advances in science that occurred in the last half century, long after Copernicus and Galileo and Newton ...made their contributions to the intellectual heritage of mankind. In particular, the critical discovery of atomic fission was not published until 1939; in 1937, no one knew how the sun produced its light and heat."

I frankly think it does as little good to *recite* some of the intellectual advances as to embellish a textbook with a few historical illustrations and dates. What we really want is to explain how science works in an evolving, self-correcting way, and a historical perspective often makes that clearer. Did no one know how the sun produced its light and heat in 1937? In 1920, Arthur Eddington wrote, "What is possible in the Cavendish Laboratory may not be too difficult in the sun," and "If, indeed, the sub-atomic energy in the stars is being freely used to maintain their great furnaces, it seems to bring a little nearer to fulfillment our dream of controlling this latent power for the well-being of the human race — or for its suicide." Or perhaps it was unknown until 1931, when Robert d'E. Atkinson argued that the combination of four protons into an alpha particle by some catalytic chain involving heavier nuclei could fuel the sun and stars. Even if we rewrite Westheimer a bit to say fusion instead of fission, and pick 1939 as the date of Hans Bethe's paper on the CNO cycle, could we say that astronomers then "knew" how the sun produced its light and heat? After all, the CNO cycle is now rejected in favor of the proton-proton chain.

History deals not with *facts*, such as who discovered something first or who "got it right," but with *historically significant facts*, which illuminate the process of creation and discovery. Used in a proper way, history can help students understand the excitement of discovery and the role that creative thinking, ingenuity, and genius play in formulating scientific ideas.

Bibliographical Postscript

For those who wish to pursue these ideas farther, here are come clues. Unit 2 of the *Project Physics* high-school physics textbook gives an excellent and accurate presentation of Ptolemy, Copernicus, Kepler, and Newton in a historical setting. The definitive account of the phony Flammarion "woodcut" is by Bruno Weber (in German) in *Gutenberg-Jahrbuch* for 1973, pp. 381-408. I have illustrated and discussed both the Holbein painting and *Nuremburg Chronicle* woodblock in "Copernicus: A Modern Reappraisal," pp. 27-49 in David Corson (editor), *Man's Place in the Universe: Changing Concepts* (Tucson, Arizona: University of Arizona, 1977). The most thorough discussion of the Holbein is Mary F.S. Hervey's *Holbein's "Ambassadors"* (London: G. Bell and Sons, 1900). Newton's path to *universal* gravitation is well told by I Bernard Cohen, "Newton's Discovery of Gravity," *Scientific American*, March 1981, reprinted in Owen Gingerich (editor), *Scientific Genius and Creativity* (New York: W.H. Freeman, 1986).

Excellent material on the Shapley-Curtis debate is found in Michael Hoskin's *Stellar Astronomy* (Chalfont St Giles: Science History Publications, 1982) as well as in Robert Smith's *The Expanding Universe: Astronomy's Great Debate, 1900–1931* (Cambridge, England: Cambridge University Press, 1982). My account, "The Discovery of the Spiral Arms of the Milky Way" is in H. van Woerden *et al.* (editors), *The Milky Way, IAU Symposium 106* (Dordrecht: D. Reidel Co., 1985), and a shorter version in *Sky & Telescope* **68**: 10–12, 1984. F.H. Westheimer's "policy forum" editorial appeared in *Science* **236**: 1165, 1987. The papers by Eddington, Atkinson, and Bethe are grouped together in Chapter IV, "Stellar Evolution and Nucleosynthesis" in Kenneth R. Lang and Owen Gingerich (editors), *A Source Book in Astronomy and Astrophysics, 1900–1975* (Cambridge, Mass.: Harvard University Press, 1979). Concerning the difference between *fact* and *historical facts*, see the introduction to Owen Gingerich (editor), *The General History of Astronomy*, volume 4A (Cambridge, England: Cambridge University Press, 1984).

Fig. 3. Holbein's The Ambassadors *(courtesy of the trustees, National Gallery, London).*

Discussion

M. Zeilik: *Do you have an appropriate place in your course for the history of astronomy outside of the western tradition?*

O. Gingerich: We assign a project or essay, which counts for a third of the grade, and we include as standard topics such questions as "Was Fajada Butte a solar marker?," "Would the Chinese have found a heliocentric cosmology?," and "How accurate was Mayan astronomy?" In the course itself, there just aren't enough lectures to touch on many fascinating topics.

THE ROLE OF ASTRONOMY IN THE HISTORY OF SCIENCE

James MacLachlan
Ryerson Polytechnical Institute, 350 Victoria Street, Toronto, Canada M5B 2K3

1. Introduction

This historian of science offers a few samples of the kinds of understandings his students will be subjected to. (a) In early times, Britons used careful observations of astronomical events to establish their calendar; (b) In the 4th century BC, Aristotle used the spheres of Eudoxus to establish his cosmological principles; (c) In the second century of our era, Ptolemy made astronomy scientific, partly for the sake of astrological predictions; (d) In the fifteenth century, Columbus used crude astronomical observations to find latitude. (e) In the sixteenth century, Copernicus revised Ptolemaic astronomy in order to improve its fit with Aristotelian cosmology, and in the process challenged that cosmology; (f) Kepler used Tycho's more precise data to destroy heavenly circularity; (g) In the early seventeenth century, Galileo based his renovation of motion studies on the investigative style he learned from Ptolemy, coupled with mathematics learned from Euclid and Archimedes.

Segments of history such as these are intended to demonstrate that astronomy has had contacts with numerous other aspects of life for many centuries. Astronomers should be able to interact with historians to the mutual benefit of both.

2. Use of the History of Science

I believe that history of science is useful in introducing certain topics of science, partly because it can pose problems in their original settings, and partly because it can often introduce complexity gradually.

Many of my students are majoring in technologies such as electrical and mechanical. They take a course in history of science and technology as an option in their third year (of four). One of my major motivations in teaching them such a course is to make them aware of the struggles innovators go through in order to produce change. Another is to try to show science and technology as intimately related to numerous other aspects of human activities. In this process I use several astronomical examples. They are helpful because of humans' eternal fascination with astronomy, and because I need them in order to account for other developments in science, particularly in physics.

The first example I introduce is based on the researches of Alexander Thom (Ruggles, 1988) in Britain over the past 50 years. Long before they could write, early Britons used stone markers to record extreme positions of the sun and the moon. I have described this in my textbook on the history of science and technology, *Children of Prometheus* (preliminary edition: Toronto, Wall & Thompson, 1988),

p. 36. I suggest that these megalithic astronomical monuments were erected for religious purposes connected with the Britons' concern for the sun's control over the seasons.

By the time of Aristotle, Greek mathematicians possessed astronomical data compiled in Babylon, and a fine set of analytical tools (*ibid.*, p. 37). After Eudoxus had done his best to use nested spheres to match the data, Aristotle adopted Eudoxus's basic model as the foundation of his cosmology. In the process, Aristotle laid down principles that influenced European thought for most of the next 2000 years. In particular, he made a sharp distinction between the sublunar and astral spheres.

Ptolemy may be called the first fully-fledged scientist for his concern to provide models of the planetary orbits that actually fitted observed positions. As opposed to being largely empirical like the Babylonians, or mainly theoretical like Aristotle, Ptolemy combined the two into an effective predictor of planetary positions (*ibid.*, p. 38).

According to Noel Swerdlow (1984), Copernicus's main motivation for altering the Ptolemaic scheme was to make an astronomical model that would be more consistent with Aristotle's principles. That is, Copernicus reluctantly adopted motions of the Earth in order to avoid the mechanical absurdity of the equant (*ibid.*, pp. 88–91).

Kepler, on the other hand, reluctantly adopted elliptical orbits in order to fit Tycho Brahe's data as closely as possible. Kepler also sought to create a physics of the heavens, but he was premature (*ibid.*, pp. 93–95). The route to a universal physics with Newton (*ibid.*, pp. 124–132) lay through Galileo's mathematizing of motion (Drake, 1978), which he did by using astronomy as his model for science (*ibid.*, pp. 108–109).

By examples such as these, I suggest various ways in which science has developed during the past centuries. In the process, there is some chance that my students may also pick up a few items of astronomical knowledge!

References

Drake, Stillman, *Galileo at Work* (University of Chicago Press, 1978)

Ruggles, Clive, ed. *Records in Stone* (Cambridge University Press, 1988)

Swerdlow, N.M. and O. Neugebauer, *Mathematical Astronomy in Copernicus's "de Revolutionibus"* (Springer-Verlag, 1984)

TEACHING HUMAN CULTURE THROUGH ASTRONOMY

N.S. Nikolov

Department of Astronomy, University of Sofia, A. Ivanov Str. 5, 1126 Sofia, Bulgaria

One of the phenomena in the development of science in the second half of the twentieth century is the appearance of complex branches of knowledge. This fact along with the accelerated increase of science information called forth well-known integrative processes in education. One of the most widely spread forms of these processes is the incorporation of one school subject in another. In such a way, astronomy in secondary school is incorporated in the subject of physics and sometimes in geography and mathematics. The argument for this, if there exists one, is that nowadays astronomy is astrophysics, *i.e.*, physics of celestial bodies, or that the cosmographical function of astronomy resembles the function of geography.

In this paper, we make an attempt to adhere to the thesis that astronomy is a school subject with wider connections in the human sphere than with only one branch of science (Nikolov, 1986). As a consequence, if the school subject astronomy is incorporated only in a specific discipline, this would limit the possibility of teaching facts or phenomena of other spheres of the human spirit.

On the other hand, people often propose to teach something of astronomy by means of, for example, science fiction (Rolewicz *et al.*, 1986; Roslund, 1986). Our basic idea in this paper is not to teach astronomy by means of something else but to learn something else not only of other sciences but also through teaching astronomy to learn widely of human culture. With reference to the integrative processes mentioned above, this is an integration in which astronomy as a school subject plays the leading role. More properly speaking, we attempt to show how some astronomical subjects are capable of being used to suggest some cultural values of mankind.

One set of subjects that supplies many possibilities in this respect is the planets and their satellites. They are more widely represented than any other astronomical subjects in the school curricula at every level. The names of the planets are taken from the ancient Roman and Greek mythology, which allows teachers to draw parallels with many of the interesting topics of human culture and its history. In the first place, these are the mythological legends and the characters included in them. In the second place, many paintings, drawings, and sculpture represent mythological subjects and/or characters in them. Thus in parallel with learning about the planets and their satellites themselves, we can learn the picturesque mythological legends that are the spiritual wealth of all humanity as well as some of the history of culture, *etc.* The lessons can help cultivate aesthetic taste.

The same possibility exists when you teach about constellations or some stars (variables, doubles, *etc.*) and other objects belonging to any constellation. Wide educational opportunities are supplied by the fact that the mythological personalities whose names we see as constellation names appear in painting, literature, and musical images by many famous artists, writers, and composers.

Here we are able to mention only some of the art masterpieces connected with the planets or with some of their satellites or with mythological persons and events connected with the constellation names.

Let us begin with Mercury. The Greek god Hermes (transformed to Mercury in the ancient Roman mythology) was one of the favorite characters of ancient mythology for many painters and sculptors. Examples in European art during the fifteenth–seventeenth centuries include paintings by Rubens, Velazquez, Rembrandt, *etc.* Also, there exist many Greek sculptures of Hermes or their Roman copies as well as newer paintings on this subject (see for instance Tokarev, 1980; Buslovich, 1971).

As for Venus, it is impossible to count not only all the artistic masterpieces but also the artists dealing with images of Aphrodite. Let us only mention that besides the sculptures and the paintings the image of Venus appears in European literature and music, including a cantata of the second half of the twentieth century.

The planet Jupiter, named after the Roman senior god primary who was also in ancient Greek mythology, gives many possibilities for aesthetic education, too. There exists numerous pictures devoted to the many stories about Zeus. Many drawings visualize Zeus as carrying off Europa but drawings with other beautiful women such as Io and Callisto exist too. So these drawings could be treated when you teach Jupiter and its Galilean satellites. It is well worth remembering that such a famous composer as Mozart named his 41st symphony "Jupiter."

The situation with Saturn and Uranus is somewhat different because of the lack of art masterpieces involving them. Nevertheless, the myth of Uranus as a primary god in the Greek mythology is very interesting in itself and can be very instructive to pupils.

Here I will stop with the planets. As for the constellations, I mention only the story in which the names of the constellations Cepheus, Cassiopeia, Andromeda, Perseus, and others are related in one legend. These mythological subjects have attracted the interest of many artists. Andromeda and Perseus appear not only in all kinds of paintings but also in some literary works as well as in many operas. We could mention here the poem "Andromeda" by Lope de Vega and the operas by Monteverdi in the seventeenth century and by Handel in the eighteenth century.

In conclusion, I think that not only the successful cosmic missions but also the nature of astronomy as a science that was created together at the same time as human culture calls forth that it is time to restore the leading role of astronomy in education. Remember that astronomy was in the "quadrivium" together with arithmetic, geometry, and music among the "seven liberal arts" that were necessary for every intellectual.

Fig. 1.   Perseus and Andromeda, a painting by Rubens.   Similar subjects from Greek mythology were painted by Velazquez, Rembrandt and others.   Gemäldegalerie Dahlem, West Berlin; Bildarchiv Foto Marburg.

References

Buslovich, D. 1971, *Mythological Subjects in Art Products*, Leningrad.
Nikolov, N.S. 1986, *Proceedings of the GIREP Conference.* ESA SP-253, 329.
Rolewicz, A., Swyslowski and Mickiewicz, A. 1986, *ibid.*, 379.
Roslund, C. 1986, *ibid.*, 141.
Tokarev, S., ed.-in-chief, 1980, *Encyclopedia "Myths of the World Peoples,"* Moscow.

Discussion

J. Fierro: *I agree with the speaker in the sense that science is part of culture and talking about mythology in class eases and enriches talks.*

M. McCarthy: *The differences between the traditional and the USA approach to astronomical education cited by Prof. Wentzel find a fine confirmation by Prof. Nikolov's paper on culture and astronomy. In Europe, the classical tradition remains strong; in the USA, ignorance of classical mythology is quite extensive. Last year, as a classroom assignment at Georgetown, I proposed that each student write an essay on the discovery, orbital features, and mythological background for 2 or 3 of the newly discovered satellites in the planetary system. Several said it stimulated them to do similar research on other satellites. It was fun; I enjoyed reading the essays and let all the students share the findings of their fellows.*

H.S. Gurm: *In reference to Professor McCarthy's remarks:*

1. *Wentzel's "traditional" model differentiates the American pattern of education from that of other countries. It is not concerned with the classical pattern in which everything was taught through language or grammar. As such, it should not be confused with the latter.*

2. *The third world has its own cultures; there is no such thing as an under-culture or an over-culture. The question is not teaching culture or "the classics" through astronomy or vice versa. Both need to be covered independently. Teaching astronomy through culture and vice versa would dilute the teaching of astronomy, and would affect the conceptual learning of it.*

B. Curran: *Should there be greater use of the humanities in astronomy classes or greater use of astronomy in humanities classes?*

N.S. Nikolov: My paper deals with the problem of how it is possible to teach human culture by teaching astronomy, *i.e.*, with astronomy classes. But I think the problem may be reversed; it is also possible to teach astronomy in the humanities classes.

INTERDISCIPLINARY APPROACHES TO ASTRONOMY

Andrew Fraknoi
Astronomical Society of the Pacific,
390 Ashton Ave, San Francisco, California 94112, U.S.A.

To understand the motivation for my talk, you must bear in mind what Don Wentzel discussed so eloquently at the beginning of the colloquium. In the U.S., the vast majority of students taking astronomy classes at the college level are *not* science majors. Many students coming into the astronomy courses are afraid and distrustful of science and often see science as a very alien endeavor, quite separate from their everyday lives and other studies.

For such students, it can sometimes be very reassuring and enlightening to show some interesting connections between astronomy and other (nonscience) fields at a few places in the introductory astronomy course. For example, many students are surprised and excited to see the inspiration that astronomy has provided for music, literature, and art and some of the interesting connections between astronomy and psychology, archaeology, and law.

But the idea behind teaching and exploring interdisciplinary connections is more than merely doing therapy for science anxiety. I also believe that there is a danger in the college or university "catalog" model of human knowledge, in which everything is divided into departments whose members speak to the members of other departments as rarely as possible. In reality, one's understanding and appreciation of science and the humanities becomes a complex mental blend, in which the pleasures of hearing a symphony and learning about a new scientific model can be very similar.

Using interdisciplinary approaches in our teaching can also serve to remind our students (and ourselves) that it is so often the cross currents and cross fertilization that lead to progress in many fields. As the great Russian-American novelist Vladimir Nabokov said, "There is no science without fancy, no art without facts."

In this brief review, I would like to mention just a few examples, both from serious and popular culture, for your edification and amusement. A reading list of useful sources for further exploration can be found at the end of this paper.

1. Science Fiction

If you think that science fiction is still about daring young men saving both the Earth and the mad scientist's beautiful daughter from ravaging aliens, it may be time for another look. Today, there are many authors writing sensitive, literate science fiction based on the extrapolation of good science. Some of their best science-related works are not only interesting in their own right but also can help students visualize and internalize some of the more abstract notions of modern science. A few of my favorite authors include:

a) Gregory Benford

Benford is a plasma physicist at the University of California at Irvine and arguably one of the finest observers and interpreters of science in modern fiction. Not only is his novel *Timescape* an exciting yarn about ecological disaster and communication backwards in time using tachyons, but also it has one of the most realistic portraits of a scientist I have ever read. Astronomers might also enjoy finding colleagues such as Fred Hoyle, Geoff and Margaret Burbidge, and others in the book.

Benford's two other recent novels, *In the Oceans of Night* and *Across the Sea of Suns*, posit a fascinating hostile universe in which powerful machine intelligences rule and organic life-forms are hunted and destroyed. A new novel, *Great Sky River*, envisions a future near the center of the Milky Way, where small bands of humans endure constant flight from hordes of destructive machines.

b) Larry Niven

Although he is not quite the stylist that Benford is, Niven's stories scintillate with ideas inspired by modern astronomy. In several of his short stories, massive or quantum black holes have been used to good effect. In *World Out of Time*, for example, a supermassive black hole near the galactic center allows the protagonist to travel three million years into the future. In the *Integral Trees* and *The Smoke Ring*, he creates a world of living beings in a gas torus in orbit around a neutron star.

c) Charles Sheffield

Sheffield has a Ph.D. in physics and many of his short stories (collected in *Vectors* and *Hidden Variables*) involve illustrating and fleshing out new ideas in astrophysics. In his novel *Between the Strokes of Night*, he suggests an imaginative form of life in intergalactic space.

d) John Varley

Varley is an excellent writer who is as comfortable describing the conditions on Venus as he is examining the social effects of genetic engineering . His first novel, *The Ophiuchi Hotline*, is a tour-de-force, with enough ideas and themes to make several books, including the sport of mini-black-hole hunting.

e) Fred Pohl

Veteran science-fiction writer Pohl has written many interesting works of hard science fiction and social extrapolation. His *Man Plus* describes how humans could be biologically engineered to survive on Mars. In *Gateway*, an award-winning novel that is a favorite with my students, Pohl describes what must surely be the first case of "black hole guilt" in literature.

f) David Brin

Brin has a Ph.D. in astrophysics from the University of California, San Diego, and writes mostly about an old science fiction theme, the problems and possibilities in contacting life elsewhere, but with stories and novels based on an understanding of the real universe. His award winning story "The Crystal Spheres" has advanced races using time dilation near black holes to bear the loneliness of a universe in which life is still rare.

Astronomers Writing Science Fiction

In the last few years, a number of working astronomers have joined a long-time science fiction author Fred Hoyle in combining science and fiction writing. Carl Sagan has written the best seller *Contact*, about the consequences of a message from an extraterrestrial civilization. (Sagan, by the way, asked Kip Thorne and his students at Caltech to design a viable black hole interstellar transportation system for his novel.)

Supernova expert J. Craig Wheeler has written a thriller entitled *The Krone Experiment*, which involves mini black holes. Donald Clayton's *Joshua Factor* interweaves neutrinos and intrigue. British physicist Paul Davies works out the consequences of the large-scale arrival of antimatter from space in his novel *Fireball*.

NASA's William Rossow has worked with M. Bradley Kellogg on two novels, *The Wave and the Flame* and *Reign of Fire*, that take place in a very alien environment. And physicist Robert Forward has invented a life-form that can live on a neutron star, in his novels *Dragon's Egg* and *Starquake*.

Although many astronomers have enjoyed and recommended Fred Hoyle's *The Black Cloud*, in which astronomers take the reins of power as the Earth faces disaster, my favorite among Hoyle's novels is *October the First Is Too Late*, a fictional working out of the consequences of the "many worlds" interpretation of quantum mechanics.

2. Poetry

Astronomy has served as inspiration for poetry for centuries and in courses tracing the evolution of astronomical ideas it can often be fascinating to see that evolution in the poetry of each era as well. (Some specific suggestions can be found in the interdisciplinary packet from the A.S.P., listed in the Resources at the end.)

Among modern poets who use astronomy and physics to particularly good effect in their work, it might be worth mentioning two of my favorites. Diane Ackerman studied astronomy with Carl Sagan at Cornell; her collection *The Planets* is a beautiful (and informed) evocation of the solar system as we understand it today. Novelist John Updike enjoys and keeps up with modern science and has written poems about neutrinos, the satellites of Jupiter, and entropy. (See, for example, his collection *Facing Nature*.)

Some anthologies and articles to make a start in this field might include:

Gordon, B. *Songs from Unsung Worlds: Science in Poetry.* 1985, Birkhauser Boston.
Marschall, L. "Modern Poetry and Astronomy" in *Mercury*, Mar/Apr. 1983, p. 41.
Phillips, R., ed. *Moonstruck: An Anthology of Lunar Poetry.* 1974, Vanguard Press.
Vas Dias, R. *Inside Outer Space.* 1970, Doubleday/Anchor.

3. Mainstream Literature

Astronomy and physics have begun to play an increasingly important role in modern literature, as the ideas from science begin to permeate other branches of human culture and leave an imprint. Through the work of popularizers of astronomy (starting with Jeans and Eddington in the earlier part of our century) a number of writers with little formal training in the sciences have learned enough to incorporate some of the most exciting ideas and themes into their work. Again, let me select just a few examples; more can be found in the sources listed in the Appendix.

The reclusive American writer Thomas Pynchon, who worked at Boeing for a while, has written a number of novels and stories using concepts from thermodynamics and other branches of physics, including *The Crying of Lot 49* and *Gravity's Rainbow*. Joyce Carol Oates has done an effective character study using black hole images in a short story entitled "Passions and Meditations" (in *The Seduction and Other Stories*).

Lawrence Durrell's *Alexandria Quartet* (consisting of the novels *Justine, Balthazar, Mountolive,* and *Clea*) uses notions from the theory of special relativity to deepen his considerations of modern love and mid-Eastern politics. Italo Calvino constructs modern fables and allegories from scientific notions in *Cosmicomics* and *Tau Zero.* Tom Bezzi describes the life and work of Edwin Hubble through the eyes of a fictional granddaughter in *Hubble Time.* And an astronomer's life and relationships form the core of the novel called *First Light* by Charles Baxter that winds backward through time.

4. Music

Over the last decade and a half, my students and I have collected over 100 examples of music inspired by astronomy. We exclude pieces whose connection with astronomy is peripheral or superficial — there are countless songs one could list extolling the romantic virtues of the full moon, for example — and concentrate only on pieces that are inspired in some way by specific astronomical ideas and discoveries. A good annotated list can be found in my two articles in *Mercury* magazine (May/June 1977, p. 15 and Nov/Dec. 1979, p. 128) but a few examples will serve to give the reader the flavor of what is now available on records:

In the 1970's Willie Ruff and John Rodgers (a geologist and professor of music at Yale) reconstructed Kepler's "music of the spheres" by converting the motion of the planets as we know them today into musical notes that could be played electronically. The full story is told in their article in *American Scientist,* May/June 1979, p. 286, and a recording was available from authors for a while. A similar piece, incorporating the motion of the Galilean satellites of Jupiter, was commissioned by the Münster Astronomical Society for the 350th anniversary of Kepler's death; entitled *The Harmony of the World of Jupiter* by Gunter Bergmann, it is available on a Schwann label recording.

In 1973, to celebrate the Copernicus Quincentenary, the National Academy of Sciences and the Copernicus Society of America commissioned a piece from astronomer Fred Hoyle and Composer Leo Smit entitled *Copernicus: Narrative and Credo.* Available on Desto records, the piece features excerpts from the life of Copernicus and a thought-provoking statement of belief from the perspective of modern astronomy.

Karlheinz Stockhausen, a iconoclastic modern composer, has put together (composed really isn't the right word) an intriguing piece called *YLEM* that is based on the oscillating universe. (YLEM was Aristotle and George Gamow's term for the primordial material from which the present universe was born.) In the piece, which is available on a Deutsche Grammophon recording, 19 players play a big bang, after which ten of the players "expand" away from the stage and move throughout the hall. Later they return, another explosion is played, and the players then leave the building entirely. (The music is — to put it delicately — not easy to listen to, particularly because specific notes have not been written down by the composer, who expects the players to be "in telepathic communication." But the idea of the piece does hold a certain fascination for astronomers.)

There are many other pieces of serious music that astronomers might enjoy: Philip Glass' opera *Einstein on the Beach* uses aspects of Einstein's life and work as mantras in a meditative piece of minimalist music; Charles Dodge's *Earth's Magnetic Field* is a piece whose notes are determined by the K_p index of terrestrial magnetism for the year 1961; and Paul Hindemith's 1947 opera, *The Harmony of the World*, chronicles the life of Kepler.

In the realm of popular music, there are many pieces based on astronomical ideas, although few of the fans who listen to the music are likely to be aware of the connections. Three songs compare the dying of a love affair or a rock career to the dying of a star: Pink Floyd's "Shine on You Crazy Diamond" (on the album *Wish You Were Here*), Labelle's "Black Holes in the Sky" (on *Phoenix*), and Rush's "Cygnus X-1" (on *A Fare Well to Kings*). Vangelis Papathanassious, who composed the theme song to Carl Sagan's television program *Cosmos*, has an album entitled *Albedo 0.39*, which includes pleasant musical pieces with titles like "Main Sequence" and "Sword of Orion."

But probably the most unusual rock record we have seen is an album with the provocative title *H to He Who Am the Only One*, by a group that calls itself Van Der Graaf Generator. The album cover explains, "The fusion of hydrogen nuclei to form helium nuclei is the basic exothermic reaction in the sun and the stars, and hence is the prime energy source in the universe." This statement is followed by the three equations of the p-p chain, certainly a first for a rock music album. (For those whose collections would not be complete without this record, it was available on the Dunhill label.)

5. Other Fields

The above are just a few areas outside science in which there are interesting connections to be made with astronomy. Connecting threads can be found to many other fields as well:

In art, many serious and popular artists have been inspired by the work of modern astronomers. Among the serious ones, few have produced work of such immediacy and interest as the Mexican surrealist painter Remedios Varo, whose images have appeared in Thomas Pynchon's novel *The Crying of Lot 49* and on the cover of at least one popular-science book. Many astronomers have used the mathematically oriented etchings of M.C. Escher in their courses. And astronomer William Hartmann, who is an excellent artist, has produced a number of informed and beautiful works that combine science and imagination.

In psychology, the moon illusion, the panic during Orson Welles' 1938 broadcast of *War of the Worlds*, and the intensity of belief in UFO's are all interesting topics for investigation. In law, students often enjoy looking at the trials of Giordano Bruno and Galileo and at the influence of Hubble's legal training on his later astronomical work. In archaeology, ancient monuments in Europe and western America have revealed a much greater astronomical interest and sophistication among earlier civilizations than standard histories give them credit for.

6. Concluding Comments

While no one is suggesting that such an approach substitute for a thorough grounding in science and scientific methodology in introductory astronomy courses, it can be very rewarding for the instructor and the student to consider the connections between astronomy and other fields on at least a few occasions and to show how our science fits into the wider range of human culture and thought. The few minutes devoted to a story, poem, or piece of music during an astronomy class could make a deeper impression on your nonscience students in the long run than the world's most exciting lecture on spectral classification.

Resources for Interdisciplinary Approaches to Astronomy

Digby, J. and Brier, B., eds. *Permutations: Readings in Science and Literature.* 1985, Morrow. An eclectic collection of excerpts and poems.

Dubeck, L., *et al. Science in Cinema.* 1988, Teachers College Press. An introduction to teaching science through science fiction films.

Fraknoi, A., *et al. Interdisciplinary Approaches to Astronomy.* 1986, Astronomical Society of the Pacific. A 32-page collection of articles from *Mercury* magazine on astronomy and music, poetry, science fiction, *etc.* Includes a 4-page detailed reading list.

Fraknoi, A. *Universe in the Classroom.* 1985, W.H. Freeman. Features suggestions for interdisciplinary paper and discussion topics and more detailed reading suggestions in every field of introductory astronomy.

Friedman, A. "Contemporary American Physics Fiction" in *American Journal of Physics*, May 1979, p. 392.

Friedman, A. and Donley, C. *Einstein as Myth and Muse.* 1985, Cambridge U. Press. A fascinating examination of Einstein as symbol for science and genius in our time, with excellent sections on the influence of Einstein's work on the humanities.

Nicholls, P., ed. *The Science Fiction Encyclopedia.* 1979, Doubleday.

Discussion

J.-C. Pecker: *I tend to be afraid of the use (perhaps abuse) of reference to science fiction although I occasionally refer to it myself in my astronomy lectures. S.F. can indeed lead to bad misconceptions! By contrast, detective stories, because of their type of strict logic, help to understand the scientific type of reasoning.*

A. Fraknoi: An interesting detective novel influenced by quantum mechanics is Stanislav Lem's *The Investigation*. A short story in the same vein is "All The Myriad Ways" by Larry Niven (available in a collection of stories with the same title).

———————————

PLANETS, PULSARS, AND POETRY

S.B. Yorka
Denison University, Granville, Ohio 43023, U.S.A.

For many of us in the United States, the majority of our students are in descriptive astronomy classes. And since these classes typically satisfy general education or core curriculum requirements that must be completed by all students, the students can range from those genuinely interested in astronomy to those who are taking the class because "it sounded less boring" than other options available. Whichever end of that spectrum the students occupy, many of them approach astronomy with quite a bit of anxiety because it is a science class. In student lore, a science class is a class that is by definition more difficult — perhaps verging on the impossible — than other classes, one that discusses totally foreign things in an arcane language and, above all, is a class that has no connection with anything else in the curriculum, except maybe another science class.

We all would probably concede that many non-science students do find science classes more difficult than their humanities or arts classes — primarily because of poor math and critical reasoning skills. And to students who do not read *Scientific American*, or *Science News*, or even the science section of magazines or newspapers, words like supernova, quarks, and quasars are indeed foreign. Nor is there a lot we can do in a semester course in astronomy to markedly improve math skills or change life-long reading (or non-reading) habits. But what we certainly *can* do is help the students to make connections — to become aware of the ways in which science courses can relate to, illuminate, and enhance material encountered in arts and humanities courses.

That is the purpose of "Planets, Pulsars, and Poetry," which is one of a series of interdisciplinary exercises designed for astronomy students in a liberal-arts curriculum. The exercise introduces the students to specific poems that can only be fully appreciated if the reader has some knowledge of astronomy. The poems — by 20th century American poets — are chosen for their use of astronomical phenomena as subjects or metaphors. The students are asked some questions about the general characteristics of each poem; but since it is desired that the students put to use their knowledge of astronomy, they are questioned more closely about the astronomical aspects of each poem. They are asked to analyze each poem with respect to its astronomical allusions: they must explain the general meaning of the whole poem and the meaning of any astronomical terms, and are usually asked to give detailed explanations of particularly interesting passages. Each poem is accompanied by a set of questions designed to direct the students' attention to specific parts of the poem, leading them to a better understanding of the poem as a whole. The exercise is scheduled about halfway through the semester so that the students have a background sufficient to allow them to understand some of the terms that they may not

yet have encountered, such as redshift, pulsar, and big bang. The poems currently
in use in this and a couple of other interdisciplinary exercises concern phases of the
moon, a lunar eclipse, planets, constellations, the night sky, pulsars, the Voyager II
satellite, and a whimsical comparison between astronomy and astrology.

Perhaps the most transparent of the poems — and a favorite of the students
— is Robert Frost's "Two Leading Lights," which contrasts the sun and the moon.
The sun is "satisfied with days" and has "the greatness to refrain" from appearing
at night while

> "The moon for all her light and grace
> Has never learned to know her place";

and while astronomers

> "Have set the dark aside for hers
> ...there are many nights, though clear
> She doesn't bother to appear.
> Some lunatic or lunar whim
> Will bring her out, diminished dim,
> To set herself beside the sun...."

The obvious concepts the students need to explain this poem are the daily
rotation of the Earth and the role of the revolution of the moon around the Earth
in producing the "lunatic" behavior of the moon. Observations of lunar phases
are carried out earlier in the semester, so the students have seen this behavior for
themselves and have had to explain the phases of the moon.

Another poem the students enjoy is "Saturn" from *The Planets* by Diane Ackerman. The students are asked to explain Saturn appearing as "An elliptical blur...'
creeping into the field of view "Bellowing light," lying "...stunned in a hall of mir-
rors, hog-tied by the cross hairs." The second stanza concerns the rings:

> "Millions of vest-pocket moons
> hang together as rings
> that loop around the planet
> like a highway skirting the golden city,
> dusky bright, and godawful sheer,
> They dog the equator (like Uranus's moons) never more than
> two miles thick: a sprawling coral reef of tailless comets
> Grinding one another
> finer and finer, lolloping boulder
> to dusty mote as, eddying
> down through the crepiest ring
> they pour into a gassy draw."

The students are asked to provide a detailed analysis of this and three other
stanzas from "Saturn." All of the stanzas have the same physical appearance, with

five short lines, two longer lines, and five short lines as the author obviously shaped the poem to resemble Saturn and its rings. Oddly enough, very few students notice that.

More subtle phrases are encountered by the students in other poems:

"...clouds of our unknowing like great nebulae"

"As through a glass that magnifies my loss
I see the lines of your spectrum shifting red...."

"Science of the Night" by Kunitz

"The tin man is cold,
the glitter of distant worlds
is like snow on his coat..."

"...the white, elusive
dandelion fuzz
of starlight..."

"Voyager II Satellite" by Kooser

"...ancient stream of starlight...has made its way to our eyes through...down-
curved ravines of space..."

"By Starlight" by Wagoner

The rich complexity encompassed by the description of stars as "taillights of the big bang..." (Saner) is rarely appreciated by students on their own, and a discussion of the phrase helps the students follow a chain from cosmology to stellar evolution.

Student reaction to this exercise has ranged from pleased incredulity (from English majors in the class) to downright hostility (from a student who was incensed that he was going to be expected to write a paper — and be graded on grammar and spelling — in a science class). The majority of the students, though, have responded very favorably, and are delighted that their knowledge of astronomy enables them to decipher poetry that they know they wouldn't have understood prior to taking the class. Perhaps one measure of the success of this exercise in reaching students in a way unusual in science classes is exemplified by the student who signed up for an astronomy class for next fall — *because* of the "poetry lab," as the students call it.

A short bibliography of poetry used accompanies this paper; a more extensive listing of "astronomical poetry," as well as copies of the full exercise "Planets, Pulsars, and Poetry" are available from the author.

Reading List

Ackerman, Diane, "Saturn" from *The Planets: A Cosmic Pastoral* (Morrow Quill Paperbacks, 1976).

Frost, Robert, "Two Leading Lights" and "Upon Looking Up by Chance at the Constellations", from *Complete Poems of Robert Frost 1949* (Henry Holt & Co., 1949).

Hollander, John, "The Great Bear" from *Spectral Emanations* (Athenum, 1978).

Kerr, Minnie Markham, "Nocturne" from *The Music Makers*, complied by Stanton Coblentz (B. Ackerman Inc., 1945).

Kirby, Inez, "Silence" and "Night Unto Night" from *The Music Makers*, complied by Stanton Coblentz (B. Ackerman Inc., 1945).

Kooser, Ted, "The Voyager II Satellite" from *One World at a Time* (University of Pittsburgh Press, 1985).

Kunitz, Stanley, "Science of the Night" from *The Poems of Stanley Kunitz 1928-1978* (Little, Brown & Co., 1979).

Rich, Adrienne, "Planetarium" and "For the Conjunction of Two Planets" from *Adrienne Rich's Poetry*, ed. B. Gelpi and A. Gelpi (W.W. Norton Co., 1975).

Saner, Reg, "That Line Drawn at the Moon" from *Climbing into the Roots* (Harper and Row Publ., 1976).

Wagoner, David, "By Starlight" from *First Light* (Little, Brown & Co., 1983).

Discussion

L.C. Hill, Jr.: *I would like your comment on a phenomenon that I have observed in student responses to astronomical materials. I have found that in student essays (and, in a few cases, in student paintings) some students show a grasp of astronomical topics and their implications that I would never have imagined simply from their responses to mathematical or more traditionally scientific examinations.*

S. Yorka: Yes, I agree that some students do show in written assignments a level of understanding higher than one would expect from the results of traditional exams.

Comment

Jay M. Pasachoff: Two noted American poets, Anthony Hecht and John Hollander, decided that it was a pity that art forms could not usually point to a moment of discovery the way science can. At what instant was the sonnet invented, for example? They decided to rectify this failing of poetry by inventing a new poetry form while reporting on the process. The result was a form of poetry called a "double dactyl," which has several strict rules: Each line is made of two dactyl (a dactyl is DAH dah dah), and there are two stanzas, each of four lines. The first stanza begins with a double-dactyl nonsense line like "Jiggery-pokery." One line in the first stanza is a single double-dactyl word. They keep an official list of such names and words, which cannot be reused in other double-dactyl poems. Their published "canon" was published as Jiggery-Pokery, A Compendium of Double Dactyls, *edited by Anthony Hecht and John Hollander (New York: Atheneum, 1967), and included (including the footnote):*

Revolutionist by Nancy L. Stark

    Higgledy-piggledy
    Nic'laus[1] Copernicus
    Looked at the Universe,
    Spoke to the throng:
    Give up your Ptolemy,
    Rise up and follow me,
    Heliocentrically
    Ptolemy's wrong.

My own effort is: (about the pulsar CP 0328[2])

    Higgledy-piggledy
    CP0329
    Pulsing its message
    To us from afar.
    Is it a message from
    Extraterrestrial
    Little green men
    On a dense neutron star?

Note from Editor John Percy: Jay (perhaps understandably) did not mention his even more famous effort, which appears as part of the Biographical Notes to his article on "The Solar Corona" in Scientific American, *October 1973:*

    Higgledy piggledy
    Jay Myron Pasachoff
    Williams astronomer
    Dabbles in rhyme
    Solar eclipses and
    Radiotelescopes
    Keep him contented
    The rest of the time.

---

[1]This is clearly cheating; on the other hand, "Nicky" would have been far, far worse. Lines 5 and 6 are nevertheless inspired.

[2](This appeared in Physics Today and was anthologized in *A Random Walk in Science*.)

THE NORTHERN LIGHTS AS AN INTRODUCTION TO THE UNIVERSE

J.E. Solheim
Institute of Mathematical and Physical Sciences, University of Tromsø, Norway

1. Introduction

The University of Tromsø is located at 70° northern latitude. It is planned for 3,000 students and was opened in 1972. Research interests in physics are centered on the physics related to the upper atmosphere. The region is coastal and mountainous, and is sparsely inhabited. People in the region are good observers, and those away from city lights watch the lively auroral displays that can be observed almost every clear, dark night. The different auroral forms — wavering structure, colors, and diffusing and changing appearance — all challenge the imagination. There are popular and folkloric descriptions as well as manifestations in art, including poetry, paintings, music, and drama.

2. The Course

An introductory course in astronomy and space-science physics has been developed — using the northern lights as the central theme — guiding us from our planet to the most distant parts of the universe.

The aurora borealis is the end effect of a solar wind blowing charged particles from the sun and interacting with the Earth's magnetic field. The magnetosphere acts like a large dynamo — producing the power to accelerate particles — just as in a cathode ray tube. The equivalent of the CRT screen is the ring around the magnetic poles where the magnetic lines open to space. The colors are due to emission lines from atomic oxygen or nitrogen at altitudes 90 km or higher. The same phenomenon can be predicted and observed on other planets that have a magnetic field and an atmosphere. The solar wind also interacts with comets and other bodies in the solar system. More energetic winds can be observed from stars of other types and stages in their development.

Finally, interacting binary stellar systems that contain a compact object — a white dwarf, a neutron star, or a black hole — as an accreter, show what happens when gravitation assists in the process of pulling off matter and dominates in the accretion process itself. We observe the high-energy version of the aurora as ultraviolet radiation, X-rays, or gamma rays from interacting compact systems.

To understand the source, the sun, we have to study the sun's interior, core, magnetic properties, cycles, twist of magnetic field lines, and how energy is transported from the interior through the atmosphere and the corona to the interplanetary medium. Coronal holes and magnetic field lines help us find the way, and we can discuss the life of a charged particle from the sun to the ignition of the auroral display.

We compare the sun with other active stars. We explain stellar evolution to show how stars interact with their environment, and lead up to the supernovae as the most dramatic space polluter. We also look at the powerful emitters in the center of galaxies producing quasars and gigantic cosmic jets. At the end of the course, we ask whether similar processes on an extraordinary scale were present in the moment of creation of the universe, when strings and superstrings may have been the prime actors on the scene.

Back on Earth, we study how auroral flames in the sky are interpreted through folkloric and scientific history. We find legends and tales, poetry and music, drawings and paintings, and certainly influence in the great epics of the people in the auroral zones. Students of art are invited to make projects concentrating on these aspects. Studying the history of science and asking how scientists interpreted the aurora during different periods is a fascinating topic by itself. Aristotle, Kepler, Galileo, Halley, and other famous astronomers and physicists had explanations of the aurora that reflected the development of science in their particular time in a very illustrative way.

In the future, we may demonstrate how it is possible from space to investigate the particles blowing from the sun and to check how the turbulence — which can vary from a gentle and steady solar breeze to a hurricane — can create problems for future space habitats. We can also study the aurora itself as seen from above, and even detect in it the daytime (in the ultraviolet). It may also be possible to find out if energy from space deposited in our atmosphere, through the action of our magnetospheric dynamo, affects the lower part of the atmosphere and the weather and climate on Earth. At this stage in the course, we have an excursion to a nearby rocket launching facility to see how rockets can be used to explore the atmosphere up to the height of satellites.

When we finish this journey searching for the explanation of the aurora, and its relatives on other energy scales, we find that we have been through most of the universe and a fair part of the physics known today.

3. Related Activities

Based on our experiences from this course, we are in the process of providing study material for high schools, in which auroral physics may be introduced as a special theme in physics. For the general public, and for the visitors to the region (who mostly come in the summer to see the fascinating midnight sun), we are in the process of construction a planetarium. In it northern lights will be demonstrated and explained, and related to the universe as described above.

Additional Reading

Seymour, P.: "Cosmic Magnetism," (Bristol and Boston, Adam Hilger) 1986.
Solheim, J.E.: "The Connection between the Earth and the Universe made visible by the Northern Lights," GIREP Conference 1986, Cosmos — An Educational

Challenge, European Space Agency SP-253, 103-111.

Video showing the real movements of the aurora. (27 minutes in color): (NTSC): Aurora color television project, Room 413, University of Alaska, Fairbanks, Alaska 99775-0800, U.S.A. (PAL): Tromsø Museum, N-9000 Tromsø, Norway.

Discussion

J.L. Dunlap: *I have wondered if the Northern Lights have been mentioned or explained in myth or legends of the northern cultures? (Is, perhaps, the description of Grendel in Beowulf an allusion to the phenomenon of the lights, and the death an attempt to end fear of the phenomenon among the people?)*

J.E. Solheim: Among all people in the north we find myths and legends related to the northern lights. In most cases it is connected with fear, but some places (Scotland, Sweden, and some Indian tribes) also connected with dances performed by heavenly creatures. It is not easy to connect the story of Grendel in Beowulf with the northern lights — since Grendel is a sea monster.

ASTRONOMY FOR BUSINESS STUDENTS: Space Industrialization and the Commercial Potential of Space Technology

James R. Philips
Math/Science Division, Babson College, Babson Park,
Wellesley, Massachusetts 02157, U.S.A.

When teaching science to nonmajors lacking an interest in science, two major goals are to stimulate their interest and to provide these students with information and scientific skills useful in their lives and careers. Business students now comprise over 23 per cent of the undergraduates in America, and they generally view science, including astronomy, as not relevant to their lives and careers. I find the students entering my introductory astronomy course for business students expect a pictorial tour of the universe, and are unhappy when asked to calculate redshifts in the laboratory or to attend an extra class meeting for telescope observing. Astronomy is not what they have come to a business college to learn, but they have a laboratory-science requirement to fill for their degree.

The challenge is to educate these future corporate leaders so that they understand how science influences business, and can deal with the problems resulting from the rapid pace of technological innovation in the marketplace. Nonmajors' textbooks in astronomy, as well as in other sciences, are written to appeal to a diversity of audiences, not one specialized group, and lack coverage of material on astronomy and business. If the key ideas in the textbook are perhaps 90 per cent of the introductory astronomy course material, then the remainder of the topics covered can be designed to interface with specific student interests. Business students can be acquainted with the ongoing efforts towards space industrialization and the commercial potential of space and space technology. Inclusion of some of this material significantly helps motivate these students to learn the fundamentals of astronomy to which most of the course is devoted.

NASA has for a number of years issued an annual publication called "Spinoff" that describes secondary uses in terrestrial manufacturing, agriculture, transportation, medicine and other areas for space technology developed by NASA. NASA's Office of Commercial Programs, however, realizes that the commercial potential of space extends far beyond secondary applications of devices NASA develops for its own needs. Private companies have Joint Endeavor Agreements with NASA, in which NASA provides shuttle flight time and the company plans the research experiment. Space represents an environment where some unique products can be manufactured, and where other products can be made purer or faster than they can on Earth. Besides space manufacturing, other areas of commercial opportunity include space telecommunications, remote sensing, and space support services such as transportation.

The shuttle disaster stopped almost all American space commercialization efforts. Space telecommunications remain today the main sector of commercial activity in space. The International Telecommunications Satellite Organization (INTELSAT) was formed in 1964 and now has a membership of 109 nations with 14 satellites carrying telephone and television signals (Pollack and Weiss, 1984). The newest series of satellites, designed and built by Hughes Aircraft Co., will each have the capacity to carry 120,000 telephone calls and at least three television channels simultaneously. Satellite-based systems are being developed for tracking and communicating with mobile receivers, such as trains and trucks (Beardsley, 1988).

The communications and direct-broadcast satellite industries are generally profitable but are nearing the saturation point (Horgan, 1988). Modern satellites can last over 10 years, and the most desirable spots in geosynchronous orbit are being taken (Gwynne, 1986). Competition from fiberoptic cables continues to increase, and point-to-point communications on the ground are getting cheaper. The explosions of the launch vehicles, and failures of boosters, have caused insurance companies to incur such large losses that very few companies will insure satellites at any price. And without insurance, you can't borrow money from banks. Claybaugh (personal communication) forecasts that the overall telecommunications satellite market will remain constant to the year 2000.

The first weather satellite was launched on April 1, 1960. Important applications to oceanography, hydrology and agriculture have been developed from weather satellite data, as well as improved meteorological forecasts (Yates *et al.*, 1986). Remote sensing of terrestrial features, such as classes of minerals and chlorophyll, biomass, and water content of forests can save industry millions in exploration costs and provide invaluable resource management information (Botkin *et al.*, 1984; Graff, 1985; Greegor, 1986). In 1985 the U.S. government turned its terrestrial remote sensing Landsat satellites over to EOSAT, a joint venture of Hughes Aircraft and RCA. Revenues in 1986 were only a little over $15 million, and there is doubt that a commercial Landsat program is feasible. The French SPOT Image Company provides stereoscopic pictures with much better resolution than Landsat, and is backed by the French government. But the biggest economic problem is the small market compared to the satellite costs.

Space-support services range from providing rockets for transportation to electrical power for operations in space or laboratory and manufacturing facilities there. The loss of the shuttle left the U.S. government with insufficient launch capacity, and private companies see the opportunity. Martin Marietta will launch a Federal Express communications satellite in 1989 (Anonymous, 1986). Orbital Sciences Co., and Hercules Aerospace Co. are developing a new commercial launch vehicle (Waldrop, 1988). Other companies are developing their own vehicles. Hawaii is examining sites suitable for becoming America's first commercial rocket-launching facility (Anonymous, 1988). The large U.S. satellite launching market may be temporary, however. China's Great Wall Missile Company promises to beat any price for launching (Gwynne, 1986).

Most other space support services remain many years away. The Industrial Space Facility, an automated materials processing center proposed by Space Industries Inc., has become a subject of Congressional debate (Marshall, 1988). Space Shuttle Co. of America wants to be the first power company in space (*USA Today*, 16 Sept. 1985).

Materials processing and manufacturing in space is extremely costly because of the very expensive hardware involved plus the launch costs. The minimum value for profitable space processing is over $10,000/lb ($20,000/kg). Space provides some unique benefits, though: containerless processing, and minimal buoyancy-driven convection and gravity induced separations or interferences with separations. Thus drugs and semiconductor crystals for electronics can be made purer in space, and new alloys can be made (Anonymous, 1983; Todd, 1985; Ben-Aaron, 1986; Boudreault and Armstrong, 1988). In fact, the first space product is on the market — latex microspheres for measurement. Specimens can be purchased from the National Bureau of Standards for classroom demonstration (Standard Reference Material 1965, $77.00).

There are many entrepreneurial opportunities in space. The Celestis Corp. offers a space mausoleum (Waldrop, 1985). We may soon need space trashmen because of the pollution (Scheraga, 1986). And, in the future, one can envision the commercial potential of mining the moon, Mars, and asteroids. Even for companies with no plans for space manufacturing or service, microgravity research can increase the understanding of terrestrial manufacturing science, by providing insight into the effects of gravity on the process. John Deere Co. has done experiments with NASA in space related to steel production — to see how terrestrial steel production is affected by gravity and if it can be improved. Although space industrialization is still in the early stages, and is crippled by launch costs and vehicle unreliability, its progress will continue. Chase Econometrics once calculated that each dollar spent in space had returned four dollars to our national economy in new jobs and new products (*Omni*, March 1987, p. 6). Business students who take an astronomy course will be better prepared to anticipate the space-related technological innovations that will affect our future, and to take advantage of the entrepreneurial opportunities that will arise in space.

References

Anonymous, 1983. For industry, it's almost lift-off time. *Business Week*, 20 June, 62-65.

Anonymous, 1985. Microgravity...a new tool for basic and applied research in space. *NASA EP-212*, 29 pp.

Anonymous, 1986. Big booster makes good. *Time*, 15 Sept., 57.

Anonymous, 1988. Spaceport Hawaii? *Sky & Telescope*, **75**, 133.

Beardsley, T. 1988. Messages from on high. *Scientific American*, **259**, 112.

Ben-Aaron, D. 1986. Growing crystals in space. *High Technology*, **6**, 11.

Boudreault, R., and Armstrong, D.W. 1988. Space biotechnology: current and future perspectives. *Trends in Biotechnology*, **6**, 91-95.

Botkin, D.B., J.E. Estes, R.M. MacDonald, and M.V. Wilson. 1984. Studying the Earth's vegetation from space. *Bioscience*, **34**: 508-514.

Graff, G. 1985. Prospecting from the skies. *High Technology*, **5**: 48-56.

Greegor, D.H., Jr. 1986. Ecology from space. *Bioscience*, **36**: 429-432.

Gwynne, P. 1986. Rethinking space business. *High Technology*, **6**: 38-45.

Horgan, J. 1988. Star-crossed. *Scientific American*, **258**: 37-39.

Marshall, E. 1988a. The space shack. *Science*, **239**: 855-857.

Pollack, L., and H. Weiss. 1984. Communications satellites: countdown for INTELSAT VI. *Science*, **223**: 553-559.

Scheraga, J.D. 1986. Pollution in space: an economic perspective. *Ambio*, **15**: 358-360.

Todd, P. 1985. Space bioprocessing. *Biotechnology*, **3**: 786-789.

Waldrop, M. 1985. Ashes to ashes — to orbit. *Science*, **227**:615.

Waldrop, M. 1988. Pentagon boosts a small rocket. *Science* **240**:1396.

Yates, H., A. Strong, D. McGinnis, Jr., and D. Trapley. 1986. Terrestrial observations from NOAA operational satellites. *Science*, **231**: 463-470.

THE THRILL OF DISCOVERY[1]

Gerrit L. Verschuur
4802 Brookstone Terrace, Bowie, Maryland 20715, U.S.A.

In the process of communicating to the student or the lay person, we often forget to point out why it is that scientists are involved in their esoteric explorations of the universe. As a result, the scientific endeavor is painted as being somewhat removed from "real life." This unfortunate misconception can be remedied by discussing the thrill of discovery.

Scientists love to "boldly go where no man has gone before."[2] The act of discovery produces a heady sensation akin to the thrill of victory for an athlete. The thrill may come as a surprise, as the result of a serendipitous turn of events, or may be experienced at the conclusion of a dedicated piece of research whose end-point was always understood. In either case the joy is associated with the revelation of what was previously unknown or not understood. It is this joy that is the reward for the researcher.

Myth would have us picture Archimedes in his bathtub solving the problem of why objects float. He rushed into the street, naked, shouting "Eureka! I have found it!" Witnesses concluded that he had "lost it," but the moral of the story is clear. This was his moment of joy that accompanied insight, a moment associated with the thrill of discovery.

Abraham Maslow, the psychologist, has described a "peak-experience" as "...the single most joyous, happiest, most blissful moment of your whole life."[3] This is precisely how the astronomers I interviewed described the thrill of discovery. Comparison of the two phenomena suggests they are indistinguishable. Maslow described peak-experiences in the context of esoteric pursuits where a deep insight, a "discovery," may sometimes even lead to the founding of a new religion. The scientist who has a profound insight cannot run outside, proclaim to all and sundry that he has "seen the light," and then expect his colleagues to "believe." He has to calm down and dispassionately report what he has just understood. It is his duty to communicate the essence of his discovery in order to show others that it was meaningful and adds to the fund of shared scientific knowledge.

The thrill of discovery can be experienced in a different context, as an astronomer/teacher once reminded me. "The first time it happened to me, even before it happened in my research career, was when I had to teach. Every time I understood another piece of what I had to teach, I felt elated."

The thrill of discovery, of experiencing insight or understanding can propel someone into a state of ecstasy, euphoria, or elation that may last minutes, weeks, or

[1]An expanded version of this talk appears in *Interstellar Matters* (Springer Verlag, 1988).
[2]Star Trek.
[3]Maslow, Abraham H., "Music Education and Peak Experiences," *Journal of Humanistic Psychology*, **2**, 1962.

months. The memory lasts forever. Once someone has tasted the thrill of discovery, he/she will devote the rest of his/her life to the pursuit of more of the same, and be willing to do so with amazing dedication. Is this not a benign form of addiction?

A heartening fact for the student who has not had the experience is that achieving this thrill does not depend on having lots of qualifications, training, or even brilliance. It is given as a reward for curiosity.

An astronomer recalls a night when he was 12 years old. He was looking at Saturn through a small telescope. The night was partly cloudy and the planet appeared but briefly. As the child gazed upon the magnificent rings, he suddenly felt that he was the only person in the world who, in that very instant, was looking at Saturn. It matters not whether he really was; what matters is that he believed it. He described the sensation. "I suddenly felt one with the Earth and the universe. There was nothing else." A wave of elation and euphoria swept over him and decided his career. This was the peak-experience in full cry. "I knew then that I would become an astronomer, no matter what it took."

Maslow reported that peak-experiences have the characteristics traditionally associated with moments of deep religious insight, as culled from the recorded history of many creeds and faiths.[4] The great difference between the scientific endeavor and the esoteric tradition is what the experiencer does *after* having the sensation. For the esoteric and his followers it is enough for him to say that he has "seen the light." He does not have to prove anything to anyone. His followers need only believe. But the scientist must explain what the light has revealed. Otherwise the experience has no significance beyond the personal thrill induced. Scientists must find the words to enable them to report what it is they understood or discovered so that others may achieve an equivalent level of understanding. Thus science as a discipline moves forward, evolving through sharing, while religions and sects remain static, rooted in the words of a founder who did not have to "prove" anything to anyone.

An astronomer recently spent years analyzing data related to the structure of the spiral arms in the Milky Way. Day after day he worked at it. Then suddenly it happened! "One day I heard the Galaxy," he told me excitedly. "I hesitate to admit this, but I heard the music." He stared at me, wondering if I would think him crazy. "I could hear the music of the spiral arms. They have motion within them and I could hear it! It was an incredible feeling. I understood!" He confessed that it was the greatest moment of his life, could not tell his colleagues, and wasn't optimistic that anything as wonderful would ever happen to him again.

Then he had to confront the astronomer's next great challenge. After the elation wore off he had to write a report on his research. Today his neatly rational paper on the structure of the spiral arms rests between the covers of a journal in astronomical libraries throughout the world. The report makes no mention of his moment of profound insight, nor of the music.[5]

---

[4] Goble, Frank G. *The Third Force.* Grossman Publishers, 1962. New York.

[5] The reader will appreciate that most of these "confessions" were obtained with the promise not to reveal names, a sobering comment on our shared attitudes toward the conduct of scientific research and the image that researchers find they need to uphold even in the eyes of their peers. This topic surely deserves further, more formal, research.

Frank Goble noted that "the peak-experience may be brought on by many causes: listening to great music, a great athletic achievement, a good sexual experience, even dancing."[6] And, we must add with considerable emphasis, by the thrill of scientific discovery, by the satisfaction of curiosity that accompanies understanding, by the completion of a piece of research. Seeking this feeling may be the strongest driving force behind the scientific endeavor, but it is never mentioned, because, as Maslow put it, the experience is intrinsically "unscientific."

Recently, astronomers at the Naval Research Laboratory in Washington, D.C., reported the discovery of a fascinating new phenomenon. A dark, unseen, mass had moved in front of a distant quasar and interfered with the radio waves on their way to Earth. A few months later everything was back to normal. The moment of recognition of this phenomenon (called an extreme scattering event) is etched vividly in the mind of Dr. Ralph Fiedler. It was a moment that culminated months of hard work for him. "I was elated and ran screaming and shouting down the corridor," he says, expressing a feeling to which many a researcher can relate. But his elation had to be tempered so that he and his colleagues could take up the challenge of explaining what had been observed. And when the explanation was found, the thrill returned; it was a little less intense perhaps, but still a delight. (The dark masses responsible for the phenomenon are suspected to be diffuse clouds of hot gas, having planetary dimensions, that are more numerous than stars in the galaxy.)

The discovery of a supernova, the explosive death of a star, has provided a rare thrill for only a handful of human beings in recorded history. Not since the times of Tycho Brahe in 1572 and Johannes Kepler in 1604 had anyone actually seen a supernova with the unaided eye; not until February 24, 1987, that is. On that night, at Las Campanas Observatory in Chile, astronomer Ian Shelton discovered a star in the act of exploding.

"I don't think anything is going to replace that night of actually seeing it for the first time. That was memorable." These were Shelton's words on the *Nova* television program describing the discovery of Supernova 1987A.

The thrills and the rewards that accompany scientific research are personal and scientists have conspired mightily to keep this a secret. They have come to be pictured as slightly inhuman beings in white coats pursuing lofty goals for reasons no one understands. Yet they are human, and the truth is that scientists "get off" on what they do.

Everyone can experience the thrill of discovery, even if you don't have a Ph.D. All you need do is exercise your curiosity and be willing to recognize that there are things in nature worth discovering for yourself. So why not say *yes* to curiosity? Embrace those moments when questions presents themselves. Seek solutions and find out how beautifully elegant the world of nature is. Prepare to enjoy the thrill, the "high" that is the reward. It could present you with the "single most joyous, happiest, most blissful" moment of your life.

[6]Goble, Frank G., *op. cit.*

Discussion

C.R. Chambliss: *It is extremely important in introductory astronomy courses to discuss scientific methodology. What is science? How does the scientific method work? The pseudo-scientific natures of such topics as astrology and creationism should be explained. Important concepts such as the falsifiability of a theory should be carefully explained. A good scientific theory, of course, is one that is falsifiable (i.e., testable) but that has not been falsified.*

G.A. Carlson: *I share with my students a comment I read that a high percentage of Nobel prize winners and of science fiction writers in past decades were "turned on" by seeing Halley's Comet in 1911.*

W. Osborn: *Do you feel that this "thrill of accomplishment" is one of the purposes of problem sets, such as in physics courses?*

G.L. Verschuur: Absolutely. Or at least it should be. Too often this is forgotten.

H. Shipman: *As you were talking, I asked myself, "how do I share these feelings with my students? Quotations from unnamed sources in a book just won't do." The American Institute of Physics (335 East 45th St., New York, N.Y. 10017) sells an audiotape, recorded at the telescope, of the discovery of optical pulses from the Crab pulsar, including the astronomers' reactions: "It's a bloody pulsar!!!" (Emphasis in the original.) This tape, and the few others like it, would be good sources.*

J. LoPresto: *Such excitement can be realized by beginning students with the use of a commercially available Hα filter (DayStar Filter Corp., P.O. Box 1290, Pomona, California 91769, U.S.A.) to be used with small telescopes to observe the sun.*

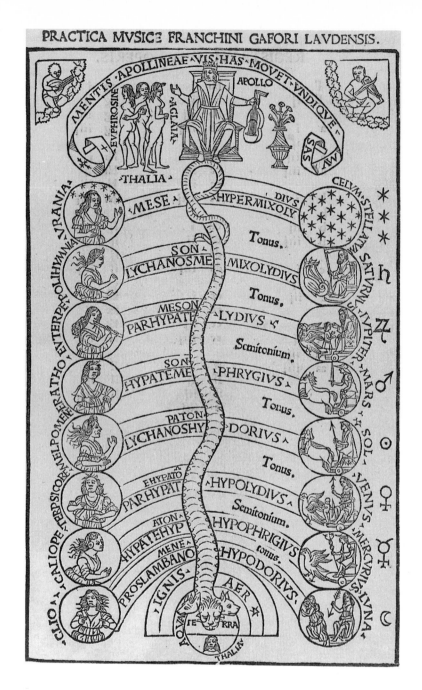

Music of the Spheres, artist unknown; Title Page of *Practica Musicae*, by Franchino Gaffurio. Chapin Library of Rare Books, Williams College. Further information on page 433.

Chapter 3

The Teaching Process

Many non-science students enjoy and benefit from astronomy courses. Papers in this chapter discuss improving teaching abilities and handling and testing students in the large general astronomy classes. Large courses for non-science students are prevalent in North America. One paper deals with a summer institute for training teachers of such courses, primarily in small institutions. We also hear of "education at a distance" — teaching of astronomy by correspondence to students in the United Kingdom and the United States. We finally read more philosophical discussions on how to structure an astronomy course.

STARTING OUT: THE DILEMMA OF THE BEGINNING COLLEGE ASTRONOMY TEACHER

Harry L. Shipman
Physics and Astronomy Department, Center for Teaching Effectiveness,
University of Delaware, Newark, Delaware 19716, U.S.A.

Most beginning college professors in the United States receive little if any help in the important task of learning how to teach well. Occasionally, a colleague or department chairman can provide some guidance, but in general most American college faculty are taught how to teach by the "sink or swim" method. There are many resources that can help the new professor swim, rather than sink, but in my case at least, I found out about them many years after I needed them the most. This paper will identify some of the literature, journals, and on-campus facilities that can help both new and experienced faculty members develop and hone teaching skills.

The literature on teaching and learning is vast. Fortunately, the new teacher need not spend years becoming an expert on this literature; much of it is like the astronomical research literature in that it is full of jargon and it requires some expertise to distill useful, easily implementable ideas from it. Fortunately, there are a few (too few!) books that distill this vast literature into practical advice from it.

Wilbert McKeachie's *Teaching Tips for the Beginning College Teacher* (McKeachie, 1986), now in its eighth edition, is a classic. I think all new faculty members should be given this book along with their first paycheck. Another book along the same lines is *The Art and Craft of Teaching* (Gullette, 1982). Chapters in these books carry titles like "Countdown for Course Preparation," "The First Day of Class," "Stimulating Discussion," and "How to Win Friends and Influence Janitors." You can read these chapters and implement the ideas immediately in your class.

Familiar physics teaching journals such as *The Physics Teacher* and *American Journal of Physics* often contain articles on astronomy teaching. *The Teaching Professor*, a newsletter, may be less familiar; like the books cited above, it contains short, easily implementable articles on generic teaching problems. For example, the May 1988 issue contains a one-page article on "Questions: Making Learners Think." It provides examples of "higher order questions," question that require students to think rather than to memorize, questions that can provide a way of keeping a discussion section moving along rather than degeneration into a small-sized lecture. A 12-issue subscription costs $39; write to 2718 Dryden Drive, Madison, Wisconsin 53704. *College Teaching* is another good general resource.

Many colleges and universities have set up local organizations to foster better teaching on campus. The University of Delaware's Center for Teaching Effectiveness has a part-time faculty director, two professional staff, and a graduate student. A major focus of our activities is confidential consultation with individual faculty members. We give grants to faculty for teaching improvement. We run a series of panel discussions, presentations, and workshops; topics range from those addressed primarily at beginning teachers (for example, "Lecturing Well") to topics of more general interest (for example, "Balancing Teaching and Research"). I find these programs interesting both because of their content and because they allow me to meet other people on campus who really care about teaching. We can make arrangements to videotape classes. I found this useful both as a beginning teacher and even after I'd been teaching for years. It's useful to watch the tape with someone else around. Another way of getting feedback is to talk with your students or ask them to give you written, anonymous feedback a few weeks into the term (not just at the end).

It helps to learn a little about students' learning and reasoning styles. Ten years ago, a number of American physicists discovered Piaget's learning theories (Fuller, Karplus, and Lawson, 1977). In some extreme cases, these ideas were overpromoted as the "magic bullet" that would solve all the problems of physics teaching. They didn't, of course. However, students don't think in the same way that their teachers do; the primary reason that universities exist is to transform people from memorizers into thinkers. Understanding the differences in thinking and problem solving styles can help bridge some communications barriers between teacher and student. Because this is an active area of current educational research, there are no authoritative, compact books like McKeachie's. Beginning and experienced teachers might find articles by Fuller, Sheila Tobias (1985) on math anxiety, Jill Larkin on problem solving styles, and a forthcoming book by Kurfiss (1988) to be good starting points.

Especially in the current student climate, the examinations given in a course communicate a teacher's expectations far more powerfully than anything that's said in class. Science teachers like to emphasize thinking and problem solving, but American astronomy teachers then proceed to give their students exams which, if the questions in instructors' manuals for textbooks are any guide, almost always ask for simple memorization and recall. We can do better. Educators have developed ways of constructing more sophisticated multiple-choice questions; Ebel's *Essentials of Educational Measurement* (Ebel, 1986) contains some hints, though it has many of the flaws of educational writing at its worst. (Skip the first half.) One tactic I use, for example, is to ask students to demonstrate that they know the connection between observations and the conclusions that are drawn from them. For example, "Most objects in the Universe contain 25-30% He by mass. This observation shows that (i) the Big Bang theory is probably correct, (ii) ...." If the multiple choice questions are machine graded, software packages like ITEMAL (available for most mainframe systems) can analyze the examination and pick out those disastrous questions that the weak students answer correctly and the stronger students answer incorrectly.

Because astronomy is such a visual science, slides, videos, and movies are an excellent resource. I usually obtain slides from the Hansen Planetarium, the Astronomical Society of the Pacific, or the European Southern Observatory. Videos are a bit harder to find; the Planetary Society (65 N. Catalina Ave., Pasadena, CA 91106) has a few. NASA maintains a number of teacher resource centers in the United States where a visiting teacher can bring a blank videotape and copy anything that is available at the Center. Movies can also be borrowed from NASA, though the use of this medium seems to be fading fast.

References

Ebel, R.L. 1986, *Essentials of Educational Measurement*, 4th ed. (Englewood Cliffs, New Jersey: Prentice-Hall, 1986).

Fuller, R.G., Karplus, R., and Lawson, A.E. 1977. *Physics Today* **30** (February 1977); 23-29.

Gullette, M.M. (ed.) 1982, *The Art and Craft of Teaching*, (Cambridge, Mass.: Harvard-Danforth Center for Teaching and Learning.)

McKeachie, W. 1986. *Teaching Tips for the Beginning College Teacher* (Lexington, Mass.: D.C. Heath Co., 125 School Street.)

Tobias, S. 1985. *Physics Today* **35** (June 1985); 61-68.

Discussion

R.J. Dukes, Jr.: *If facilities or time do not permit video taping, audio taping using an inexpensive recorder can prove helpful in analyzing what has happened in your class.*

O. Gingerich: *Videotaping class sections has some unanticipated advantages over audiotaping. For example, one of our best teaching fellows discovered she had id-iosyncratic habitual motions that she had been entirely unaware of. Another was very surprised to see the expressions on the students' faces when he was writing on the blackboard with his back to the class.*

H. Shipman: I agree; videotaping should be done if available. Audio tapes are better than nothing.

TEACHING 7000 STUDENTS PER YEAR

R. Robert Robbins,
Department of Astronomy, University of Texas, Austin, Texas 78712, U.S.A.

The undergraduate program at the University of Texas has grown into the largest astronomy teaching program in the world, with some 7000 students per year (almost 20,000 credit hours). The department has 22.5 Ph.D.-level teaching faculty, about 45 graduate students, and about 40 pre-professional undergraduate majors. But most of the enrollment is in courses that satisfy the science requirements of students in liberal arts and non-technical majors. In 1985–86, 96.4 per cent of our undergraduate credit hours taught were in such classes. It is instructive to examine the historical reasons for our growth and its educational consequences, and to draw some conclusions from both for other programs.

The Early Growth of the Department: "Reclaiming" the McDonald Observatory

Up until 1958, there was no astronomy department at all at the University of Texas — only the McDonald Observatory, named for the Texas banker who left the money to build it in 1926. It was run by Yerkes Observatory, but when efforts began to bring the University of Texas into the first rank of academic institutions worldwide, it was decided to develop a competitive, major astronomy department in Austin and "reclaim" the observatory from the University of Chicago. To this end, in 1963, Dr. Harlan Smith was brought in to be both the director of the observatory and the chairman of the new department; shortly thereafter, the 2.7-m (107") telescope was begun at the observatory. (For more details, see *Big and Bright: A History of the McDonald Observatory*, by David Evans and Derral Mulholland (University of Texas Press, 1986)).

The original goals called for the new department to grow to about 12 faculty members, a number judged appropriate to support a major observatory. However, it has now grown to almost double this targeted size, and we will see below that undergraduate teaching has been responsible for this.

The "Post-Sputnik" Degree Plan at the University and the Influence of the "Space Age"

The figure shows that student enrollments did not begin to increase significantly until the school adopted a new degree plan in 1967, requiring *15 hours of science* in every major at the University. This is one of the strongest science requirements in the country, and it immediately began to generate growing enrollments in the service courses of all the science departments. Astronomy became the science department that was most successful in acquiring new enrollments, no doubt partially aided by the well-publicized space program of the United States in the 1960's and the 1970's, which directed the attention of much of the world towards astronomy. (Note: the University grew by about 50 per cent over this period, not enough to explain our huge growth. And grade statistics show that the enrollments cannot be attributed to easy grading.)

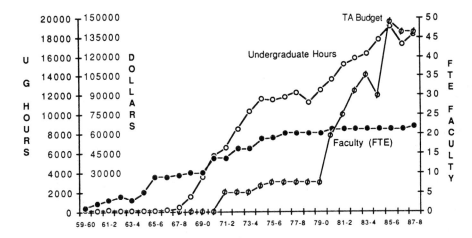

Fig. 1. Credit hours, teaching assistant budget, and number of faculty in the astronomy department at the University of Texas.

"Intangible" Factors Creating Sustained (*i.e.*, Subsequent, Continuing) Growth

It must be noted that many other large universities also had the potential to use the widespread interest in astronomy to build up sizable departments, but in reality it did not happen very often. To explain our continued growth over two decades, I feel that it is necessary to turn to an *intangible* factor.

Many astronomy departments concentrate their resources almost exclusively at the graduate level. At the University of Texas, it was decided early on to also put significant effort and funding into undergraduate programs for non-science majors. It began with a "critical mass" of the early faculty members who were interested enough in undergraduate education to influence the subsequent development of the department. In the late 1960's, this group included Dr. Smith and me, and also Frank Bash and Bill Jefferys. An overall commitment of the whole department did not occur all at once, but developed over time. But one can isolate some of the positive factors that grew from this departmental commitment:

1. *All the faculty teach classes for non-science majors* at least once a year. All the faculty take the needs of the program seriously, since they are all involved in it.

2. The first class in astronomy (for non-science majors) is a *very accessible, general non-mathematical survey of the universe*, covering the contents of the universe and the astrophysical models that modern astronomy constructs from its observations.

   The lack of math is simply an unfortunate necessity, because American college students are typically not prepared to handle any level of mathematics at all. We have decided that we cannot right all the wrongs of the American secondary school system in one course, but we can give the students a powerful contact with the modern universe, an exposure that in many cases may significantly affect their lives. The lack of math does *not* necessarily imply a substandard course, because the high degree of abstract thinking and reasoning skills that are demanded by the sophisticated syllabus of modern astronomy guarantees that the course will be one of the more challenging in the undergraduate program (See Robbins, *Annals of Engineering Education*, Dec. 1981, p. 208, for a study of the factors that determine student success in introductory astronomy).

   The introductory course is presented in three modes: (i) A conventional lecture with an optional 1-unit lab. However only about 5 per cent of the students elect the lab. (ii) A lecture class with observing activities blended into it, that is, it has a lab component integrated into the big class itself, and (iii) a large self-paced (no lectures, Keller-method) reading class taken by about 900 students per year. These options provide variety for students with different types of study skills.

   The first class is then followed by an attractive smorgasbord of followup classes. About 70 per cent of our introductory students go on to a second semester of astronomy, so this approach is obviously successful. These classes are also non-mathematical, but they develop a particular area of astronomy in more depth. Some examples are: *Man in the Solar System; The Search for Extraterrestrial Life; Stars and Stellar Evolution; Astronomy in Science Fiction; Galaxies, Quasars, and Cosmology, The Milky Way Galaxy; Archaeo- and Ethno-Astronomy of the Americas;* and *History and Philosophy of Astronomy*. These courses are designed by the various faculty in their areas of greatest interest, and this freedom seems to result in follow-up classes taught with enthusiasm.

3. *We have consistently given time and money to a vigorous outreach program* that produces brochures, film series and star parties, and other programs of high visibility, and funds an educational services office with a staff of 3 people to

handle public relations and assist with the audiovisual and other needs of large classes.

A central lesson to be drawn from our Texas experience is that to develop a significant undergraduate enrollment, the decision must be consciously taken that undergraduate education is important and should be encouraged and supported. It doesn't happen by itself.
The graph demonstrates that our growing enrollment did create a momentum that carried the department past its targeted size of 12 faculty members to almost twice as large. But it also shows that since the mid-1970's, increased enrollment has not generated new faculty positions.

Our "Mature" Department and Its Natural Enrollment Limit

Today, most of the students who are allowed to take astronomy by their degree programs do so. We are quite close to the maximum possible enrollment. But we are also up against another limit, which is (1) the number of classes we teach *times* (2) the number of students that will fit into our largest lecture hall (230). The only way to expand further would be to acquire more faculty or to seek out larger classrooms. I do not know of any research that settles the question of what is the optimum size for a class, but by anyone's standards, ours are all quite large already!

Does the increased enrollment inevitably result in increased budget and faculty lines? The Texas experience suggest "Yes, but..." since there may be a considerable time lag (*i.e.*, years or decades; see the graph) before the support appears. There will typically be no immediate reward for "overteaching."

Some Educational "Pros" and "Cons" of Our Size

An advantage of large programs is that they have a great diversity of offerings to select from, to appeal to the wide range of interests that students can exhibit. But many of the drawbacks to a very large program are fairly obvious to any experienced educator.

1) Evaluation and Grading: Evaluating huge classes can become very time-consuming. Most instructors of very large classes eventually turn to multiple-choice exams. I do not intend to present a point of view on what form of testing is superior; I am simply pointing out consequences.

2) Labs and Observations: Although a "lab" experience is probably the best way to teach astronomy, at Texas fewer than 5 per cent of the students get one. This is the principal source of disappointment noted on student evaluations. It is simply not possible to offer an intensive laboratory class for as many students as we have; we do not have the rooms, equipment, or instructors. One lecture section does offer measurement experiences to its students, but the instructor teaching that class must replace conventional homework assignments with observing activities, and many instructors prefer not to do this.

3) Less personal contact with the instructor: Students will most commonly interact personally with a Teaching Assistant. The graph shows that funding for TA positions is the only factor that has tried to keep pace with the growing enrollments. We generally employ about 20 TA's per semester, at 20 hours per week (half-time), which still only results in about 1.5 TA's for each class of 230 students.

A considerable problem for the teaching program is that the research programs at Texas will often hire most of the astronomy graduate students as Research Assistants. Thus, we must hire some of our TA's from other departments (physics, engineering), and to try to train them for the job. Clearly, such TA's will be much more limited in the duties they can carry out. Also, many of the astronomy TA's today are foreign students who often have such a poor command of the English language that they are quite restricted in their usefulness.

Summary

Whether or not you feel that the Texas experience is directly or only partially applicable to your institution, it does indicate the ultimate *potential* of an undergraduate program. If all U.S. colleges would develop programs comparable to ours, the job market for astronomy teachers would increase by at least a factor of 10! Imagine the effect of that on astronomy as a career.

In researching this paper, I personally was surprised by the importance of intangible factors, and most particularly, the importance of the *belief* that undergraduate education is important and worthy of effort. This factor explains why growth was allowed to continue, even when the tangible rewards from it were not always evident.

WORKSHOP FOR TEACHERS OF INTRODUCTORY ASTRONOMY

George S. Mumford
Department of Physics and Astronomy, Robinson Hall, Tufts University, Medford, Massachusetts 02155 U.S.A.

As interest in astronomy develops through missions to Mars, SETI, and heaven-only-knows-what earth-shaking new discoveries in the future, demand for astronomy courses at all levels will increase. Without adequate numbers of professional astronomers to teach them, persons from other fields will be thrown into the breech. Already a significant number of college students in the United States are receiving instruction from persons not trained in astronomy. I suspect that this is currently true world-wide, especially as physicists who adopt our field for their research on neutrinos or cosmic strings are assigned or volunteer to teach elementary courses.

Comprehensive universities in the United States, some private (such as Harvard or Princeton) and others public (such as the University of California system or the University of Michigan) generally have several astronomers on their faculties who provide an extended program of courses. At another level are the four-year colleges. Some, such as our host for this colloquium, Williams College, have well-established astronomy programs; however, many others do not. Finally, in the American scheme of higher education, there are the junior or community colleges — two year institutions offering programs to a mixed clientele ranging from eighteen-year-olds to senior citizens. Seldom does one find an astronomer here. But, astronomy courses are popular and in demand and many of these institutions offer them at the introductory level. Since there are few astronomers associated with these places, often a volunteer versed in another discipline assumes the teaching function.

With funding from the National Science Foundation's program for the Enhancement of Undergraduate Faculty, a two-week workshop was run this past June at Tufts University for persons with a minimum of three years of college-teaching experience who had not had graduate training in astronomy but found themselves teaching introductory courses. Of particular concern were persons from institutions that likely did not have access to modern observational facilities to train students. Despite the fact that announcements of the program did not go out until early April 1988, by mid-May there were 25 completed applications for 20 slots, plus inquiries from another dozen persons. The program was repeated in 1989.

Eleven of the applicants were from community colleges; five from public four-year colleges; and the rest from private four-year colleges. Thirteen of the applicants held Ph.D.'s: eight in physics; two in physical chemistry; and one each in chemistry, education, and science education. These degrees were from universities that are recognized sources of research-oriented doctorates. Three additional doctorates, not Ph.D.'s, were in education, while the remainder of the applicants held masters' degrees, including five in education or teaching. All degrees had been earned rather uniformly throughout the period 1960 to 1982. There was a comparable even distribution in the ages of the applicants.

In general, members of the group were conversant with modern textbooks, used many laboratory exercises, and had access to portable telescopes. Most, however, needed instruction in how to set up, orient, and use such an instrument; many were unaware of resources for slides, computer software and other materials. At a more mundane level, several consistently inverted the magnitude scale, leading to some peculiar looking color-luminosity diagrams; most had, at best, only a cursory knowledge of various astronomical coordinate systems.

Our intent in the workshop was to minimize lecturing and maximize hands-on activities that simulate some of the steps an astronomer makes in reducing data and obtaining a result. To accomplish this, a variety of projects were used including many of the *Laboratory Exercises in Astronomy* from Sky Publishing Corporation. Further, beyond having participants work their way through the various exercises, we hoped to get them to think critically about what they were doing. While some exercises are quite complete, many can be improved. Not only are data frequently out of date, but also accidental and systematic errors in the measurements are sometimes ignored. Participants were asked to think about such matters. For each of the activities he or she completed a brief questionnaire evaluating the materials used and suggesting improvements.

We are teaching young astronomers; we are teaching teachers; but this group has apparently escaped the net. How many does it contain? It's difficult to say until someone makes a comprehensive survey, but my guess is several hundred in the United States alone.

It seems to me that there are two rather different ways to approach the problem of having elementary astronomy courses taught by nonastronomers. On the one hand, a statement from the profession urging the employment of only those with graduate training in the field to teach in these programs would increase employment opportunities in academia for many graduates who cannot otherwise find jobs related to their discipline. But, while this might be possible in the long run, because of the economics of the situation an alternative solution is required. Simply stated, this alternative is to identify, instruct, and support those persons who are obviously interested in learning about their adopted field. The structures and mechanisms to accomplish this exist, and we should use them.

Discussion

R.R. Robbins: *In the mid-1970's, I presented NSF-Chautauqua workshops on the teaching of astronomy to just the type of participants you had, and the problems were the same. But I gave these to hundreds of teachers at 12 to 15 education centers across the country, and the demand was so great I could have done it over and over again at every location. Fifteen years later the problem has not changed at all! We must find some organized mechanism to update our teachers, or we will be meeting again around the year 2000 and noting that the problem is still with us.*

G.S. Mumford: This is the very reason why I presented this talk.

R.J. Dukes, Jr.: *I teach in a physics department of 3 astronomers and 4 physicists. All of the physicists have taught the introductory astronomy course. After 10 years of subtle observation, I have found that the physicists generally do as good a job on this course as the astronomers!*

J.C. LoPresto: *Many Planetarium directors have little or no background in astronomy.*

MULTIPLE-CHOICE QUESTIONS

Roy L. Bishop
Department of Physics, Acadia University, Wolfville, Nova Scotia, B0P 1X0, Canada

Although few people enjoy tests and examinations, these aspects of education appear to be essential to the maintenance of standards, and they provide a vital learning environment for both students and instructor. Opinions regarding "objective," or "multiple-choice," questions range from very positive, which is my own qualified opinion, to very negative. In the latter instance, some claim that such questions should *never* be used, and refer derisively to multiple-choice questions as "multiple-guess questions."

I shall first address the disadvantages of multiple-choice questions. I begin with a disadvantage that is not well appreciated:

1. It is difficult to make *good* multiple choice questions. The difficulties include the following:

   (a) One must try to avoid standardized questions and, especially, trivial questions.

   (b) It is not easy to arrive at unambiguous, concise wording, including all the necessary caveats.

   (c) Even in questions that are otherwise good, numerical answers that are incorrect can lead students astray.

2. Multi-step problems are difficult to present in a multiple-choice format.

3. Assuming the usual 5 possible answers, random guessing will guarantee, on average, a grade of 20 per cent. (However this grade is too low to be of great concern. Also, a marking scheme that eliminates this possibility would result in some students receiving negative grades! Further, guessing based on elimination of incorrect answers or on a "feeling" for what is correct, deserves some credit.)

What are the advantages of multiple-choice questions?

1. Good multiple-choice questions confront a student with several plausible and often subtle choices. Such situations are more like real-life than those proposed in the usual, essay-type questions where just enough information has been provided to allow the answers to be deduced. The necessity of sorting through plausible sounding alternatives probes a student's understanding in a special way. A good multiple-choice question goes beyond mere regurgitation. It can stimulate critical thought and/or require an awareness of the real world.

2. Multiple-choice questions make efficient use of testing time. Nearly all of a student's time is spent reading and thinking, rather than writing, and writing, ...and writing. Not only is this more efficient in probing a student's knowledge, but also more material can be examined in the time available.

3. There are some important practical advantages for the instructor:

   (a) Marking is easy and quick, a tremendous advantage when dealing with large classes. The total time saved is not as great as might first appear, however, for it takes much *more* time to make up good multiple-choice questions than essay-type questions. Nevertheless, making up questions is an interesting, creative endeavor compared to the drudgery of marking hundreds of essay-type answers!

   (b) The grade for each student is quantitative and unambiguous (provided the questions are good ones).

   (c) The instructor does not have to decipher students' writing! (Some, however, will rightfully bemoan the loss of a chance for students to practice written expression.)

4. Analysis of incorrect answers is relatively easy and can be quantified. Such analysis yields insights into the reasoning processes of students and suggests ways that teaching can be improved. For example, in the paper: "Constructing Objective Tests" by Aubrecht and Aubrecht (*American Journal of Physics*, **51**, *pp.* 613, July 1983), two indices of question quality are defined:

   "Difficulty Level": The average grade on any one question should be about 65 per cent. An average of 100 per cent or 20 per cent (in the case of a question with 5 possible answers) indicates a question useless for testing purposes.

   "Discrimination Index (DI)": Pick the top 27 per cent (=N) of papers ("highs") and the bottom 27 per cent (=N) ("lows"). For a given question, the number of highs that get it correct = H, and the number of lows that get it correct = L. DI is defined as (H-L)/N and ranges from +1 (perfect discrimination) to -1 (poor question!). DI should be at least 0.3.

I conclude by presenting 6 examples of multiple choice questions at the introductory astronomy level. The first 3 are poor, and the second 3 are good:

P1 Space debris is of most interest to astronomers for which one of the following reasons?

   1. It contains much gold and silver.
   2. It all came from the Moon.
   3. It all came from comets.
   4. It tells us about the early Solar System.
   5. It tells us about the last ice age on Earth.

Comment: The question itself is not good because the term "space debris" is not defined (fragments of satellites in near-Earth orbit? Interplanetary dust and meteoroids?). Also, answers (1), (2), and (5) should be obviously incorrect to even a relatively poor student, so a guess at the remaining answers would mean a 50% chance of choosing the correct answer.

P2 During which season of the year is Earth closest to the Sun?
(1) winter (2) spring (3) summer (4) fall (5) Earth is always the same distance from the Sun.

Comment: The instructor who composed this question has ignored both the opposite seasonal pattern in Earth's other hemisphere and the absence of mid-latitude seasons in the tropics.

P3 The surface temperature of Venus' atmosphere is most accurately described as:

1. as hot as the surface of the Sun.

2. twice as hot as Earth's.

3. about as hot as Earth's.

4. half as hot as Earth's.

5. less than a tenth as hot as Earth's.

Comment: This is an almost trivial question. Answer (1) is ridiculous, and a knowledge of the order of the planets from the Sun (which is taught at elementary school) would rule out answers (3), (4), and (5).

G1 Suppose you are in Williamstown. If, toward the northwest, you see a first quarter Moon near the horizon, what month is it?
(1) August (2) March (3) December (4) June (5) September

Comment: This question is not trivial, and it does not involve regurgitation of standard material. The student must know about: (i) the phases of the Moon and the associated Sun-Earth-Moon geometry; (ii) the inclination of the ecliptic; (iii) that the Moon's orbit lies (approximately) near the ecliptic; (iv) the possible range of compass directions of the intersection of the ecliptic with the horizon at mid-northern latitudes; and (v) how the Sun's position on the ecliptic is related to the time of year. Furthermore, the student must be able to correlate these various things and fit them to the question. In brief: A low Moon in the northwest means that the summer solstice point (most northerly point) of the ecliptic must be in that vicinity. The first quarter phase means that the Sun must be about 90 degrees further westward along the ecliptic ...near the spring equinox. Therefore the answer is option (2).

G2 In which stage of the life cycle of a single, isolated, massive star does it have the least mass?

1. a dilute cloud of collapsing gas and dust.

2. a T-Tauri type star.

3. a main sequence star.

4. a red giant star.

5. a black hole.

Comment: The wording of the question rules out significant accretion of material by the star from a dense region of the interstellar medium after it has formed. The answer (5) is obvious provided the student knows about energy/mass conservation and has an appreciation for a star: namely, that it continually loses photons, protons, electrons, helium nuclei, *etc.*, to the dark, cold void surrounding it. Vague misconceptions about black holes (5) can lead the less able student to avoid this answer.

G3 Most locations in space (*i.e.*, locations selected at random anywhere in the universe) receive an amount of light

1. much greater than daylight on Earth.

2. comparable to daylight on Earth.

3. comparable to moonlight on Earth.

4. comparable to night-time starlight on Earth.

5. much less than night-time starlight on Earth.

Comment: Although intentionally not obvious at first reading, this question is designed to see if the student has an appreciation of the large-scale structure of the universe: that it consists basically of dim galaxies scattered here and there in the darkness of intergalactic space, and that Earth is located well inside of a galaxy. If this appreciation exists, then the locations specified in the question will clearly be in intergalactic space, and so the answer can only be (5).

In conclusion, I should add that only the best students can handle questions like the last three. A test composed entirely of questions of this level would be a disaster (pun intended) for a majority of the students in an introductory, university-level astronomy course. Tests must have a mixture of simple factual questions, questions that require one step (or two) in reasoning, and a few questions that require more sophisticated patterns of reasoning.

Discussion

M. Zeilik: *Do you do an item analysis on your tests, and if so, what kind of reliability index do you get?*

R. Bishop: I determine the success rate for each question; that is, the fraction of the total number of students who answer correctly. A success rate near 6/10 or 7/10 indicates a good question: neither too difficult nor too easy. Questions below 5/10 are examined for clarity and level of difficulty. If they still appear reasonable, then they indicate topics that need extra emphasis.

I also examine the frequency distribution of the incorrect answers selected for each question. This reveals the sort of errors or misconceptions that students have (and will also reveal inadequate questions — fortunately this does not occur too frequently!).

In addition, for questions that are disasters (success rates of 2/10 or 3/10), I make particular note of the answers given by a few of the best students in the class. Their answers usually indicate that the question was not totally unreasonable, but sometimes their answers will reveal flaws in the question that had missed my attention.

Barrie W. Jones:

1. *At the Open University (UK) we carry out item analysis on every multiple-choice question and find now (with time and care spent on the question setting) a good correlation between a student's overall performance and his/her performance on individual questions.*

2. *We also correlate performance on multiple-choice questions with performance on conventional tutor-marked assignments. Again (with care and time spent on the question setting) we find a good correlation between the two performance indicators.*

William Waller: *I would like to point out the virtues of true/false questions. Like multiple-choice questions, they have the potential of being unambiguous, analyzable gauges of student understanding. The restriction on possible answers, however, helps to focus on what is important in the course. To my surprise, I have found that the true/false format can efficiently probe student understanding — even for astronomy majors in their senior year of study.*

Lon Clay Hill, Jr.: *How confident are you of the actual validity of test items when judged solely by intrinsic internal consistency tests (I have some data that suggest that regard for important naked-eye phenomena is* negatively *correlated with success with the more academic foci of the classroom.)*

R. Bishop: I am not certain what you mean by "intrinsic internal consistency tests." I try to examine *all* aspects of the course — theoretical and observational. Particular attention is given to questions that good students cannot answer "correctly," and to questions that *few* students answer correctly. Only after I have examined the answers of many students to a particular question can I rate that question as being "good" or "bad."

———————————

INNOVATIONS IN ASTRONOMY AT THE OPEN UNIVERSITY

Barrie W. Jones
Physics Department, The Open University, Milton Keynes MK7 6AA, U.K.

1. Introduction

Dreary correspondence colleges that send tatty typewritten notes to their suffering students are gradually being eclipsed by full-fledged open-learning institutions of wide educational significance. The Open University is the premier open-learning institution in the U.K., perhaps even in the world. It already offers an astronomy course as a significant part of a science major degree, and further courses are planned, including an in-service course for school teachers.

In this article I discuss:

1. open learning systems, including the ways in which various educational media can be used to good effect

2. some features of the existing astronomy course of The Open University, and some possibilities for future courses.

2. Open Learning Systems

There are still comparatively few open-learning institutions in the world, though their numbers are growing. They all differ strikingly from conventional educational institutions but nevertheless the materials they produce have been used successfully by conventional institutions, and many of their educational strategies and techniques have been put to good effect very widely in education, including within the third world. Let's look at some of this in more detail via the specific case of The Open University.

2.1 The Open University

The Open University (OU) was founded in 1969 for the specific educational purpose of providing adults with the opportunity to study at tertiary level in their own homes in their own time. It offers fully recognized bachelor's degrees, higher degrees, and a great variety of short courses. It accounts for nearly all of the open learning at tertiary level in the U.K., and currently it has 68,000 undergraduates. Over half of those who start on degrees obtain them.

The OU is *open* in four distinct ways.

(i)   To qualify for entry, students need only be 21 or over and be resident in the U.K.

(ii) Students can study almost exclusively in their own homes or workplaces.

(iii) Students can study in their own time and, within loose constraints, can progress towards a degree at their own rates.

(iv) Student fees cover only a small fraction of the tuition costs: an honors degree costs a student on average, about $4,000 at current prices. (The difference is provided by the U.K. government.)

The OU uses a rich array of media to educate its students, and a great deal of thought and production resource goes into each medium. *Table 1* lists most of the media that we use, and any particular course will use most of these.

Table 1. Media Used

Printed materials	- the main texts - supplementary materials	Face-to-face teaching	- small group tutorials - residential schools

Audiovision	- broadcast
	  television
	- videocassettes
	- audiocassettes
		Assessment
		of students
			- continuous assessment
			  - computer marked
			  - tutor marked
			- final examination

Computer-assisted learning (including videodisc)		Feedback from students and tutors to the course team.	
Home experiment kits			

A course is produced by a team of people. The team defines the content and the general approach, and for each medium the team discusses the material at each of various stages in its production. For example, in the case of the main texts, an author, who is normally a member of the team, has to write to a fairly tight brief, and has to suffer course-team discussion of several drafts. This is ideal for those who like delivering public whippings (and also for the smaller number who like receiving them), but nevertheless we believe that the course-team mode of production improves the quality of all of the various media that we use.

The courses are designed and produced at the main campus in Milton Keynes (where the research laboratories are also housed). In addition there are 13 regional offices, the main job of which is to sustain the regional support system, notably the appointment of part-time tutors and the organization of the non-academic aspects of tutorials and residential schools.

2.2 The Use of Various Media

From the long list in Table 1, I'll examine the use of just a few.

First consider, the *main texts*. At the OU these are the main teaching vehicle, and this will remain the case until almost everyone in the U.K. has easy local access to videocassette players, microcomputers, or whatever else is required for a medium that replaces the printed page. This means that for the distance learner, even in the U.K., the printed page will be predominant for the foreseeable future, the other media finding a more prominent place at residential schools or in tutorials.

We write our own main texts and make relatively little use of available textbooks. This is because for the distance learner it is particularly important that a text is more "user friendly" and student-active than the usual textbook. Note that any student can benefit from a text designed initially for the distance learner.

Second, consider *broadcast television* and *videocassettes*. Broadcast television programs are made on the assumption that students do not record them, whereas videocassette programs are either loaned to students on already recorded videocassettes or the assumption is made that the students will record them off the air when the programs are broadcast (which can then be in the middle of the night). Thus, broadcast television should be used in a different way from videocassette because of the ephemeral nature of television compared to the replay and freeze facilities available with videocassette.

In particular, broadcast television programs should have a soft pedagogical role: low density teaching; showing students places and situation beyond normal reach; and telling the story of a discovery, an experiment, or a theory. By contrast, videocassette programs can have a *hard* pedagogical role: high-density teaching; interactive/student active teaching; and visuals with intricate details.

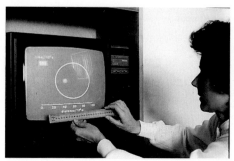

Fig. 1. *A student working with the videocassette supplied as part of The Open University course Matter in the Universe.*

Fig. 2. *A student (in 1978) making a hologram at home with the home-experiment kit supplied with The Open University course Images and Information.*

Fig. 1 shows a student working with the 120-minute videocassette supplied as part of the OU astronomy course (see Sect. 3.1). In this particular example the student is observing the motion of a small body around a far more massive body: by taking off-screen measurements of the orbital semi-major axis and orbital period, the mass of the larger body will be obtained. More complicated cases then follow. This videocassette contains much else, including a simulation of the spiral density-wave in spiral galaxies.

Finally, consider *home-experiment kits.* These provide the student with the opportunity to do experimental and observational science at home, though the lack of supervision can be a problem. Nevertheless it is surprising what can be achieved: in Fig. 2, a (1978) student of our optics course is shown making a hologram at home — yes that *is* a laser. However, a good deal about experimental and observational science, including astronomy, can be learned with far simpler equipment, indeed with little if any more than the sort of things that can be found around the average household, even in the less wealthy parts of the world.

3. Astronomy Courses at The Open University

3.1 Matter in the Universe

Matter in the Universe is the OU astronomy course, and it comprises 1/16 of a general bachelor's degree with honors. Table 2 lists the major topics covered ,with the approximate fraction of the course devoted to each of these. The course has a few unusual features, as follow.

Table 2. Matter in the Universe

Subject area (in order)	% of course
Introduction and overview	6
The interstellar medium	25
Stars	32
Galaxies	12
Planetary systems	25

First, it is apparent from Table 2 that the interstellar medium is prominent in both position and fraction of the course. This induces culture shock in many of our students, who come to the course thinking that astronomy is only about stars, planets, and galaxies. However, we believe that the interstellar medium deserves its prominent role.

Second, we use a story line to link together much of the course. We took as our story line the evolution of matter in the cosmos, not only within galaxies — from the interstellar medium to the stars and planets, with the subsequent return of some of this material, somewhat modified, to the interstellar medium — but also the evolution of the galaxies themselves.

Third, students are not shielded from the uncertainty that still pervades much of astronomy, and linked to this we have a yearbook to update the course.

Fourth, the course shows that astronomy both draws on and feeds into other subject areas — a two-way process.

Finally, and related to the previous point, we teach the underlying physical and chemistry not taught elsewhere in prerequisite OU courses. The physical examples will probably be familiar to you (spectroscopy, nuclear physics, *etc.*), but the chemical examples, relating mainly to the interstellar medium (ISM), will probably be less familiar:

- the ISM is far from thermodynamic equilibrium, so we teach some non-equilibrium chemistry and chemical kinetics
- the ISM contains appreciable concentrations of ions and these react readily, so we teach something of ion-molecule reactions
- the best way to make H_2 in the ISM is probably on dust-grain surfaces, so we teach something about reactions on solid surfaces (once we have H_2, gas-phase reactions seem able to provide the other molecules).

The course is now in its fourth year of presentation. It is a rather high level course, but nevertheless attracts over 400 students per year, over 60 per cent of whom successfully complete the course.

3.2 Possible Future Courses

The possibilities for future OU courses in astronomy are legion, but life is short and resources are limited. There is however a pressing opportunity arising from the recent specification in the U.K. of a national science curriculum for schools, to be implemented starting in autumn 1989. This curriculum includes some astronomy, and so there is now an urgent need to provide teachers with the necessary astronomical background via short in-service courses. Development is under way.

Less pressing, but highly desirable, is an undergraduate astronomy course at a lower level than *Matter in the Universe*, and somewhat broader. This would be a general-appeal "astronomy for poets" course similar to the many liberal arts courses taught in the U.S.A. and Canada.

Astronomy is certainly alive at The Open University, but there are many opportunities for us to do a lot more, both within the U.K. and beyond.

Discussion

S. Torres-Peimbert: *For how long has the astronomy course been given? How many students have taken it? What are the results?*

B. Jones: The astronomy course has been running each year since 1985. Over 400 students start it each year, and over 60% of these achieve a full pass.

R. Ramkarthikeyan: *What do you do for practicals in the "open university"?*

B. Jones: We send out home experiment kits (one contains an optical bench, laser, *etc.*) for *unsupervised* use at home. We also have day schools around the U.K., and summer schools, for supervised laboratory and field work. Overall we have somewhat less practical work than at a conventional university.

A. Cool: *How is The Open University funded, and how much does it cost a typical student to complete the degree?*

B. Jones: The Open University is largely funded by central government, with much smaller contributions from other bodies, including foundations and industry. The students on degree courses pay only a small fraction of the true costs.

Roughly speaking, it costs about $4,000 to obtain a full honors degree, and this will be spread over, typically, six years.

A SELF-PACED MASTERY-ORIENTED INTRODUCTORY ASTRONOMY COURSE

John L. Safko
Department of Physics and Astronomy, University of South Carolina, Columbia, South Carolina 29208, U.S.A.

I will be describing a self-paced mastery-oriented introductory astronomy course that has been offered at the University of South Carolina since 1972. This course is a three-semester sequence with total enrollments of 800–1000 students per term. Most of the enrollment is for the first semester of the sequence. Although these particular courses are designed for non-science majors, I believe that the methods developed in these courses can be applied to courses for astronomy majors. These methods could provide a solution to the problems of teaching classes and, at the same time, traveling for research and meetings.

A mastery-oriented course involves developing a list of learning objectives that are described in terms of performance of given tasks. That is, words such as *understand* and *appreciate* should not be used. Performance objectives involve words such as *recognize, calculate, measure,* and *describe.* Any set of learning objectives must be tied to a set or readings, a textbook, and/or supplementary notes. These items are chosen or written by the instructor(s) to provide the background necessary for the students to master the objectives.

At the University of South Carolina, our learning objectives were developed by examining the test questions that had been collected from 5 years of a regular lecture course. This allowed us to determine what performance we were expecting from the students, rather than guessing what the learning objectives should be. In a few cases it became obvious that additional objectives and/or additional questions were needed. In some cases it was clear that questions should be eliminated. Item analysis and student questions told us what notes in addition to the text were needed. We combined the objectives, the notes, and the laboratory materials that we use into a single volume.

A regular text publisher would want modifications in order to sell sufficient copies to justify publication. Rather than remove all references to particular procedures at our institution, we chose to go to a publisher who prints and warehouses sufficient copies for two-year intervals. A sample of the volume, published by Kendall/Hunt Publishing Company, was on display at this meeting.

As an aside, this method of publishing can provide texts for small enrollment courses such as graduate astronomy courses. With today's microcomputers and word processing programs, an author can prepare camera-ready copy for printing that reduces the cost of small press runs to a reasonable level. In those countries where regular publishers are not interested in publishing astronomy textbooks, you could prepare a camera-ready copy and take it to a local printer, and arrange for the students to buy them. If no other option is available, you can even pay for the press run and hope to recover your investment from sales. All questions of ethics can be resolved by selling at a price that will only recover costs and a modest payment for the preparation time. Thus, I believe, students in any country can have texts in their native tongue.

A self-paced course allows students to progress at their own rate, subject to the scheduled completion time of the course. We have an area that is open about 40 hours per week. During the open time, students can come at times of their own choice to use our study materials and to show mastery of portions of the material. A majority of the students spread their work out over the semester, but, as might be expected, some put the course off until the last days of the term. These are the same students who wait until the night before a scheduled exam to study.

We have an audio-visual room with computer programs, video tapes, and slide material covering selected material from the course. Many of the video tapes and some of the computer programs were made at the university specifically for this course. All students are encouraged to try the material in the audio-visual room. Those students who are having difficulty with particular concepts are directed to the appropriate tape or program. The video tapes thus take the place of lectures with the advantage of being available and repeatable whenever desired by the student rather than only at fixed times.

Most of our evaluation is carried out with multiple choice questions, although some of the material also has brief paragraphs expected. Answer keys are provided for the staff to use. This evaluation process allows us to use staff who may not be able to answer all the questions themselves. The mastery evaluations are computer-generated from a large field of questions that are subdivided into the learning objectives. For any given subject a large number of different evaluations are available; hence, those students who need to repeat receive a completely different test on the next try. We require 13 correct responses out of 15 questions for credit. If mastery is not achieved, immediate feedback is provided by reference to the learning objectives the missed questions were testing. If that is not sufficient and if the staff in the testing center is not able to help, the student is referred to one of the senior staff or to me.

Most of the staff and laboratory instructors are recruited from students who have successfully completed the course. Many of these are not science majors.

This procedure has the advantage of having a staff who have seen the course from the students' viewpoint. We have also found that many non-science students are, unreasonably, intimidated by science students. Thus a non-science student staff can often provide aid and a friendly ear where a physics or astronomy major could not (note, I said "could not," not "would not").

Although the course I have been describing was designed for non-science majors, I present it as a possible model for undergraduate and graduate major astronomy courses. By freeing ourselves from a fixed schedule of lectures, we allow greater flexibility in travel and observation. There are also educational advantages in forcing the student to take more responsibility for his or her own learning process. Texts, articles, and prepared notes can provide most of the aids that a student needs. Individual consultation and student study groups can fill in any remaining gaps. Feedback from the students will show you where the notes need further detail.

Our self-paced astronomy has a large enrollment, so we can offer the students a large number of possible testing and audiovisual hours and still be economical. Smaller enrollment courses would, probably of necessity, have fewer hours available to the student. This handicap can be overcome, in part, by computer testing and the minimal need to proctor students in advanced courses. We have also found that several courses can be offered with the same testing and audio-visual center.

There are several *caveats* you must consider before embarking on this method of teaching:

1. Don't consider a self-paced course to save yourself time. You will spend a larger amount of time in preparation, organization, and dealing with individual students than you would in a normal lecture course. This time spent, however, will be more productive since much of it will be spent with individual students.

2. Self-paced, mastery-oriented courses require careful preparation and more attention to detail than you would spend with a normal lecture course.

3. If several faculty share a single course, one person must be clearly in charge and responsible for all exceptions to the rules.

4. If commercial texts do not meet your needs, you will have to write extensive notes, which will take additional time.

5. You must be philosophical about what students do to themselves. It is their responsibility to study and read the material, not yours.

I hope this brief introduction can give you an idea of the nature of self-paced, mastery-oriented instruction and that it may inspire a few of you to experiment with this method.

Discussion

P.P. Saxena: *This type of teaching is only useful for those institutions where there is a scarcity of astronomy teachers and/or where the number of students is very large.*

J.L. Safko: The particular course that I described is offered at such an institution (scarcity of astronomy teachers and a large number of students); however, this is not the only situation where the method would be useful. Observation schedules and shifts in telescope scheduling may pull astronomy faculty from their home institution at odd times in the year causing a disruption of normal classes. Then another faculty or graduate student must cover those lectures in addition to their normal classes. Proper scheduling of self-paced duties can eliminate this problem.

The efficiencies of size can be gained by combining the evaluation process for several courses into one administrative structure, open hours, and location. This would allow a few classes of 5 or 6 students to be incorporated with several introductory classes with minimal or no extra staffing over the introductory class needs.

The main difficulty with self-paced courses involving several instructors is reaching an agreement on the wording of the learning objectives for the course. This problem must be resolved to everyone's satisfaction for the method to be successful.

A. Fraknoi: *I may be the Neanderthal of astronomy teaching, but I have one concern about the very interesting program you have described. I remember vividly that during my college education, some of the most exciting and valuable experiences involved watching and hearing a brillant scholar/teacher interact with others students, respond to questions in class, digress about his own work or recent discoveries, and share his or her personal love and excitement about the field. Is there a role for an inspiring live lecturer in a course such as the one you described?*

J.L. Safko: Many students do not agree. They choose astronomy because they won't have to go to lectures 2 or 3 times per week. I do offer special opportunties for credit for attending lectures, such as those given by visiting astronomers, special lectures at the local astronomy club, *etc.* Educational psychologists say that the only factor that correlates with a student's grade and learning is how much work he/she puts into the course.

A CORRESPONDENCE COURSE IN ASTRONOMY

Michael Zeilik
Department of Physics and Astronomy, University of New Mexico,
800 Yale Blvd., N.E., Albuquerque, New Mexico 87131, U.S.A.

1. Introduction

As do most universities in the United States, the University of New Mexico (UNM) offers a one-semester introductory course in astronomy aimed at the vast audience of non-science majors. UNM is a non-selective, minimally-funded state research university with an undergraduate enrollment of about 20,000 students — with a median age of 26. Many of the students have jobs and families and attend on a part-time basis. The Physics and Astronomy Department usually offers three sections of Astronomy 101 to accommodate students' schedules: in the morning, afternoon, and evening. Total enrollments per semester range between 400 to 600; the course is also regularly taught during the summer. The class has *no* prerequisites of any kind, is advertised as "non-mathematical," and does *not* have a required lab for observations. (A related lab course, Astronomy 111, *is* available, and is taken by about 100 students per semester.) The course attempts to cover "all" of astronomy with a special emphasis on the theme of cosmic evolution. Instructors tend to use visual aids (slides, videos, films, laser disks, and microcomputers) heavily. Students usually rate the instructors and the course in the top 10 to 30% of all instructors who choose to use a university-wide class evaluation instrument, compared to a group of other lower-division courses.

When I teach the course, I use my textbook (*Astronomy: The Evolving Universe*, 5th edition, 1988, John Wiley and Sons) and typically cover, in all or part, 18 out of 22 chapters. I require four one-hour-long tests plus a two-hour-long final exam and also provide 8 practice tests during the semester (Folland *et al.*, 1983). All tests are in a multiple-choice format; my reliability coefficients are greater than 0.8, and the standard error of measurement averages 2%. Students are given reading assignments but no homework as such. The four term tests count 80% of the final course grade; the final exam is worth 20%.

The characteristics of the students in Astronomy 101 are typical of those found in the United States in non-selective state universities in science courses with no prerequisites: about 1/3 of the students are formal reasoners in the Piaget sense and the other 2/3 transitional or concrete. Their fear of math is so strong that arithmetic and graphs mystify many of them. Some 90% or more have never had a physics or astronomy course; 50% have never had algebra/trig/geometry. Astronomy 101 is probably the only science course most of them will take at the university. One redeeming point: 2/3 say that they are taking the course to satisfy their curiosity about astronomy rather than only to satisfy a science requirement for graduation.

2. Correspondence Course Description and Operation

I developed the correspondence course from a Personalized System of Instruction format of a Harvard University astronomy course (Zeilik, 1974; Bieniek and Zeilik, 1976), which also formed the basis for my textbook. (I have also used this PSI format for upper-division undergraduate courses; see Zeilik, 1981.) These materials were revised and used with a special PSI section of Astronomy 101 at UNM; these were streamlined into the correspondence course materials and the study guide to my textbook (by John Gaustad and me) that was available with the second and third editions. (It is no longer published.) The correspondence course, called Astronomy 101C, is offered for academic credit through UNM's Division of Continuing Education at the same cost as a regular in-state tuition for a 3-credit hour course. Continuing Education handles all logistics and records; the course can follow me around by mail.

The student has 22 written lessons to complete (one for each chapter of the book); each lesson asks the students to answer *in writing* three to six questions per chapter in an open-book environment. In addition, the students have a Lesson 0 that requires naked-eye observations. The complete lessons are mailed to me via Continuing Education, and I grade each individually with comments, usually within a few days after I receive them. The lessons are then returned to the students; turn-around is typically one week. The grades on the lessons count 80% of the final grade for the course. The final exam counts for the other 20%; it is the *same* final given by me to the Astronomy 101 students in the most recent term. It consists of 100+ multiple-choice questions given in 2 hours with a proctor (provided or paid for by Continuing Education). I grade it with the same scaling as the regular term exam. Because the correspondence students have not seen multiple-choice questions on their lessons, I provide them with sample tests upon request (the same sample tests given to the lecture students).

Officially, the students are given 12 months in which to complete the course. If they have not finished in that time, they can request two 90-day extensions for a total completion time of 18 months. Students are formally not supposed to complete more than one lesson a week; in fact, some do; I have never seen this rule enforced. I have noticed that students who rush through the material simply do not give in-depth answers on the lessons and tend to do poorly on the final exam. I announce to students that I am available by telephone, but have received very few calls from them.

Since 1976, 73 students have enrolled in Astronomy 101C for an average of 6.1 per year. They included 31 residents of New Mexico; the rest tended to reside in the western part of the United States, though one student lived in France and another in Hong Kong. Occupations ranged from "student" to "housewife," and "alcohol counselor" to "airplane pilot." Of those *enrolled*, 55 had *completed* the course before Fall 1988 for an average of 4.6 per year; 31 were women and 24 were men. The average completion time was 10.3 ± 5.3 months; note the large standard deviation. The fastest completion was 2 months (by three students); the slowest,

22 months. Of these students who did complete the course, the average age was 32.3 years; the average grade on the lessons was 89.8%; and the average final exam grade 75.3%. Of the students who did *not* complete the course by not doing all the lessons, the average number of lessons completed was 9. They started out well but could not keep up their motivation.

3. Results

First, I compare in Table 1 the final grades of the Astronomy 101C students to those in Astronomy 101 in Fall 1980, Spring 1982, and Fall 1984, which I also taught:

Table 1

Grade	Astro 101	Astro 101C
A	23%	33%
B	28%	33%
C	33%	10%
D	4%	0%
F	9%	0%
Other[a]	3%	25%

[a]Note: "Other" includes incompletes and *recorded* withdrawals; term-time students have six weeks to unilaterally withdraw from a course and leave no record of having been enrolled. For Astronomy 101C, the "other" category includes all students who failed to complete the course.

The trend is clear: term-time students who had to finish the course by a deadline either withdrew early or stuck it out after six weeks had passed. In contrast, Astronomy 101C students who finished the course tended to do well, as shown by the lack of D's and F's. In a sense, the correspondence students had more control over their destiny with a flexible, extendable deadline; the course was student-paced to a much greater extent than the regular one. (The grading scales were the same for both courses: 90% and above for an "A"; 80 to 89% for a "B"; 65 to 79% for a "C"; 50 to 64% for a "D"; below 50% for an "F.")

Another basis of comparison is the final-exam scores. The average for the term time students was 69%; correspondence students, 75%, for a statistically-significant difference (about three standard errors). You might argue that this comparison is biased, because the correspondence students who would score low simply did not take the final. That is true; however, it is also the case that most of the students who received "D's" and "F's" in the regular course did so because they did not take the final exam, which is the main reason their final grades are so low. So the comparison is apt.

From my twelve years of experience with Astronomy 101C, I conclude the following:

1. A correspondence course is an effective means to deliver an introductory astronomy course to students far from the instructor and home university; a "plus" for New Mexico, which has a low population density and a wide geographic spread.

2. Such a course helps wheelchair-confined students who still find it a great effort to attend classes on campus.

3. It is a low-cost instructional process relying mostly on the textbook; the lack of a rich visual classroom environment and a live instructor does not hinder the learning of the key concepts for mature, motivated students.

4. Acknowledgment

I thank Connie Terry of Continuing Education at the University of New Mexico for tracking down old student records, which were essential to the preparation of this paper.

References

Bieniek, R.J. and Zeilik, M., "Follow-up Study of a PSI Astronomy Course," *American Journal of Physics,* **44,** 695-696 (1976).

Folland, N.O., Manchini, Robert R., Rhyner, C.R., and Zeilik, M., "Report on Using TIPS (Teaching Information Processing System) in teaching Physics and Astronomy," *American Journal of Physics,* **51,** 446-449 (1983).

Zeilik, M., "A PSI Astronomy Course," *American Journal of Physics,* **42,** 1095-1100 (1974).

Zeilik, M., "Flexible, Mastery-Oriented Astrophysics," *American Journal of Physics,* **49,** 827-829 (1981).

ASTRONOMY AS A LIBERAL ART

Tom R. Dennis
Mount Holyoke College, Five College Astronomy Department,
South Hadley, Massachusetts 01075, U.S.A.

Every year in the United States, many thousands of university and college students take a one-semester course in astronomy as part of a "Liberal Arts" curriculum. At many institutions these courses are the *raison d'etre* for the astronomy program, and at most other institutions they are viewed as an important term in calculating an astronomy department's contribution.

I have often had a sense that what we are doing in these courses[1] is counterproductive to any serious understanding of the goals of liberal arts curriculum. Over the last six or seven years, as I have been participating in my college's efforts[2] to introduce a year-long, interdisciplinary survey of the humanities into our curriculum, this hunch has been confirmed: in our efforts to fit astronomy into the liberal-arts curriculum, we have ironically been undermining the enterprise.

The problem originates when teachers (and textbook authors) quite accurately perceive that the students in these courses are not scientists: few of them have taken science during the last two years of high school, and have certainly not taken chemistry and physics; and while most have taken (and passed) 11 or 12 years of math, they appear to have learned almost nothing, and are often hostile to the subject. Most of them consider themselves "weak in science [and/or] math."

As a response to these deficits, we decide to teach a "descriptive" course in astronomy. And at that moment, we inevitably commit ourselves not only to reinforcing their negative self-evaluation but also to misrepresenting the character of science.

Negative Self-evaluation: By presenting these students with a mass of "facts" about the universe, we essentially assign them a substitute task, which they can pursue *instead of* taking a "real" science course.

This works — insofar as it works — because most students who are sufficiently disciplined to have reached college can ingest a few "facts" and regurgitate them at the appropriate time.

But they still "do not understand." Indeed, one suspects that the mass of "facts" they "do not understand" has only increased, and so their impression of being intellectually inferior is unmodified at best, or perhaps proportionately increased. Certainly, by the very act of making a special course available, the instructor and the institution are reinforcing the judgment that some students cannot be expected to do "real science."[3] Many of them feel consigned to live in a universe that defies

[1] And I have taught such a course at least 33 times.
[2] Funded by a grant from the Mellon Foundation.
[3] In my institution students call these "baby" courses — "baby astro." There are no "baby"

comprehension, but which certain untrustworthy wizards (scientists and engineers) seem to be able to manipulate almost at will. But at least the science requirement is out of the way.

Misrepresenting Science: Consider an anthropologist's view of a professor lecturing to a descriptive "Astro 101" class. There is an authority figure describing almost unbelievable things: holes in space-time, or ionization at the edge of infinity.[4] The homily is reinforced by catechism: students learn responses to questions, and their perseverance and skill are taken as a measure of their good works.

What is wrong here is that the uniqueness of science is missing: by ignoring the revolutionary, anti-authoritarian epistemology of science, we present something indistinguishable from magic, superstition, political ideology, or theology. As Neil Harris, a historian at the University of Chicago,[5] has put it, "Contemporary science is believed in by a credulous public rather than understood by an informed one."

This mind-set leaves the public open to all sorts of pseudosciences, from creationism to astrology. And the evidence is that the pseudosciences are winning: according to Martin Gardner[6]

> A 1984 Gallup poll showed that the number of teen-agers who say they believe in astrology had risen from 40 percent in 1971 to 55 percent. Among teens who claim to believe in astrology, those from well-educated families outnumber those from untutored classes, and whites outnumber blacks.

Even in the community of intellectuals — professors of the humanities and social sciences — there are individuals who fail to understand the epistemological differences between physical science and their own disciplines, and consequently foist a 20th–century version of scholasticism upon Harris' "credulous public."

An effort at reform: For the last several semesters I have been addressing these issues in my teaching, using my own text materials. The hallmarks of my approach include:

- *A Conceptual Outline*, which begins with a discussion of the spatial and temporal structure of the universe, and builds to a treatment of cosmic evolution informed by extensive use of the principle of conservation of energy;

- *Quantitative, but Elementary Treatment*, utilizing elementary algebra and "scientific notation," but no magnitudes, logarithms, or trigonometry;

- *Regular Problem Assignments*, emphasizing problems that illustrate and reinforce the argument of the book, rather than problem-solving ability *per se*;

courses in the humanities.

[4] To borrow a phrase from a literary friend, in a conversation about the origin of the cosmic microwave background.

[5] In a review of *How Superstition Won and Science Lost*, by John C. Burnham, in *Science* **240**, 1552, 1988. Harris continues: "Today's popularizing processes, compared with some earlier versions, seem condescendingly simple-minded or narrowly opportunistic."

[6] New York Review of Books, 1988 June 10.

- *Laboratory Work*, including naked-eye and telescopic observations (for which they must set the telescope themselves, using coordinates and sidereal clock);
- *Explicit Connections with the Humanities*, in an effort to develop an appreciation of science that goes beneath the "gee whiz" level.

This is still in an early stage of development, but I am truly encouraged with how well the experiment is working out. Perhaps the most encouraging aspect of it is that students do not really consider it nearly as different as I do — they are accustomed to taking math courses, and this therefore looks very familiar to them. I therefore suspect that as a community we have simply overreacted to the declining quality of mathematics instruction in the schools; rather than opting to eliminate the quantitative content from our courses, we can instead adapt our teaching materials and syllabus to provide the increased support these students require.

COMPARATIVE METHODS FOR TEACHING AN ASTRONOMICAL TOPIC

Syuzo Isobe
National Astronomical Observatory, Osawa, Mitaka, Tokyo 181, Japan

Astronomy has greatly progressed in this decade. The number of people who are interested in astronomical phenomena is certainly increasing as the number of IAU members increases. These people are divided into several categories, depending on their degrees of interests. For Japan, these are as follows:

Categories	Rough Number of People	Definition
A	10^2	produce useful observational data
B	10^3	observe frequently
C	10^4	observe several times per year
D	10^5	read astronomical magazines
E	10^6	read general science magazines
F	10^7	read scientific articles in newspapers
G	10^8	no interest in science

The situation is changed when some such special event, such as a return of Halley's comet, occurs. We can say that our present target in teaching astronomy is to increase the number of people in each category except the last one.

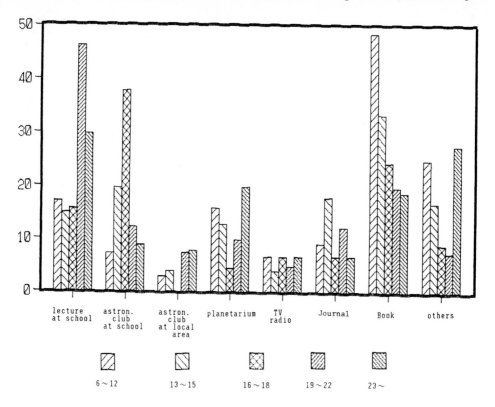

Fig. 1. Main triggers by which one started to be interested in astronomy.

Considering the above situation, astronomical and physical knowledge is also increasing for the general public, and it has become a popular idea not only that the Earth revolves around the sun, but also that stars form, evolve, and die. We sent a questionnaire to all the members of the Astronomical Society of Japan, and received 300 answers out of 700. One of the results is shown in Fig. 1.

The number of people within each age group whose interest in astronomy was triggered by different items is shown in percentages. It is clear that the main trigger for children 6 to 12 years old was books that explain new discoveries in astronomy. These children sometimes know more about astronomical topics described in newspapers and magazines than do their teachers!

There are many branches of astronomy. Since a Hertzsprung-Russell (H-R) diagram is one of the most popular astronomical concepts, is very important to understanding evolution of stars, and has a secure position in astronomy, we compared methods for teaching about the H-R diagram. We compared the treatment of the H-R diagram in 8 textbooks for Japanese senior high schools. Two examples are given in Fig. 2.

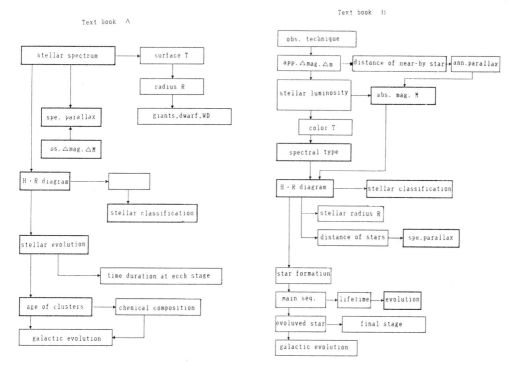

Fig. 2. Flow of teaching astronomy in two examples of textbooks.

All the textbooks have similar structure. First, stellar spectra and absolute magnitude are explained, an H-R diagram is drawn, and then a story of stellar evolution follows. This is a historical and orthodox sequence.

We also sent similar questionnaires to all the members of IAU Commission 46 (The Teaching of Astronomy) and obtained 32 answers (Table I). First, we asked what should be the order of astronomical topics to be taught at school and university. Most of the answers were in order of distance from Earth, that is, solar system, stellar system, and galactic structure and galaxies. This is also the historical sequence. About the H-R diagram, although nearly all the answers stressed its importance for studying stellar evolution, they used different parameters to draw an H-R diagram. For considering stellar evolution, one should use effective temperature and luminosity for the axes.

We also compared 11 English textbooks for undergraduate students available at our observatory. As shown in Fig. 3, different parameters are used from page to page in the 10 textbooks. Only one uses effective temperature and luminosity for the axes consistently throughout the book.

		A	B	C	D	E	F	G	H	I	J
consistency		×	O	×	△	△	×	×	△	△	×
T	L		O	O		O					
T	M										O
T	L + M							O			
T + S p	M				O		O				
T + S P	L + M									O	
T + color	L + M					O					
T + S p + I	M	O									
T + S p + color	M							O			
S p	L								O		
S p	M	O		O	O	O	O			O	O
S p	L + M							O			
S p + C I	M			O							
C I	L								O		
C I	M					O					O
C I	m										O
color	M					O					

Fig. 3. Parameters used for both axes in 11 English textbooks. T: temperature, Sp: spectral type, c: color, and CI: color index. L: luminosity, M: absolute magnitude, m: visual magnitude.

From Fig. 1, we see that many children already have some ideas on astronomical topics. It is good for them to understand the theoretical stellar evolutionary sequence. When this sequence is taught, theory and observation are compared. Such a comparison is also true for planetary motions. Children are taught those apparent motions in the sky assuming that they know the heliocentric motions of the planets.

Since the theory of stellar evolution is now settled, it may be the time for us to teach the H-R diagram by first considering theoretical stellar evolution. It is interesting for me that one textbook is arranged in the order stellar system, galactic structure and galaxies, and solar system. The author intended to arrange it that way on the grounds that the theory of stellar evolution is settled. However, to teach galactic structure and galaxies, we should follow a historical sequence because these branches of astronomy are not completely settled at present.

Table 1 appears on page 108.

Discussion

J.-C. Pecker: *Your diagram about the triggering of interest for astronomy show that TV is very inefficient in this respect. This is, I believe, a world-wide truth. I feel that we should put strong efforts in getting through TV programs that could give good ideas about astronomy — not only about the objects or the facts, but also about the ways that astronomers use to reach knowledge.*

T. Clarke: *In response to Professor Pecker regarding the low level of TV in stimulating interest in astronomy, since television programs are selected by the viewers, those not having a prior interest in astronomy simply do not choose a program on astronomy. They select something else. Hence, we get a low indication of TV's value on the graph.*

S. Isobe: One should remember that most of those who replied are the members of the Astronomical Society of Japan. I think that TV programs help the general public to enjoy astronomy and science.

D. McNally: *If we are to take your survey at face value, then for greater effect we should advise that authors write only for 6–12 year olds and that lectures should only be given to 19–20 year olds.*

S. Isobe: To simulate interest of pupils and students in astronomy, all kinds of the efforts contribute up to some level. I would like to say that each kind of the efforts work efficiently with the help of the other kinds of effort.

Table 1. Answers from 32 members of IAU Commission[a]

No	Country	1 a	b	c	d	e	2 a	b	c	d	e	f	3 a	b	c	d	e	4 a	b	c
1	Egypt	3	1	2	4	5	y	y	y	y	y	y	y	y	y			y	y	y
2	India	4	1	2	3	5	y	y	y	y	y		y	y				y	y	y
3	Poland	y	y	y	y	y	y		y				y					y		
4	U.S.A	2	1	3	4	5	y	y	y	y			y					y	y	
5	Yugoslavia	1	2	3	4	5	y	y	y	y		y	y	y	y			y	y	y
6	Germany	1	2	3	4		y		y									y	y	
7	Bulgaria	y	y	y	y		y		y				y		y			y	y	
8	Australia	4	1	2	3	5	y		y				y		y			y	y	
9	Poland	2	1	3	4	5	y		y				y		y			y	y	
10	Finland																			
11	Indonesia	y	y				y		y				y		y			y	y	
12	U.S.A.	y	y				y		y				y		y			y	y	
13	Australia	1	4	2	5	3	y	y	y	y	y	y	y	y				y	y	y
14	U.K.	2	1	3	4		y	y	y		y		y	y				y	y	y
15	Malaysia	y																		
16	U.K.						y	y	y	y	y	y	y	y	y	y		y	y	y
17	U.S.A.	1	2	3	4	5	y	y	y	y	y	y	y	y	y			y	y	y
18	Switzerland	y	y	y	y		y	y	y	y	y	y	y	y	y			y	y	y
19	U.S.A.	1	2	3	4		y	y	y	y	y	y	y	y	y			y	y	y
20	U.S.A.	2	1	3	4	5	y	y	y	y	y	y	y	y	y			y	y	y
21	South Africa						y	y	y				y	y				y		y
22	Canada	y	y	y	y	y	y	y	y	y	y	y	y	y	y			y	y	
23	Czechoslovakia	y	y	y			y	y	y	y	y	y	y	y	y			y	y	
24	U.S.A.	1	2	3			y		y	y			y					y	y	
25	U.S.A.	1	2	3	4		y	y	y	y								y	y	
26	Sweden						y	y	y	y	y	y	y	y	y			y	y	y
27	France	y	y	y	y		y	y	y	y	y	y	y	y	y	y		y	y	y
28	Hungary	y	y	y	y		y	y	y			y	y	y	y			y	y	y
29	Portugal	y	y	y	y	y	y	y	y	y		y	y		y			y	y	y
30	Mexico	2	1	3	4	5	y		y				y					y	y	
31	Australia	1	2	3	4	5	y	y	y	y	y	y	y	y	y	y	y	y	y	y
32	China	1	2	3	4	5	y	y	y	y	y	y	y	y	y	y	y	y	y	y
	Total						26	19	29	11	14	13	25	16	22	4	1	27	25	16

[a]Question 1: In what order of astronomical thema are pupils taught? a: Celestial dynamics, b: Solar system, c: Stellar physics, d: Galactic structure, e: Cosmology.
Question 2: For what purpose is the H-R diagram taught? a: Classification of stars, b: Spectral type, c: Stellar evolution, d: Stellar mass, e: Spectroscopic parallax,

Chapter 4

Student Projects

Involving students in laboratory and observing projects is often one of the most effective ways of teaching astronomy. The first two papers deal with involving students in projects of original research. The following pair of papers discuss using existing data of research grade. Next, three papers discuss student observing with telescopes, for visual observing and using CCD's and photometers. Finally, we read two philosophical statements about the nature and value of student laboratory and observing work.

THE ROLE OF PROJECTS IN ASTRONOMY EDUCATION

D. McNally
Department of Physics and Astronomy, University College London, Gower St., London WC1E 6BT, U.K.

1. Varieties of Project

At first sight, there may appear to be a very wide range of project situations that are in use. Students may seem to be doing the equivalent of postgraduate work, or may seem to be working on contract basis, or may even seem to be doing rather long term laboratory exercises, or some mixture of all three. However, there are in essence two major types of project situations — the free-scope project and the set-piece project.

The free-scope project attaches importance to the selection of project topics by the student. The student may be told to select a problem from the entire spectrum of problems offered by the science of astronomy, and to get on with it, or there may be guidelines to restrict the choice. Nevertheless, the choice of topic is the prerogative of the student, and it is believed that the effort of making that choice is important in the teaching/learning process. Because of the range of formats for free-scope projects, it is a convenient form to adapt to meet varying local circumstances.

The set-piece project depends on there being a set of specific projects available, and choice is restricted to that set only. It is usual, because of investment in necessary equipment, for such projects to be available for several years as a minimum. Projects that decline in scope can be phased out and replaced by fresh examples. Such a system is time-consuming to devise and set up. It has advantages in that by having well-defined boundaries, the assessment and inter-comparison of students is easier than in free-scope projects. Economies can also be effected by assigning students in pairs.

The boundary between free-scope and set-piece projects is not sharp. Free choice can be limited by availability of supervisors, equipment, availability of data, or accessibility of celestial object; the set-piece project need not inhibit enthusiastic student initiative. Often a menu of possible projects is predetermined, and the student is given a choice within the menu. Other students will shop around members of staff to see if they have an interesting problem on offer. The mix can become so blurred that it may not be immediately clear to which class any project actually belongs.

The mode of operation of the projects also varies considerably. In some schemes, each project student is heavily supervised, while in others, the student sees the supervisor only if in difficulty. Some schemes involve the student with, and absorption into, a research group: in others the student works completely alone. Again, the wide range of modes of operation of project schemes allow ready adaptation to local conditions and university regulations.

Projects must also be assessed. There are many mechanisms ranging from continuous assessment by the supervisor, evaluation of a written report, or a single short talk on the work done or a combination of several such elements.

Adequate time must be allocated for project work and its assessment. It is clear that it will be rare for an undergraduate project to break new ground. Yet it is the essence of project work that it is new to each undergraduate, and should represent something of value when complete, whether it be a new item of equipment, a set of complete programs for data analysis, or the analysis of a spectrogram that might otherwise not have been analyzed. Assessment is a considerable problem in its own right, and will be returned to later in this paper. At this point it is sufficient to say that any assessment scheme must give the student credit for adequate work done, ability to draw conclusions from the work, and to see how the project fits in with Greater Astronomy. A student should also receive credit for being able to present a clear, well-set-out report.

2. The University College London Astronomy Project

Perhaps the best way to illustrate project work is to take a particular example. I shall take the Project Course that forms a part of its final year program in the B.Sc. Degrees in astronomy, astronomy/physics, and mathematics/astronomy. The Project course is compulsory in the degrees in astronomy and astronomy/physics (in the latter, the students have the option of electing to do either the Astronomy or Physics Project), and represents 25 per cent of the final-year workload. In mathematics/astronomy, the project is optional, and has a reduced demand being equivalent to a 12.5 per cent of the year's workload.

The Astronomy Project operates in the free-scope mode (the UCL Physics Project operates in the set-piece mode). The student is free to choose a topic that the course directors consider has sufficient astronomical flavor. Choice ranges from cosmology to Earth resources, and even astronomical history or philosophy are not excluded. The student is asked to consider a project topic during the vacation before the final year and this topic is discussed with the course directors at the beginning of the final year. Students are simply advised to consider any logistical constraints, such as accessibility of objects to be observed, availability of data, and computational requirements. Many students make a tour of the research groups to see what may be on offer (thereby introducing an element of the set-piece project) and indeed the majority find a topic this way. Nevertheless, a number of students each year arrive at a choice of topics themselves, or at the very least have identified an area of interest in which to define a project topic. Since the project is compulsory for astronomy students, the course directors keep a number of possible projects on hand for the few who seem unable to reach a choice for themselves. (It is surprising that by the final year of a degree course, there are students incapable of identifying an area of interest.) The UCL astronomy project, while free-scope in essence, has set-piece elements within it.

The project course is run by two course directors who are available for discussions with the students on any matter concerning projects from science and supervisors to logistics. The course directors discuss the proposal for projects with the students at the beginning of the final year, ensuring that the proposal is possible and represents a reasonable astronomical endeavor. This discussion is fraught with pitfalls — clearly one cannot allow a student to proceed with a project that demands observations of objects that will not be available until nearly the time when the project must be concluded. It is discouraging for the student to be told that his or her proposal is not good science (one needs to be wary of dismissing a proposal by a colleague for the same reason!), but with a little exploration, amendment, reconsideration, and enterprise it is usually possible to find a reconciliation of student expectation, a scientifically worthwhile project, and a willing supervisor by the end of the first four weeks of the Autumn term. At this stage, the student submits a short statement of the project topic. The supervisor countersigns the proposal to show that the logistical implications are accepted and can be met as of the date of signing. This means that, for example, data are available then and are not awaited from some outside source. The students are interviewed about 2/3 of the way through the Autumn term to ensure that they have begun work and are not procrastinating. Problems are identified, clarified, and, we hope, sorted out. On return from the Christmas vacation, the student presents a short report on work done and plans for the upcoming Spring term. This report is again signed by the supervisor to indicate its accuracy. The report is assessed by a course director and credit can be lost for failure to meet the submission deadline. The students accompanied by their supervisors if they (the students) wish, are interviewed by the course directors in the middle of the second term. This is a crucial interview. If it is felt that the student has made insufficient progress with the project, the student is not permitted to proceed with a full project, but is told to halt that work, and must submit an extensive essay review. Students are also questioned to ascertain that they are not devoting time to the project that should be spent on other courses. Finally, on return from the Easter vacation, the students submit a report on their project for assessment, to be followed two weeks later by a 20-minute project talk, given before a small audience of lecturers and other students. Each report is assessed by the supervisor and another academic close to the topic of the project. All reports and talks are assessed by a course director, so that the final grade awarded is a moderated amalgam of all three assessments.

In assessing the project, all three assessors should use similar criteria, though they may weight these criteria differently. One for example will give a substantial weight to the project talk — another will give it almost no weight. However, all assessors are looking for a capable piece of well-presented scientific work. It is surprising how little disagreement in mark differing philosophies achieve — a spread of 5 per cent is common, and less than a degree classification band (about 10 per cent) will account for all but exceptional cases.

It is hard to say whether or not the students like projects — in all but a few cases, most like doing their project work. The report is normally regarded as a chore and the talk is universally disliked. The project report, even though it will be available for consultation by future generations of project students, retains some vestigial remains of the confidentiality of the confessional whereas the project talk exposes the soul to the gaze of the masses. The project talk is a nerve-wracking and not-to-be-repeated ordeal for most.

In the UCL Astronomy Project, choice of project topic is seen as an essential part of the training. A final-year student should be capable of choice of topic and be able to justify that choice. It is also regarded as important that the student should be able to write a report understandable beyond immediate practitioners of the project topic, which demonstrates clearly the work the student has done and the conclusions which can be legitimately drawn. The student should also be able to communicate the essence of the work and conclusions verbally. These are skills that are featured nowhere else in the UCL Astronomy-based degrees. The scientific quality of the project should demonstrate the student's capability in following an astronomical problem, or skill in practicing an astronomical technique. Over the years, the project course has brought forward projects of great professionalism, and have been published in the refereed astronomical literature. Most are simply good reports of good work done. Some are admittedly awful — being inept, inaccurate, innumerate, and illiterate. Yet even that rump end have probably learned something useful in the process. The project may bring forward students who have not shown distinction in any other way. Conversely, the project may undermine students whose previous examination records have been impeccable. It is clear that the project is a different measure of a student from standard essay-type examinations.

3. Arguments in Favor of Projects

The argument most often advanced in favor of projects at the undergraduate level is that they offer students the opportunity to attempt to study a real problem in the raw, and not as predigested in textbook and lecture note. This is a compelling argument for most academics, since that is the essence of research. The project, they contend, is a valuable way of assessing students as possible research workers. Standard examinations offer little scope for real assessment of research potential, other than the broad assumption that he or she who does well in examinations is likely to be a useful research worker. By and large, such an assessment is adequate. The project supplies further evidence for research potential, even though the range of that evidence is perhaps more limited than most are willing to admit.

That the project offers a chance to get to grips with a problem could be an illusion. Even in a free-scope project, the opportunity to attempt something really new is unlikely. The important part of the well-organized project is to remove the student from the lecture/textbook situation. The student has an opportunity to see real astronomical data, to really see what noise means, to appreciate how data must be handled to extract meaningful information, and above all, to appreciate what a difficult and demanding task an original investigation can be. The pace of information transfer of a three-year degree course, becomes abruptly slowed in the project through the demand on student energy is increased. Some students do not survive the spartan exposure by project.

The students who do survive (*i.e.*, the majority) learn in varying degrees about the difficulty of making sound astronomical progress. They learn that a consideration of error bars is as important as actual measured values. They learn about judgment. They begin to understand techniques in a way that no number of set experiments in laboratory classes will teach. They make a start in learning to communicate and to realize the value of the astronomical literature as a resource. They will learn about themselves. They will begin to realize what type of problem drives their enthusiasm, and what may cause them to procrastinate. They may even realize that an ambition that sustained them through three years of university study is not practical, but that its likely lack of realization is not a disaster either. Projects help people mature and gain experience. The gain in confidence between the first- and second-term interviews is often seen as a step function.

The argument in favor of projects should be not that they give students a taste of research but that they give the students freedom from the predigestion of textbook, lecture, and formal laboratory classes. They give students a chance to try for themselves at their own pace, and to learn how to handle their acquired knowledge in attempting to understand a particular problem or technique.

The project is not a wonder antidote for standard examinations, but is another measure of particular individuals. The project is only one among a number of the measures.

4. The Argument Against Projects

While projects have advantages, there is no doubt that projects have serious disadvantages and irreducible drawbacks. The first objection is that a project — for all its apparent similarity to a research situation — is essentially contrived, and, in order to be scheduled within the constraints of an academic year, simplistic. It could be argued that undertaking project work only gives students a false impression of real scientific work so that it would be best were projects avoided. These arguments have force and sensibility.

The UCL Astronomy Project must be started, read about, worked on, concluded and reported upon, within the space of 30 weeks of which 7 are vacation. It represents 1/4 of a full year's work, *i.e.*, approximately 70-75 equivalent lecture hours (*i.e.*, 70-75 actual hours in the lecture rooms, plus time spent additionally on that lecture course). A realistic estimate would be 150-200 actual hours, or not more than 1 day's work per week (on the basis of a 30-week period). Clearly, any project that can be started and concluded in such a period must offer only a cursory introduction to scientific investigation. Any project scheme must suffer such limitations of timetable. Again, the time given to the mechanics of understanding the problem, actually executing an adequate amount of data reduction, and writing the final report must take up so much available time that there is little time to think the problem through — surely the most valuable part of the exercise. Unfortunately, there is ample evidence from many reports that this has been the case. Indeed there is considerable evidence from the project reports that many students fail to appreciate the astronomical context of the work, as well as failing to draw adequate conclusions from what they have done. The impression gained is "I was told to do this, that way, and I did it, but I don't know way I did it, or why I did in that way." Such a student has gained little scientific insight from the project. Other students will give a confident report of what they did, yet under questioning reveal little grasp of the underlying science. At best, such students may have learned a technique, but little else.

The scientific value of the work done is often questioned. Why should a student be permitted to waste time and resources to produce results that are not even publishable in the scientific literature? Is it not more valuable for the student to take extra courses, from which something useful can be learned? Since pressure on research resources is severe, is it not more sensible to reserve resources for the productive scientific workers? These are arguments which are not easy to answer, because they are true, although in a limited sense.

Of more importance is the psychological trauma inflicted by projects on the student. Many students will claim they are disadvantaged by examinations. In the case of standard examinations, where the situation is well understood by examiners and examined alike, the number of people actually disadvantaged is about an order of magnitude less than those who claim such disadvantage (on the grounds that the greatest number end up with lower 2nd-class degrees, though the argument does have an element of circularity!). In the case of projects, a significant fraction are actually disadvantaged by the project situation. This is in very large part because of reliance on standard essay-type examinations almost exclusively in the school and university system. The project is a new experience and most are not equipped to meet the challenge. The mere fact that a project cannot be resolved like a homework question is unsettling to many. To some, the sheer challenge may be completely overwhelming. The project route is littered with those who have effectively given up somewhere along the way. Some of these may achieve a poor passing grade, others fail unequivocally, yet others do not complete a report, and some do not attend for their talk. The project does impose additional strains on the student. They feel the project format does not allow them to acquit themselves in the way they feel does them justice — a justice they are perfectly capable of achieving in the standard examination format. The project talk often produces refusers. It is frightening to stand up and present ideas to an audience (an audience often more sympathetic than the speaker imagines) and then to respond to the challenge of being questioned by acknowledged leaders in their subject. It requires a cool head to appreciate that the renowned world authority is seeking a simple and not a sophisticated explanation.

The project is often critized on the grounds of subjectivity in assessment. It is argued that no supervisor will want to fail his or her own student — it would be a reflection on the research capability of the supervisor. Again, two assessors could disagree in their respective marking of the same project. The reasons could range from peer rivalry to diametrically opposed expectations of the student. It has been alleged that some project assessments are assessments of the supervisor. It is clear that assessments of projects will have a subjective element of necessity. A project that fulfills the aims and achieves the result sought is necessarily pleasing to a supervisor. A project that does not achieve the expected result, or falls short of expectation, may not please, but gratification of supervisors is not the object of undergraduate work. The value of a project to science may not be seen in the same light by two different assessors. Assessment problems will not arise with assessors of integrity, but any scheme must have very clear aims what a project examination is trying to measure about the student and the criteria on which that judgment should be made. Each assessor may weight the criteria differently, but the criteria themselves should be visible and agreed.

5. A Justification of Undergraduate Project Work

I am a believer in the opinion that undergraduate project work is a valuable teaching/learning experience and a useful measure of potential that is clearly differentiated from standard essay-question examinations. I recognize all the objections against projects and I feel particular concern for the stress that project work places on students. Each year produces a fresh crop of project difficulties related to student stress (some of the stress is self-induced through poor judgment on the part of the student) but learning to cope with stress in its many manifestations is part of learning to cope with life. Therefore, an element of stress is necessary in the project situation. However, assessment strategies must be flexible and sympathetic to cope with the effects of excessive stress. The form of the project must be well defined, so that students, supervisors and assessors are clearly informed on what is expected and the schedule that is being followed. But in order to cope with excessive stress (affecting 10-15 per cent of the students) the rules may need flexible interpretation, which is both fair to the stressed student and the remaining unstressed (!!) students.

To maximize the benefit of the project a fair assessment scheme needs to be devised. A very clear but very fine distinction has to be made. The project assessment *must not judge* the quality of the science implied by the project — it is to judge the quality of the science displayed by the student within the context of the project. This is the crucial distinction that has to be kept clearly in mind at all stages of the assessment. It is a distinction that is crucial to the success of the free-scope project. Set-piece projects can be set up in such a way that they all contain about the same amount to work even if some set-piece projects have more student appeal than others. The distinction just made has reduced force in the set-piece project area. In the free-scope project, the course directors must ensure that each and every accepted project has sufficient substance. Beyond this, however, the assessors must look for:

a) student awareness of the scientific context of the project;

b) the understanding the student shows of the problem to be tackled;

c) the contribution of the student to a solution of the problem;

d) the logistical approach of the student

e) the ability of the student to draw conclusions from the work done;

f) the ability of the student to report scientific context, the problem to be tackled, the methods used, and to present data and draw conclusions both in writing and orally (including the quality of the final report);

g) the breadth of the student's reading around the project topic.

Other factors may enter into particular assessments. The weight given to factors (a) to (g) will vary with assessor. There is no realistic way to enforce uniformity, and indeed it would be counter to the spirit of student projects to seek such uniformity. However, all assessors should agree on the fundamental basis of the assessment scheme. In the assessment, supervisors will place emphasis on (c), (d), and (e). A moderator might put more emphasis on (a), (f), and (g). Clearly moderator and supervisor will be looking at the student from well-separated points of view. A second assessor, neither supervisor nor moderator, may wish to emphasize (b), (e), and (f). It is interesting to reflect that assessors will sometimes agree to within 1 per cent, often to within a degree class, but occasionally vary by several degree classes in their assessments. The experience of normal agreement to within a degree class is encouraging and removes much of the criticism of project assessment as subjective. To achieve this objectivity, a minimum of 3 assessors for each project is essential — the supervisor, an assessor close to the science, and a moderator. The moderator must assess all the projects presented in a given year, and have sufficient authority to decide the relative standing of a project among its peers. How the moderator achieves this goal is quite open. I have found discussion with the other two assessors normally resolves any difficulty. In the last analysis, the moderator must have final responsibility for the assessments.

The assessment scheme must also be made clear to the students. The students must know the time schedule for the projects and be aware of the important deadlines at the start of the project period. They must be told of the mechanics of the assessment and the basis on which it is made. The students should also know who their assessors will be (from the beginning the supervisor and moderator are clearly identified — only the second assessor in the UCL Astronomy scheme is not selected until the project reports are in the writing-up phase). Students can then see that the system is fair to them and be aware of what they have to achieve and when. A fair scheme that is well known to all involved goes a long way to resolve most of the criticisms that assessment is subjective, that one assessor marks another's science, and that students receive only rough justice. Nevertheless, the scheme only works optimally if all assessors approach their task with integrity.

The criticism that projects are a waste of scarce resources is much harder to counter. Projects do consume resources. One can always restrict the maximum expenditure for projects in actual cost terms; *e.g.*, at UCL no project can exceed an expenditure of £200 on materials. However, there is no quantifiable estimate of what each project costs in terms of computer resources, technician time, other support staff time, and supervisor's time. There is no doubt that such costs would be substantial. The criticism of diversion of resources has foundation — projects are costly. The argument must be based on balance. Degree-level students represent a substantial investment in educational resources. In U.K. terms, a current estimate would be that £20,000 has already been invested in each undergraduate by the time they enter the final year if only recurrent expenditure is considered. Therefore their final year should offer the best opportunities to maximize the return on outlay. It could be argued that the best also includes a range of experience. There is no doubt that project work represents a very different learning process from attending lectures. Students are placed in a situation where they must use their accumulated knowledge and experience to tackle a problem whose solution may not be obvious. Most science graduates are going to meet exactly such situations for the bulk of their working lives and a brief introduction is therefore of value. If you can afford it, the project is a useful luxury.

But even if one is willing to devote resources to the project, should strain on students be considered a major disadvantage? There is no doubt that project work enforces strain on students. While most students enjoy the actual project work and research group contact that goes with it, there remains the deadline of the report submission date and the implication that something must be achieved in a finite time. Those requirements can cause stress, sometimes severe stress. There are students who do not enjoy the freedom of the project work — they are discouraged by the amount of information to be acquired, new techniques to be learned and the sheer exasperation of failure to get the problem to yield. They often become depressed by how little they appear to have achieved and can magnify out of all proportion the alleged progress of their peers. Such depression can have disastrous effects not just for the project course but also for other more conventional course work as well. The project leaves its mark and is not neutral — it can both make and break. This is why at UCL we have the safety net of a half-unit essay review type of project. There must be a safety net to catch those who are not coping in the full-blown project situation. Those who do not cope must not be regarded as failures, otherwise injury to self esteem may injure their chances in standard course work. The project situation should be tough and be accepted as such, if it is to have educational value, but should always be engineered with loopholes so that those who do not cope can attain a recognized and respected fall-back position. The value of the fall-back position should be made clear to all students at the outset. With such safeguards, the strain on students can be kept within acceptable limits.

It has been found in practice that the most stressful part of the UCL Astronomy Scheme is the project talk — given before an audience of assessors and students. For most students this is their first experience of speaking in public. They must put their ideas on the line and defend what they have done. All students are nervous, some excessively so. Only a few manage a good performance striking the right balance between context, conclusions, and methods with appropriate illustrations. Most stagger through, and their inexperience in knowing how best to illustrate their work shows up painfully. But they are not being assessed for their elegance of presentation and the experience is an invaluable learning experience even if a hard one. But worse is to follow — questions. About half the students cope with questions. The other half stagger blindly from straw to straw and some are reduced to silence by what they perceive as sophisticated questions from the internationally famous in their subject. But the more internationally famous the questioner is, in my experience, the simpler the answer that is required. Although told of this, it does not conform with undergraduate perception and mythology! However, all students after the talk would agree it was not as bad as anticipated. It is the anticipation that drives some to hide in their lodgings or the Union Bar, or anywhere they think they will not be found. In extreme cases, some students are allowed to give the talk only before their assessors. I would have doubt about the value of the project talk — were it not for the fact that talking about one's work is such an integral part of scientific life.

Conclusions

The project serves a purpose in science education — it allows the students freedom from the textbook or lecture room situation and gives them time to investigate for themselves. They learn about actual data as they come from real instruments raw and uncorrected. They learn about signal/noise in a direct way and the weight that should be assigned to conclusions. A few students take to projects like a duck to water, most learn to live with them, a few are disadvantaged by them. Projects are expensive in the time of academic staff who act as supervisors and their research groups. Projects are also time consuming to assess adequately and fairly. Without the investment of adequate manpower resources, projects are likely to be a disaster. With proper investment, projects can be a valuable learning strategy in which students learn perhaps more about themselves and their abilities than in any other learning situation. While recognizing the difficulties and dangers, it would be my view that projects win by a small but significant margin.

Discussion

L.C. Hill: *Have you found students who performed at a significantly different level (either better or worse) in their project as opposed to their performances in other academic work?*

D. McNally: By and large, good academic students do well on their projects. However, some 2rd and 3th class students have turned in professional, polished projects. You also find that some 1st class students (on examinations) find projects frustrating because progress is slower than they are accustomed to.

Comment

Jay M. Pasachoff: We have a similar program of research projects at Williams College, and we view it as a major enticement for students to come to Williams and to major in astronomy. Still, the students often come up against the same problems so perceptively listed by Professor McNally. I think that giving Professor McNally's article to new undergraduate thesis students, as I shall surely do, will help them greatly.

MAXIMIZING THE EDUCATIONAL VALUE OF STUDENT RESEARCH PARTICIPATION

Emilia Pisani Belserene
Maria Mitchell Observatory, 4 Vestal Street,
Nantucket, Massachusetts 02254, U.S.A.

1. Introduction

The Maria Mitchell Observatory monitors variable stars photographically and has lately begun to receive photometric data remotely. The staff consists solely of the director and undergraduate student assistants during summer and January vacations. Research topics are chosen for both their scientific interest and their educational potential.

The scientific goal is to improve variable-star statistics by answering any unresolved observational questions. Lately the emphasis is on pulsators in the Cepheid instability strip. Can we watch the stars grow older? We look for deviations from a single, constant period.

The educational goal is to give the student assistants a realistic sense of the research process, beyond what is usual in early undergraduate years. The selection criterion is a university record such that a career in astronomy is a realistic possibility. The observatory is a memorial to America's first woman astronomer. In the selection of students and in the assignment of tasks, the program tries to be especially careful to give women equal opportunities and equal encouragement.

How, then, to divide the work among the students? Not necessarily for the greatest astrophysical efficiency. No darkroom specialists, programming specialists, *etc.* They should all sample a variety of tasks. Each student is put in charge of at least one star.

Typical tasks:

> Inspecting plates. Measuring magnitudes.
> Computer processing.
>
>> Data entry.
>> Period search (Fourier, *etc.*).
>> Graphic output (Light curves, O-C diagrams, ...)
>> Least Squares.
>> Data-base management.
>> Computer-program revisions.
>
> Astrometry.
> Library search for other observations, theory.
> Writing and presenting a paper.
> Blinking for additional variable stars.

All students also work cooperatively at some of the tasks, such as taking and developing plates, presenting colloquia to each other, and running programs for the public. The key feature of the division of labor, however, is that each student is, in effect, the principal investigator for a portion of the project.

What does science gain?

> Improved variable-star statistics
> Potential talented recruits.

What do the participants gain?

> Exposure to research methods and the professional literature.
> A preview of a potential career.
> Contact with students from other campuses and with visiting astronomers.

Does the research suffer from the educational emphasis?

Not much. There is some loss of efficiency, since all assistants are learning all parts of the work. There may be a slight observational bias in the variable-star statistics, since variable stars are put on the program at least in part for their educational usefulness.

How do former participants view their research experience? A survey elicited these comments:

The Maria Mitchell experience was ...

...research on a level suitable for undergraduates ....

...our first shot at publishable science ....

...a perspective on the scientific process ....

...not just data entry and number crunching ....

...real research with all its joys and discouragements ....

...nurturing and encouraging to budding scientists ....

It was valuable to ...

...get experience with real data ....

...be able to complete an entire project ....

...work for a woman ....

...become familiar with important reference works ....

...collect and analyze data when the results are not known in advance ....

...follow a project through from taking plates to publishing results ....

...see how real research is done ....

...realize early that research isn't quick and glamorous all the time ....

We learned to ...

...do literature searches ....

...write up results for journals ....

...use computers, view plates, look up sources ....

I work on Crays now but first became comfortable with computers on the little one at MMO.

We were exposed to ...

...the need for careful, meticulous work ....

...the satisfaction of getting scientific results ....

...a valuable historical perspective ....

...an opportunity to try a research job ....

...the example of scholarship set by Dorrit Hoffleit.

The MMO experience was an important factor in making the decision to become a professional astronomer.

Appendix: The Survey of the Participants

The responses are from a survey that attempted to reach as many as possible of the alumnae and alumni of the program. Responses came from nearly one-third of the 100 or so students, all but three of them women, who worked under Dorrit Hoffleit during 1957–1978 and from nearly half of the 58, 60 per cent of them women, who worked under the present author from 1979 to date.

The respondents did, as was to be expected, include a large number working or studying in astronomy and related fields. The percentages were 68 per cent working and 7 per cent studying among the students from 1957–1978; 14 per cent working and 46 per cent studying among the students from 1979–present. For the purposes of these statistics, "related fields" are defined to include physics, mathematics, and history and philosophy of science. The percentage of respondents who identify themselves specifically as astronomers or as graduate students in astronomy or astrophysics are 26 per cent and 14 per cent, respectively.

Under both directors, most of the work was photographic research on variable stars, usually leading to the presentation and/or publication of a paper. Some of the students' time was devoted to running open nights at the telescope, giving lectures for children, and hosting a series of lectures by visiting astronomers. There was also a weekly seminar on variable-star astronomy at which each participant presented at least one paper.

The questionnaire asked about present and former occupations and invited answers to four questions:

1. What aspects, if any, of the Maria Mitchell student internship program in your day were most useful to astronomy, in your opinion?

2. What aspects, if any, of the Maria Mitchell program in your day were most educational for the student participants, in your opinion?

3. Which part(s) of the Maria Mitchell program would you encourage me to change to increase the benefit to science?

4. Which part(s) of the Maria Mitchell program would you encourage me to change to increase the educational benefit to the assistants?

The summary given above and presented in the poster version of this paper drew on those portions of the answers to question 2 that addressed the educational benefit of the research portion of the work. Many of the respondents also mentioned the educational benefits of meeting the visiting astronomers, working with the public, and giving seminar papers.

The other questions, while not strictly related to the topic of this paper, elicited interesting responses that seem worthy of being summarized are here for use of astronomers planning out-of-the-classroom activities for students.

In addition to improved data on variable stars, two other benefits to astronomy appeared among the responses to question 1. The Maria Mitchell program was viewed as aiding astronomy by encouraging potential recruits to the profession, and also, through the public programs, by increasing the appreciation of astronomy among members of the public.

Among the suggestions for increasing the benefit to science (question 3) were to introduce newer technologies and/or topics closer to "cutting edge" astronomy than survey projects on variable stars. Respondents whose data reduction had been done in the years of desk calculators were pleased that the Observatory now has computers.

Newer technology also occurred among the suggestions to increase the educational benefit (question 4). Other requests: lessons in Fortran; more understanding of what the computers are doing; more opportunity to write programs; the preparation for publication of joint papers on larger aspects of variable star astronomy; reading assignments on the topics presented to each other at the seminars; more guidance about choosing graduate schools and advisers; hard information about being a woman in astronomy.

The suggestions elicited by question 4 raise two interesting questions. How much are we losing, in terms of student understanding of the research process, when we use the power of computers in an educational setting? And, considering that some of the requests are for activities available on the campus, are the usual academic techniques, after all, more useful to students than simply doing research tasks as research assistants? I answer confidently, from the enthusiastic response to the research program, that a taste of research is a very valuable addition to classroom and laboratory study. But we should be careful to consider the educational aspects of our use of student assistants and work hard to encourage understanding of the techniques to which we introduce them.

What does our project suggest to others employing students in research? If you are working on a hot topic, use your student assistants in any way at all; they will be glad to contribute in no matter how routine a way. Otherwise be willing to sacrifice some efficiency. Set up the tasks to maximize educational benefits. Your data will be useful even if unspectacular, and your assistants will have a realistic, early taste of research. Whether or not they become astronomers, they will carry this knowledge of scientific processes into their future lives. The value of the program should not be measured solely by the number who go on in astronomy, I heard from some of the respondents who are not astronomers. The lessons of scholarship are useful to parents, educators, and others.

Finally, what did the survey reveal about the later career paths of women who were students of astronomy in 1958–1987? Some revealing comments:

...astronomer 100 per cent, housewife 40 per cent — we all work 120 hr weeks ....

...managed to have a career in astronomy while being a mother to three children

— it has been difficult, however ....

...engineer 40 hrs/week (=50 per cent of time) + mother (more an art than a
science) ....

...worked half time for 12 years while children were small ....

And, from someone identifying herself as a housewife/programmer analyst,
"my husband's job and the two children have taken precedence."

Acknowledgments

Variable-star research at the Maria Mitchell Observatory has received funding from the National Science Foundation, currently under AST86–19885, which
includes an allotment from the program Research Experiences for Undergraduates.
Since 1979, one of the students each summer receives support from the Dorrit Hoffleit Assistantship Fund of the Maria Mitchell Association. The use of remote photoelectric observing was funded by a Theodore Dunham, Jr., grant for research in
astronomy.

THE EDINBURGH ASTRONOMY TEACHING AND EDUCATIONAL PACKAGES

M.T. Brück and S.B. Tritton[1]
Department of Astronomy, University of Edinburgh
[1] U.K. Schmidt Telescope Unit, Royal Observatory Edinburgh, Blackford Hill,
Edinburgh EH9 3HJ, Scotland

1. Introduction

In this paper we describe our experiences with film copies of original astro-
nomical photographs taken with the 1.2-meter U.K. Schmidt Telescope (UKST) in
Australia, and used as teaching material in the Department of Astronomy of the
University of Edinburgh. Two packages are intended for undergraduate use; the
Education Packages are designed as visual aids for colleges, schools, and amateur
groups.

The original purpose of the telescope (which was commissioned in 1973) was
to carry out a Southern Sky Survey to match the Northern Survey done by the
Palomar 48" Schmidt Telescope. The telescope has a very wide field — 6.5 x 6.5
degrees, or equivalent to over a hundred and fifty full moons, and the photographs
reach objects of 23rd magnitude: they record stars like the Sun to the very edges

of the Galaxy and galaxies to a thousand million light years. Each photograph has an area of 356 x 356 mm and records between 100,000 and 1 million stars and galaxies. The survey photographs are the deepest available maps of the sky and are indispensable tools for astronomers in searches for unusual objects, for investigation of the distribution of galaxies and many other tasks. The UKST is equipped with objective prisms that are capable of producing low dispersion spectra of stars and galaxies. A prism of very small angle is placed in front of the aperture of the telescope so that each individual image is drawn out into a tiny spectrum on the photograph. The dispersion is very small, only a few millimeters from the red to the ultraviolet, but the dominant features in the spectra are recognizable and have many applications.

The films that we use in our Teaching Packages are, like the original photographs, negatives, with the stars as black images on a transparent sky background. The original photographs are taken on glass; the copies are contact reproductions on film (as is the Southern Sky Survey) so that every detail on the original is retained. The use of these films for teaching purposes has many advantages. On the purely practical level they are relatively cheap; they are also unbreakable and, if kept enclosed in their transparent envelopes and treated with common-sense care, have an indefinite lifetime. A major attraction is that they constitute the highest quality modern observational material such as is currently used for advanced astronomical research.

Much of the research carried out with Schmidt photographs involves measuring dimensions of features, counting numbers of objects, and classifying objects. Our exercises also involve measuring, counting, and classifying. Some exercises are elementary and can be carried out in about one hour by first-year students, many of whom are non-physicists. For these, the only equipment required is a millimeter ruler, a light box, and a hand magnifier (preferably fitted with a graticule). For some of the other exercises a light table with a high-power microscope is recommended and a simple photometer allows one or two additional exercises to be done. At Edinburgh, we also use the films for more advanced projects that are in the nature of research projects occupying up to thirty hours for final year honors students of astrophysics: students can carry out genuine projects that yield answers to real problems and in which they encounter real difficulties.

The photographs need to be accompanied by other information and by some level of theoretical knowledge. We have included the relevant additional data with each exercise to make it self-contained. The exercises have become part of the practical course in astrophysics in the University of Edinburgh and have been found very successful not only for the experience of carrying them out but also for the results.

In our Packages, we have chosen sets of films that include examples of typical objects of astrophysical interest ranging from the solar system to distant galaxies. Worked solutions for four of the simple exercises and a summary of one more advanced exercise are given in the Appendices at the end of this paper.

2. The Solar System

Our first example (Appendix 1) is the simplest, a study of images of asteroids on photographs of fields in the plane of the ecliptic. During the one-hour exposure time, asteroids that are in motion relative to the background of fixed stars leave trails rather than dots on the photograph. The actual calculation in the exercise is in principle simple: to deduce, from the length of the trail, the distance of the asteroid from the Sun assuming circular orbits and Kepler's third law. The practical matter of converting trail lengths to circular angular measure and of realizing that the final answer has to be obtained by trial, makes this an excellent puzzle, which takes about an hour to solve.

The exercises on comets (Appendix 2) are at a similar level of difficulty. The student is asked to demonstrate that the comet's tail points away from the Sun and to calculate the unforeshortened length of the tail, given appropriate information about the comet's path. The two films of Comet Halley taken on successive days show moving knots of material in the comet's ionic tail. The speed of the knots, which are being swept along in the direction of the solar wind, can be calculated.

3. Galactic Objects

Moving outwards among the stars, there are examples of various typical constituents of our Galaxy — star clusters, dark dust clouds, reflection nebulae, and a supernova remnant. The Vela supernova remnant is a luminous shell of ionized gas that outlines the boundary between the material bursting outwards from the explosion of a star in the final stages of its evolution, and the interstellar medium through which it moves. By chance the whole of this shell fits into one Schmidt photograph. The distance of the remnant is known from radio observations. Radio observations have also discovered the pulsar that is the remains of the exploded star, located about one degree away from the center of the shell, having recoiled at the explosion from its original position. The practical part of the exercise, in which the linear dimensions of the shell and the linear displacement of the pulsar are calculated, can be done within one laboratory session. The exercise also introduces the student to the theory of shocks in the interstellar medium.

So far the exercises involve only linear measurement. We now describe some that involve the counting of images and which are intended for longer projects. We have an example of a globular star cluster (Appendix 3). The question is how massive must a cluster be in order to withstand what is in effect the tidal force of the Galaxy. The mass is calculated indirectly by observing the apparent distribution of stars in its outer regions. The task set in the exercise is to count the numbers of stars per arbitrary unit of area radially outwards from the center of the cluster; at least several hundred images have to be counted. The exercise is a test of care and patience; it is incidentally also valuable practice in dealing with number statistics. An advanced exercise involves matching the distribution of stars in the cluster to a mathematical model.

We have also chosen a typical field in the galactic plane that very strikingly illustrates the enormous numbers of stars in the Milky Way and the presence of dark clouds of interstellar material. In fact, the apparent dearth of stars in some patches of sky is caused by loss of light from the stars behind, through scattering by small particles of dust. By counting the relative densities of stars inside and outside the area of a dark cloud, it is possible to calculate the fraction of light being removed by the dust. From the known properties of scattering grains, the total number of grains in the line of sight may be calculated and an estimate made of the mass of the cloud. The actual star-counting part of this exercise is not difficult in principle, but the result is exciting and the student is introduced to the theory of the scattering of light.

Two films taken with the objective prism are provided (Appendix 4); one is of a Milky Way field and the other of a high galactic latitude field. At these dispersions only the dominant features in the spectra can be distinguished. Spectral classification into the main classes is an obvious exercise that is particularly instructive if it is performed on contrasting regions in the sky. The differences are reflected in the spectral types of the stars, and it is very satisfying for a student to discover this.

4. Extragalactic — The Magellanic Clouds

In the extragalactic sphere there are many possibilities for student exercises to be found on photographs of the Magellanic Clouds. On deep photographs they extend over a large area of sky and appear on several Schmidt photographs of which a few have been selected for our Packages. One exercise is a survey of star clusters in the Large Cloud. The Large Magellanic Cloud has an irregular spiral structure and is fortuitously oriented in such a way in space that from the point of view of the observer on Earth its disk is seen almost face-on. Several hundred clusters, easily recognizable under low magnification, are identified. Their average number density plotted as a function of radial distance from the center gives the distribution of mass within the disk. The distribution turns out to be exponential, in accordance with standard models of galactic disks.

An important use of objective prism spectra is to search for objects with unusual characteristics, in particular objects with emission-line spectra that show up conspicuously on photographs by contrast with the normal absorption spectra of the vast majority of stars and galaxies. Emission lines are found in very young stars with hot gaseous envelopes and in planetary nebulae. The spectra of these types are distinctive and a student soon learns to tell the difference. An objective prism film of part of the LMC where there are large numbers of emission line objects of all kinds has been chosen for this exercise.

5. Extragalactic — Clusters of Galaxies and Quasars

One of the most useful films in the Package is a photograph of the Virgo cluster of galaxies. The most obvious use of the photograph is for the classification of galaxies according to the "tuning-fork" scheme that distinguishes elliptical from spiral galaxies and further subdivisions. The photograph shows such fine detail that an exercise can also include measuring the diameters of galaxies, estimating their linear dimensions and observing the number of spiral arms in the giant ones. The same photograph shows many very distant background galaxies whose distances may be inferred from their apparent diameters.

Another example is Abell 1060 (Appendix 5), a cluster of galaxies rather similar to the Virgo cluster but at a considerably greater distance. This advanced exercise leads to quite a sophisticated analysis of the internal dynamics of the system when the spatial separation of the galaxies as deduced from counts is combined with motions of individual galaxies, which are known from spectroscopic Doppler shift observations. The final result is a good estimate of the total mass of the cluster. There is a considerable amount of recording and computation and, if carefully done, the result obtained bears comparison with the research investigator's. It is especially instructive to compare the mass calculated by this method with that expected from the total sum of individual galaxies. The student, taking all the uncertainties into account, finds himself debating as to whether there is or is not evidence for "missing" or invisible mass in the cluster.

An exciting exercise is to search for quasars that are identified from their peculiar emission line spectra. There are thousands as yet undiscovered on UKST objective prism photographs. A student, with no more equipment than a hand lens, can expect to find 30 or 40 on each of the films which we have included in our Package; with rather more labor he can measure their redshifts and, adopting a suitable model for the expanding Universe, estimate their distances.

6. Educational Packages

These packages have been made up from film copies originally made for the ESO/SERC Southern Sky Survey but which failed the very stringent quality control criteria. Each package contains films of different areas of the sky and contains examples of most types of astronomical objects. Care has been taken to include films in each package which show the high density star fields of the Milky Way and also areas of the sky rich in galaxies. Comets, asteroids, star clusters, planetary nebulae, dust clouds, reflection and emission nebulae, and spiral and elliptical galaxies are all represented. Each film is accompanied by a sheet listing some of the interesting objects which can be seen. No attempt has been made to provide exercises for these films which are therefore mainly useful as visual aids.

7. Details of Packages

1. Edinburgh Astronomy Teaching Package for Undergraduates (M.T. Brück 1983, revised and expanded 1988) (£100): Ten film copies together with a set of notes suggesting a variety of exercises. The notes are intended for teachers who can then choose the most suitable exercises for their particular class. General solutions and references, rather than complete detailed workings, are provided for numerical problems. The exercises are on Asteroids, Comets, Globular Clusters, the Galactic Plane, the Vela SNR, the LMC, and Clusters of Galaxies.

2. Edinburgh Astronomy Educational Packages for Schools (£50 per set): These packages are mainly designed as visual aids. Each package contains 10 films of different areas of the sky and each film is accompanied by a brief sheet describing the most interesting objects that can be seen. Five separate packs are available.

3. Edinburgh Astronomy Spectroscopic Teaching Package (M.T. Brück and S.B. Tritton, 1988) (£100): Eight film copies (5 from plates taken with the objective prism and 3 from matching direct plates) and a set of notes suggesting exercises as in Package 1. The exercises include spectral typing, searching for emission line stars, and searching for quasars. The package also contains the Atlas of UKST objective prism spectra (Savage *et al.*, 1985).

Acknowledgments

The authors thank Mrs. E. Gibson for her patient typing and the U.K. Schmidt Telescope Unit and Photolabs, Royal Observatory, Edinburgh.

References

M.T. Brück, 1986, "Photographic Teaching Packages in Astronomy," in Proceedings of the GIREP Conference 1986: *Cosmos — an Educational Challenge.* ESA-SP-253.

S.B. Tritton, 1986, "Educational Packages for Schools and Amateur Groups," in Proceedings of the GIREP Conference 1986: *Cosmos — an Educational Challenge.* ESA-SP-253.

A. Savage *et al.*, 1985, "The UKST Objective Prisms." Royal Observatory, Edinburgh.

Appendix 1: Asteroids or Minor Planets

Plate 1: Several trailed images of asteroids in the ecliptic plane photographed at opposition.

Exercise: Measure the length of the trails and calculate, assuming circular orbits, the distance of the asteroids from the Sun in Astronomical Units. The measurements are made with a graticule with 1/10 mm divisions, viewed on a light box with a x10 hand magnifier. Trail lengths are estimated to a 1/4 of a division.

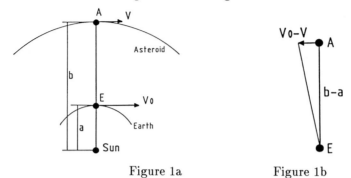

Figure 1a   Figure 1b

Let   a = distance of the Earth from the Sun
         b = distance of the asteroid from the Sun
         V_0 = Earth's orbital velocity
         V = asteroid's orbital velocity
         ω = angular motion of the asteroid as seen from Earth
            = $(V_0 - V)/(b - a)$ radians per second

By Kepler's third law, circular velocity is inversely proportional to the square root of the radius of the orbit, therefore

   V = $V_0(a/\sqrt{b})$, which gives
   ω = $\omega_0 a(1 - \sqrt{a/b})/(b - a)$ where $\omega_0 = V_0/a$, the angular velocity of the Earth in its orbit around the Sun.

If a and b are in astronomical units, $a = 1$.
   ω_0, the Earth's angular velocity, is 360 degrees per year (3.16×10^7 seconds) or $360 \times 3600/3.16 \times 10^7$ seconds of arc per second (or minutes of arc per minute). Substituting the numbers and simplifying the algebra by factorizing $(b - 1)$ into $(\sqrt{b} - 1)(\sqrt{b} + 1)$, one gets

      ω = $4.10 \times 10^{-2}/(b + \sqrt{b})$ minutes of arc per minute for the
      rate at which the asteroid moves as seen from the Earth.

A typical trail on the film is 0.62 mm or 0.70 arcminutes (given the scale of the film of 1.12 arcminutes per mm). The exposure time was 70 minutes; therefore the observed motion ω of the asteroid is 0.01 arcminutes per minute.
   The problem is to find the value of b that will make $\omega = 0.01$, that means solving $b + \sqrt{b} = 4.10$. It is quicker to find b by the trial and error using a hand

calculator than to solve the quadratic equation for $\sqrt{b}$. Try whole numbers first and then smaller steps: for $b = 2$, $b + \sqrt{b} = 3.41$; for $b = 3$, $b = \sqrt{b} = 4.73$; for $b = 2.5$, $b = \sqrt{b} = 4.08$.

Now estimate the accuracy of the result. The uncertainty lies in the measurement of the trail length (all other numbers used in the calculation are very accurate). The trail length was estimated to a quarter of a division, *i.e.*, 0.025 mm in a length of 0.62, or 1 part in 25. The calculated motion of the asteroid and also the number $b + \sqrt{b}$ have therefore the same uncertainty of 1 part in 25. If the calculation for b is repeated using the upper and lower limits (4.10 ± 0.16) for $b + \sqrt{b}$, the limits for b are found to be 2.4 and 2.6. Thus the result for the distance of the asteroid from the Sun, assuming circular orbit, may be given as 2.5 ± 0.1 Astronomical Units.

Appendix 2: Halley's Comet

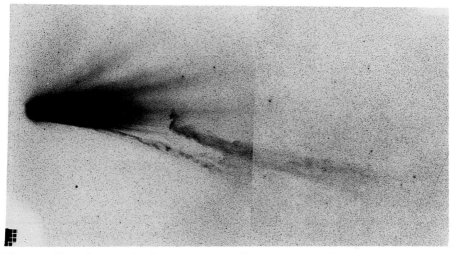

Plate 2: This photograph shows Comet Halley on 10 March 1986. It is made from two original photographs of adjacent areas of the sky to enable the complete tail to be seen.

This photograph is provided together with two films showing Halley's Comet on consecutive days, March 9 and 10, 1986, and taken exactly 24 hours apart. The coordinates of the comet and the Sun and also the distances of the comet from the Sun and from the Earth on the two dates are given. The photographs show a discontinuity in the plasma tail where the tail or part of it has broken away from the head of the comet and is separating from it. The actual break or "disconnection event" is known to have occurred on March 8.

Exercise 1. Measure the length of the plasma tail on the composite photograph and calculate the true length in Astronomical Units.

From the coordinates given (see the notes) for March 10, mark the positions of the Sun and the comet on a globe and find the angle between them.

Measure the length of the tail on the composite print with a millimeter ruler. Measure from the sharp front of the comet's head; the end of the tail, which is faint, is not so easy to estimate. Find the scale of the print by comparing the distances apart of some bright star images on the print and on the film which has a scale of 1.12 arcminutes per millimeter. Convert the length of the tail to degrees.

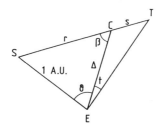

Figure 2

The diagram (Figure 2) shows the configuration of Earth, Sun, comet, and tail. The distance of the Earth from the Sun is 1 Astronomical Unit (A.U.); the distance of the comet from the Sun is r A.U. and of the comet from the Earth δ A.U. The length of the tail is s A.U. The angular distance between the comet and the Sun as seen from Earth is θ; the angular length of the tail is t.

Solution by trigonometry: Solve the triangle SCE to obtain the angle β. In the small triangle CET the angle at T is $(\beta - t)$; the sine formula in this small triangle gives

$$s/\sin t = \delta/\sin (\beta - t)$$

Substitute

$\delta = 1.05$ A.U. (given)

$\theta = 49°$ (measured on globe)

$t = 10°$ (measured on photograph)

$\beta = 62°$ (from triangle)

The result is $s = 0.23$ A.U.

Solution by construction: The solution may also be found without the use of trigonometry. On millimeter graph paper draw the triangles to scale, starting with the line SE and using a suitable scale (*e.g.*, 10 cms to 1 A.U.). Mark the angle θ with a protractor; draw a line along this direction and mark off the position of C to scale. Mark also the angle t and the direction to T. Join S to C and continue until the line crosses the other lines at C and T. Measure the distance s with a millimeter ruler to give the length of the tail on the scale of the diagram.

Exercise 2. Calculate the average velocity of the disconnected tail relative to the comet in meters per second, assuming that the motion is in the direction of the tail.

Measure the distance of the disconnection from the head of the comet using a millimeter ruler on each of the films and convert these distances to degrees. The subsequent calculation is the same as in Exercise 1, replacing t in the diagram by the difference between the first and second positions of the disconnection. The angle θ in this case is the angle between the Sun and the comet plus the angle between the comet and the position of the disconnection on March 9. The distance moved should be given to decimals of a degree.

Angle β (from solution of triangle or from construction) = 60°

Angular distance moved by the disconnection between March 9 and March 10 (t) = 102 mm on film = 1.9°

Δ (mean for March 9 and 10, given) = 1.06 A.U.

Linear distance moved $(s) = \Delta \sin 1.9°/\sin 58° = 0.041$ A.U.

Interval between the two photographs = 24 hours; 1 A.U. = 1.5×10^{11} meters. Therefore rate of motion of disconnection = $0.041 \times 1.5 \times 10^{11}/24 \times 3600 = 7.1 \times 10^4$ meters per second.

Appendix 3: Globular Cluster

Globular clusters, which may contain 100,000 to a million stars, have very high densities of stars in their central regions; their centers are therefore quite unresolved on photographs. Globular clusters are held together by their own gravity, which is able to withstand the disrupting gravitational influence of the Galaxy as a whole. An estimate of the total mass of a globular cluster may be made by assuming that the cluster is tidally limited by the Galaxy, that is, that the differential gravitational force between the extremities of the cluster exercised by the Galaxy is balanced by the self gravitational force of the cluster itself. The effect of the differential gravitational force is similar to the tides on the Earth's oceans caused at opposite sides of the globe by the Moon.

Let r be the distance of the cluster from the center of the Galaxy, looked upon as a point where the mass of the Galaxy interior to the cluster is situated. Let R be the radius of the cluster. One side of the cluster is nearer to the Galaxy and the opposite side more distant from it by an amount R. This differential or tidal force is not capable of disrupting the cluster because the cluster has sufficient mass to hold itself together. The balance depends on the ratio of the masses of the cluster and the Galaxy, and the ratio of the cluster's radius R to its distance from the center of the Galaxy r. The formula connecting these ratios, assuming that the cluster is in a circular orbit around the center of the Galaxy, is:

$$m/\text{M} = 3.5 \ (\text{R}/r)^3$$

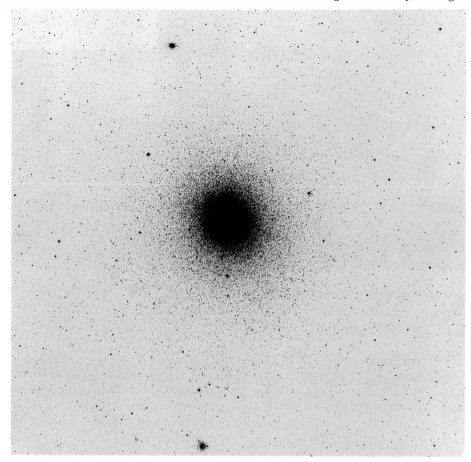

Plate 3: The large globular cluster 47 Tucanae (NGC 104) as it appears on one of the films. The small clusters seen belong to the Small Magellanic Cloud, which also appears on the film.

Exercise 1. Find the radius of the globular cluster from radial star counts. The number of stars per unit area is to be plotted as a function of distance from the center of the cluster. It is unrealistic to attempt to count all the stars; counts in one (or more) radial directions should be made, using a transparent grid laid over the film on the light box (Figure 3a). A grid with 1 mm squares is very suitable and viewing should be done with a x10 hand magnifier.

Choose a diameter of the cluster which is free of large star images which might interfere with the counting. Lay the grid very carefully so that the row of squares to be counted goes through the very center of the black core of the cluster. Attach the grid firmly to the protective envelope of the film. Count and note down the number of stars per square and distance of each square from the center, continuing the counts until well beyond the obvious termination of the cluster.

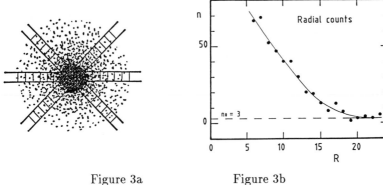

Figure 3a   Figure 3b

Many very faint stars will be seen. It is not necessary to include the faintest ones, but it is important to count to the same limit of brightness throughout. As the center of the cluster is approached the stars become very numerous, with many overlapping and blended images. However, these regions need not be counted; the purpose is to find the outer limit of the cluster.

Plot the numbers as a function of the distance from the center. The limit of the cluster is the point at which the numbers level off to a constant number, representing the foreground and background stars in the general field. This limit is not always easy to assess because of statistical fluctuations in the numbers of field objects. The cluster may not be perfectly symmetrical, and the field objects not uniform; therefore the more sets of counts made, the better the estimated result.

Figure 3b show one sample radial scan. The background count is 3 per unit area and the limit is estimated at 22 mm on the original film.

The distance of the cluster (given) is 4.1 kiloparsec. The radius is 22 x 1.12 arcminutes. Convert to radians and multiply by 4.1 to give the radius in linear measure. (1 radian = 2.06×10^5 arcseconds). The result is 0.029 kiloparsecs or 29 parsecs.

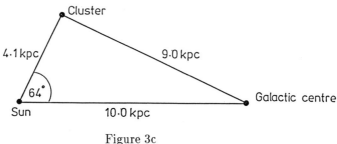

Figure 3c

Exercise 2. Calculate the mass of the cluster from the tidal radius and other data.

The coordinates of the cluster (in galactic coordinates) and the distance of the Sun from the galactic center are given.

From the coordinates the angular distance θ between the cluster and the galactic center (coordinates 0,0) are found either by trigonometry or by using a globe. In the triangle in the diagram, the distance r of the cluster from the galactic center is found by trigonometry or by drawing the triangle to scale (Figure 3c).

Coordinates of the cluster (306°, - 45°). Angle $\theta = 64°$ (from globe or by trigonometry). Distance to galactic center = 10 kiloparsecs (given). Distance to cluster = 4.1 kiloparsecs (given). Distance of cluster from galactic center (r) = 9.0 kiloparsecs (from triangle). Radius of cluster (R) = 29 x 10^{-3} kiloparsecs (exercise 1). Mass of Galaxy (M) = 2 x 10^{11} solar masses (given).

$$m/M = 3.5 \times (29 \times 10^{-3}/9.0)^3 = 1 \times 10^{-7}$$

$$m = 2 \times 10^4 \text{ solar masses}$$

Accuracy of the result. The value of the tidal radius obtained from star counts probably tends to be underestimated because in the outer zones a very low level of counts is not distinguishable above the background ("noise") level. The distance to the center of the Galaxy and the mass of the Galaxy are quantities which are not yet agreed on with certainty.

Appendix 4: Spectral Classification

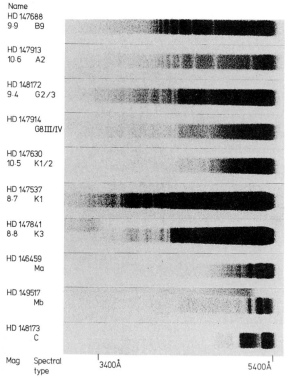

Plate 4: Examples of standard stars of different spectral types as seen on the objective prism film copies.

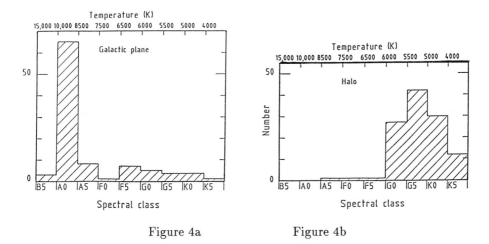

Figure 4a   Figure 4b

With a little practice it is easy to classify stars to one spectral class (or better). The exercise is to classify stars in similarly sized small areas of one photograph taken in the plane of the galaxy and of another taken perpendicular to the plane. The difference in stellar populations between the disk and arms of the galaxy (which generally contain young, hot stars) and the halo (which contains older stars) is easily seen in the histograms (Figures 4a and 4b).

Appendix 5: Cluster of Galaxies

There are some hundreds of galaxies in the field, made up of the cluster members and also the individual lone galaxies which are scattered more or less uniformly over the sky. The student's task is to decide, by careful analysis of counts over the whole film, where the center of the group lies and to find the average density of background galaxies. At high magnification it is possible to classify the galaxies according to Hubble's "tuning fork" scheme. The classification, together with the apparent diameter of each galaxy, gives an estimate of its distance. The distance of this cluster, inferred from the redshift of its members, is 45 megaparsecs. The spatial distribution of galaxies in the cluster is found by counting and fitting to models; from this the average separation of galaxies is found ($\sim$ 3 Mpc). This information combined with the internal velocity dispersion, known from spectroscopy, gives the cluster's total mass (7.5×10^{14} solar masses).

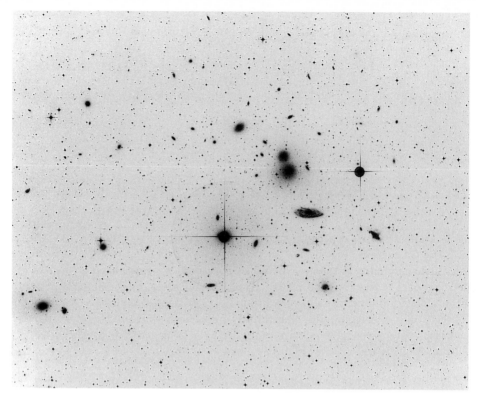

Plate 5: Abell 1060. The photograph shows only the central part of the cluster and covers only a small part of the provided film.

CLASSIFICATION OF STELLAR SPECTRA

M.M. Dworetsky
University of London Observatory, University College London,
Mill Hill Park, London NW7 2QS, U.K.

Summary: *The objectives of this exercise are to gain familiarity with the appearance of the spectra of stars, to understand some of the temperature and luminosity criteria used to classify spectra, and to gain insights into the relationship between the appearance of stellar spectra and physical conditions in stellar atmospheres.*

Our experience with exercises in which beginning undergraduate astronomy degree students are challenged to classify "real" stellar spectra according to Harvard or MK types has not been entirely satisfactory. These inexperienced students lack appreciation of the subtleties involved and are sometimes unable to explain the physical phenomena underlying the appearance of stellar spectra. For this reason, we decided some years ago to construct a "cloudy night" experiment, based on high-quality slit spectrograms taken at the University of London Observatory, which promotes familiarity with the appearance of spectra of different stellar types, and also illustrates the underlying physical basis of spectral classification (ionization and excitation). The equipment required is rudimentary: aside from the set of high-quality photographic prints, all that is needed is a good transparent plastic ruler, an identification chart for the Cu-Ar comparison lamp, a table of identified lines in the star γ Pegasi (Aller 1949; Aller and Jugaku 1956), some references and background reading (*e.g.*, Zeilik and Smith, 1987; Roy and Clarke 1982), and a calculator for evaluation of the Saha and Boltzmann formulae.

The photographic prints are first-generation enlargements of (mostly) IIIa-J spectrograms of 13 standard stars, taken with an original dispersion of 140 Å/mm using the Cassegrain spectrograph described by Dworetsky (1980). The widening, resolution and quality of these prints conforms reasonably well with the original MK specifications. The exercise itself is divided into three sections, of which the first two could serve some instructors as independent "mini-experiments" in their own right.

In the first section, students measure the positions of several comparison lines and as many stellar lines in the spectrum of a B3V star (Fig. 1) as they think they can detect. Experience has shown that, if the measurements are made with care, stellar wavelengths can be determined with an accuracy of the order of 1 Å, sufficient for the student to identify the visible stellar lines such as H, He, C, and Mg. In the second section, students make eye estimates (on a scale from 1 to 10) of the strength of Hγ in the seven spectra from B3V to F9V. They calculate, using the Saha and Boltzmann formulae, the population of H atoms in the $n = 2$ state relative to the total number of H atoms and ions. This ratio, raised to the 0.4 power and scaled so that $I(max) = 10$, should provide a close fit to the estimated strengths of the Balmer lines if the calculation has been correctly carried out (Aller 1963). To assist the students, we provide a table of stellar effective temperatures and electron pressures, as well as a tabulation of the partition function of H I. In the final part of the experiment, students examine the spectra of F to M stars and study the systematic behavior of several prominent spectral features.

We assess students on their ability to present their results accurately and clearly, and on their answers to "thought" questions that are asked at each stage of the exercise.

Space limitations preclude a full presentation of the written script here, but the Observatory is able to provide a complete set of the photographic prints of spectra, a copy of the script we have prepared for the students, and instructor's notes, at a low cost (at this writing, approximately US $18 for the first set and US $10 for each additional set.) that covers the printing and postage expenses. This also includes free replacement of lost or damaged prints, future prints of additional spectra, and future revisions to the script. A substantial discount is available for additional sets. Enquiries should be addressed to "The Director" of the Observatory at the author's address given above.

References

Aller, L.H. 1949, *Astrophys. J.,* **109**, 244.

Aller, L.H. 1963, *Astrophysics: The Atmospheres of the Sun and Stars,* Second Ed. (Ronald Press: New York), 124.

Aller, L.H. and Jugaku, J. 1956, *Astrophys. J.,* **127**, 244.

Dworetsky, M.M. 1980, *Q. J. Roy. Astr. Soc.,* **21**, 50.

Roy, A.E. and Clarke, D. 1982, *Astronomy: Structure of the Universe,* Second Ed. (Adam Hilger: Bristol), ch. 7.

Zeilik, M. and Smith, E.v.P. 1987, *Introductory Astronomy and Astrophysics,* Second Ed. (Saunders: Philadelphia), ch. 13.

Fig. 1. The spectrogram of the B3V star ι Herculis used in this experiment (original dispersion 140 Å/mm). University of London Observatory photograph.

TELESCOPES FOR STUDENTS

Roy L. Bishop

Department of Physics, Acadia University, Wolfville, Nova Scotia, B0P 1X0, Canada

An astronomy class typically is composed of twenty or more students, one instructor, and one telescope. For observing sessions, these numbers mean that at any one moment most of the students are standing, slowly freezing in the dark, waiting their turn at the eyepiece. Moreover, when they do have their 20-second peek, moist breath, a bump by the last observer, or unfamiliarity with the instrument often mean that the object will look hazy, be out of focus, or even be out of view. A couple of sessions under these circumstances can dampen the interest even of keen students. Also, the instructor may be reluctant to let individual students have unsupervised access to the one, expensive, fragile telescope.

Ideally each student would have a telescope during supervised observing sessions. Each telescope should be easy-to-use, rugged, equipped with setting circles, and mechanically and optically designed so as to minimize conceptual barriers between its operation and the geometry of the sky. As a step toward this goal, for the introductory astronomy course at Acadia University we have designed and constructed a set of six telescopes (Bishop, 1986).

Each telescope is a modest, wide-field, equatorially-mounted refractor. The objective is an f/5 achromat with a clear aperture of 74 mm and is combined with a 16 mm eyepiece having an apparent field of 69°. This gives a magnification of 23× and an actual field of 3°. Thus a finder telescope is not required and the field is large enough (and the aperture sufficient) to provide striking views of several objects that cannot be encompassed by the narrower fields of larger telescopes (for example: the Pleiades, the Praesepe cluster, the North America Nebula, the entire sword of Orion with the Great Nebula, the Andromeda Galaxy together with its two companion galaxies, *etc*).

With these telescopes, angles as small as 10 seconds of arc can be resolved, yet atmospheric turbulence is not a problem. The disk of Jupiter, the rings of Saturn, the double star Mizar, and the Trapezium in the Orion Nebula can be seen. Thus, in addition to providing interesting wide-angle views, the telescopes provide an an introduction to sights that are displayed to better advantage in larger instruments. Even for some objects that are generally considered to require larger instruments (for example, the Triangulum galaxy, and the Crab Nebula), these telescopes provide pleasing views.

The usual neck-saving, 90° bend in the optical path is provided by a roof (Amici) prism. The *two* reflections ensure that the image is both upright and correct left-for-right, and thus has the same orientation as the naked-eye view of the sky when students raise their heads. Also, the telescopic view may be compared directly to star charts. This eliminates a source of confusion for students, and, even for experienced observers, is a convenience that must be experienced to be fully appreciated.

Each telescope is mounted on a substantial, portable, three-legged pier and is designed to be used while the observer is comfortably seated on a low stool. There are no clamps or slow-motion controls to bother with: all motions are smooth and restrained by friction in Teflon-lined adjustable bearings. The right ascension and declination dials are large and easy to read, and, to minimize conceptual difficulties, are oriented so that their pointers are parallel to the optic axis of the telescope. Thus, the calibration marks on both dials, extended radially, coincide with the imaginary grid of right ascension and declination on the sky.

Aside from providing more time at the eyepiece for students, these telescopes provide experience that is missing at observing sessions involving one, larger, motor-driven telescope that necessarily has been pre-aligned on objects of interest. When they sit down beside one of these basic instruments, students are literally cast adrift among the stars: they *have* to study the star patterns, think about coordinates, and allow for the turning of the planet beneath their feet. This experience will probably stay with them long after other details learned in class have been forgotten.

Fig. 1. Telescopes for students. Left: six telescopes mounted on the observing deck. Right: close-up view of one telescope.

References

Bishop, R.L. 1986, *J. Roy. Astron. Soc. Can.*, **80**, No. 4, 211-215.

USING CCD'S IN INTRODUCTORY-LEVEL COLLEGE ASTRONOMY LABORATORIES

Thomas J. Balonek
Colgate University, Dept. of Physics and Astronomy, Hamilton, New York 13346, U.S.A.

1. The CCD and Interactive Image Display Systems

Our university recently purchased a liquid-nitrogen-cooled CCD camera system (from Photometrics Ltd., Tucson, Arizona) which has been installed on our campus' 40-cm multiple access (Cassegrain/Newtonian) telescope. Images are reduced on-line at the observatory using Photometrics' microcomputer-based analysis software package, which includes operations for standard data acquisition and initial stages of data reduction — including corrections for bias, dark current, and flat fielding. Images are displayed on a 256-level-gray-scale black and white monitor. Additional post-processing can be done either on the CCD system's computer at the observatory or on IBM-AT/PC's located both at the observatory and in the laboratory.

Because of the ease of use of the PC-based program and the ready availability of PC's, we are using this system to develop laboratory exercises to be used in introductory astronomy courses. The eight-color image-display program, developed by Michael Newberry and Stephen Gregory (Institute for Astrophysics, University of New Mexico), has many features similar to other well known astronomical CCD image-analysis packages. For our research work, analysis is done on an IBM-AT clone, while IBM PC's and PC clones (equipped with a mouse and an EGA color graphics card with 250k RAM) will be used for introductory laboratory exercises. Images can be stored either on a hard disk or on a floppy (some images need to be "trimmed" in order to fit on a 360k floppy).

2. Laboratory Exercises Under Development

Laboratory exercises for use in our introductory astronomy courses are being developed by Colgate students (astronomy/physics majors), who are responsible for defining the scientific and educational goals of the project, obtaining the CCD images with our system, and writing the laboratory exercises.

Operation of the CCD system is demonstrated as part of the laboratory exercises, so that students experience the many steps involved in making accurate astronomical measurements. In addition to analyzing prints of several images, students use the microcomputer-based image-display program to deduce properties of the objects being studied.

Exercises under development include:

Enhanced images of galaxies in the Virgo Cluster. This exercise will contain images of several galaxies that are studied in the University of Maryland's laboratory exercise on galaxies[1]. We plan to have images of elliptical, spiral, and interacting galaxies that the students can study with the PC image display system. This exercise will require the student to learn techniques used in extracting quantitative results from high-dynamic-range images. Using these enlarged images (compared to the scale of the Palomar Sky Survey print images used in the Maryland exercise), students will be able to study the structure, brightness distribution, and density profile in different galaxy types, and to test conclusions that they had made in the earlier lab exercise.

Colors of stars and the H-R Diagram. Students will perform photometric reduction of an open cluster, determining the magnitude and color of stars. These results will be used in constructing a cluster H-R diagram. This exercise will demonstrate the procedures involved in accurate calibration of images.

Planetary-satellite orbits. Students will investigate Kepler's laws of motion by measuring satellite movements. Images will be obtained of the Jupiter or Saturn systems during several satellite revolutions. By accurately measuring the positions

---

[1] Reprinted in *Teacher's Guide for Contemporary Astronomy*, 4th ed, by Jay M. Pasachoff (Saunders College Publishing, Curtis Center, Independence Square West, Philadelphia, PA 19106, U.S.A.; 1989).

of the moons relative to the planet, students will be able to use real data to deduce Kepler's laws.

These exercises are being developed by our astronomy students under the direction of faculty. Thus, not only will introductory students use observations obtained with our CCD system, but also the astronomy majors will learn how to obtain accurate CCD images and how to develop useful laboratory exercises. Motivated students will be encouraged to undertake an observing project of their selection — in which they participate in all the steps involved in selecting a suitable object to observe, followed by obtaining, calibrating, and analyzing the CCD image. It is our hope that students will be excited by using results obtained from a research-quality instrument that is located on our campus.

3. Acknowledgment

Funding for the purchase of the CCD system was provided by a National Science Foundation-College Science Instrumentation Program grant and by Colgate University. Support for student participation in developing laboratory exercises has been provided by a Sloan Foundation grant to Colgate University. Once development of these projects is completed, we expect to make our images and exercises available to interested educators. We welcome any comments or inquiries about our CCD system or laboratory exercises.

MODERN PHOTOMETRY LAB EXERCISES FOR STUDENTS IN INTRODUCTORY ASTRONOMY

Daniel B. Caton
Department of Physics and Astronomy, Appalachian State University,
Boone, North Carolina 28608 U.S.A.

Introductory astronomy lab exercises have traditionally been rather simple in nature — observations are often limited to visual observing, sketching, and perhaps photography. These types of labs are more like 19th-century astronomy than the techniques used by modern astronomers. Students should also be introduced to *quantitative* work in astronomy, in the same way that they are in equivalent introductory physics and chemistry labs. A typical solution to some of these problems has been to use published data (for example, the *Sky & Telescope* lab exercise series) to provide numerical work. This is less satisfactory for the student than obtaining his or her own quantitative data to reduce and analyze.

At Appalachian State University we have been working on the development of a set of modern lab exercises that address the needs discussed above. The first of these are photometry exercises that use a readily available photometer, the Optec SSP-3 (Optec, Inc., 199 Smith St., Lowell, Michigan 49331, U.S.A.). This photometer is ideal for student use in that it is compact, durable, user-friendly, and relatively inexpensive. Its PIN-diode detector does not require the high voltage that a photomultiplier needs. The photometer operates off a rechargeable 9-volt battery that will run it for a typical lab session. The controls include a two-position filter slide, flip-mirror knob, and switches for scale and integration time. The output is displayed as a four-digit number on red seven-segment LEDs (for advanced student projects, the output is also available as a 0-10 kHz TTL-level pulse train that may be integrated by a microcomputer).

The photometers are used on Celestron C-8 (20-cm) Schmidt-Cassegrain telescopes. Our observing lab/deck is equipped with twelve of these telescopes and photometers, each telescope used by a pair of students. Two dozen students seems to be the maximum that may be easily managed by an instructor and student-helper.

Before using the telescopes with the photometers, the students are introduced to the instrumentation indoors and then learn to use the telescopes, visually, outdoors. Then the students are introduced to the photometers, again indoors. They learn the basic techniques of photometry by observing artificial stars. The "starbox" is simply a cardboard box with a couple of light bulbs mounted in it, and a metal plate with various size holes drilled in it. Colored plastic over the holes provides a variety of star colors (about 0.5 magnitude range in B-V). The students learn to take data, remove the (artificial) sky from the readings, and calculate magnitudes and colors. These computations are done by hand, using calculators, to provide them with some knowledge of what magnitudes and colors mean.

After the indoor introduction the students use the photometers and telescopes outdoors to observe a variety of bright Johnson standard stars. They learn that the effects of our atmosphere must be taken into account. This requires that the time of each observation be recorded. For this purpose we provide some digital clocks that display the time on 2-inch tall LEDs. The clocks are set to the sidereal time before the lab begins — the sidereal/solar rate drift during the lab session is unimportant for their work.

Doing quantitative work outdoors requires some precautions by the instructors as well as some care by the students. The instructor must be sure that the telescopes do not dew up during the evening, and that the observing conditions remain good enough to get meaningful results. The students must learn to center their star in the photometer's reticle carefully, and to handle the photometer gently so as not to cause the instrument to drift. Centering should be checked after each filter change, and at the end of the last measurement on an object. The students must learn to note whether their numbers seem reasonable, or whether a sudden change indicates drift, clouds passing, or that someone walked in front of their instrument. One should note, though, that some of these are the same concerns of professional astronomers, and that the students will learn to appreciate the difficulty involved in obtaining good scientific data!

Having obtained their data, the students use the next cloudy lab night to reduce their measurements to outside-atmosphere magnitudes and colors. The extinction coefficients for the night are predetermined by the instructor, using data from early and late lab sessions held each night. Instead of doing the calculations by hand the students use microcomputers, running a photometry reduction program written for them in BASIC. They enter their measurements, times, photometer settings, and star identifications, and the program computes and displays the outside-atmosphere magnitudes and colors. The student then plots these results *versus* the Johnson values, providing a calibration for his/her instrument. This calibration may be used to identify the Johnson magnitude and color of an observed "unknown" star, and may be used later in the Pleiades project.

After learning to use and calibrate their instrument, the students are ready for an interesting exercise in observational astrophysics: obtaining their own color-magnitude diagram for the Pleiades, and using it to estimate the cluster's distance. They are provided with a finder chart of the Pleiades and a star identification list containing the spectral types of the numbered stars. It is explained to them that they are responsible for picking a range of spectral types to get a good color-magnitude diagram. They are to make two measurements in each filter (B and V) of as many stars as possible in the two-hour lab session. In practice, the best students can measure 15-20 stars; the least capable students get about half of this number. The practical limit to the photometer on the C-8 is about visual magnitude 10. In some ways it is an easier project to star-hop in the Pleiades than it is to set the telescope on the several standard stars used in the first outdoor project. The field of view in the Optec viewer, on the C-8 telescope, is about 20 arc-minutes.

Again, the next cloudy lab night is used to reduce and analyze the data using a modified version of the microcomputer program. The reduced data are transformed to the Johnson standard system using either the previous calibration lab results or the Johnson values provided for one of the Pleiades stars. The color-magnitude diagram is plotted on one set of axes, alongside a standard H-R diagram. The vertical shift between the C-M and H-R diagrams is measured and used as a distance modulus to compute the distance to the cluster. Students who worked carefully to obtain good data are able to get results that are in good agreement with published values of the distance. Figure 1 shows one student's results. The few stars that fall far from the main sequence are in most cases known to be field stars or double stars. The increased scatter at the lower right is due to noise in the photometer near its sensitivity limit. The students are not forewarned about these problems since the discussion of these problems is a good lesson in itself!

A package of sample lab exercises, computer listings, instructor suggestions, and other helpful materials for performing photometry labs is available upon request from the author. This project was supported by grant # CSI-8551003 of the National Science Foundation's College Science Instrumentation Program.

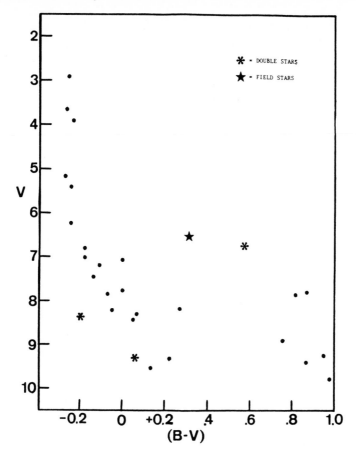

Fig. 1: The color-magnitude diagram of the Pleiades.

Discussion

M. Dworetsky: *A possible improvement of the indoor exercise might be to use a photofloodlamp rather than an ordinary light bulb. This could provide more blue light. Disposing of the heat generated could be a problem, of course.*

L.C. Hill: *Do you find that you have to hold the students' hands while working through the calculations of logarithms and magnitudes? Do some students get lost in the calculations?*

D.B. Caton: This is always the question of how much cookbook work should be used and how much the students should thrash out the questions for themselves. Some just don't make it — and that's life.

H. Shipman: *This exercise requires relatively good weather. I realize the weather on your campus is pretty good, but on how many nights is it good enough to work outside but not good enough to get a decent main sequence? Also, you mentioned that once they get data, they're set for 3–4 nights inside reducing their data. There is still a problem: you have to decide five minutes before lab starts whether the weather is OK.*

D.B. Caton: Out of 15 lab nights per semester, we will have perhaps five that are photometric and two or three more that are good enough for visual or photographic work. The problem with last-minute weather changes *is* indeed still a problem since it takes about 30 minutes to set up the telescopes. We always have both an indoor and outdoor lab exercise ready each lab night.

H.S. Gurm: *How about using a 6" or 8" Cassegrain with a larger f-ratio instead of a Celestron, particularly when doing photometry?*

D.B. Caton: I see no problem with using a six- or eight-inch Cassegrain to do photometry. Our Celestrons were chosen for their reliability, availability at competitive prices, and wide range of accessories available from both Celestron and second-source manufacturers.

L.A. Marschall: *1) How many labs does this exercise take; 2) and how many students do the lab each semester?*

D.B. Caton: 1) One lab indoors and one lab outdoors. Good students get about 20 Pleiades stars in two hours. Average students get about half that number. 2) About 75 students per semester, in three groups of 24. The students work in pairs.

CRATERING IN THE CLASSROOM

Walter Bisard
Central Michigan University, Mt. Pleasant, Michigan 48859, U.S.A.

The SEMS Project (Science Education in Middle School) in Michigan is a cooperative science education improvement effort by several universities, hundreds of school districts, and the state department of education. Upper elementary and middle/junior high school teachers have been targeted for improvement of science instruction. Although the initial area of focus included teachers of grades five through nine, the project has been successfully extended to early elementary grades. While there are many excellent teachers at these grade levels, research findings indicate that 49 per cent of all middle school science teachers are teaching without a major or minor in science! The situation is even more critical at the upper elementary levels.

In an effort to improve science teaching at the upper elementary and middle/junior high school levels, over 300 well-qualified teachers were selected to be trained as resource leaders for their respective school buildings and districts. Upon completion of their training in 1986, these teacher/leaders returned to their home school districts and made efforts to assist out-of-field and inadequately-prepared science teachers in their schools.

This technique of using local teachers as resource leaders afforded the greatest capability to reach the largest number of teachers possible. A resource training manual and model lesson plans were developed for the resource team training and utilized the goals and results of several sources for the improvement of science education for all students: 1) develop scientific and technological process and inquiry skills; 2) provide scientific and technological knowledge; 3) use the skills and knowledge of science and technology as they apply to personal and social decisions; 4) enhance the development of attitudes, values, and appreciations of science and technology; and 5) study the interactions among science-technology-society in the context of science-related issues; (National Science Teachers Association, *Science-Technology-Society: Science Education For The 1980's*, 1982).

Over seventy "ideal or model" lessons were developed and field-tested in classrooms to assist in the training of resource leaders and for utilization with their out-of-date/field teaching colleagues. One very popular lesson is Appendix 1. This activity proved to be one of the most successful for all teachers regardless of academic background. It is called "Formation of Impact Craters" and is found in the Space/Astronomy section of the *SEMS Resource and Training Manual*. The SEMS lessons contain specific sections on motivators, processes, teaching strategies, careers, extensions, and resources in addition to the usual science activity sections.

This experiment basically has the student teams investigate the relationship between the height of a dropped marble and the corresponding crater width in a box of sand. Impact craters are a significant part of our science in astronomy, planetary geology, and the evolution of life. Correct experimental techniques, the transformation of one form of energy to another, and the relationship between drop height and crater size are emphasized in this activity.

After initial introduction and motivation with visuals, the student teams, composed of three students, plan the experiment with the available materials and conditions. The experimental procedure is to first brainstorm (*i.e.*, think freely and wide-rangingly) about the possible variables and then to design the experiment to control the variables as the team tries to gather the relevant data. A marble is dropped into a sandbox from successive heights in 10 centimeter increments starting from an initial height of 10 centimeters up to as high as possible — perhaps several meters. The crater position in the sandbox should be the same location and the identical techniques/measurer should be used for each drop. Different masses and types/depths of sand may be used.

Analysis of the data via careful plotting of the height *versus* the crater width should follow the experiment. A smooth relationship is rare because of sound reflections off the container sides. This will form the basis for more experimentation by the teams. Extensions of this simple activity extend to astronomy, dinosaur extinctions, volcanic eruptions, and nuclear winter.

Formation of Impact Craters	Lesson 33 Space Astronomy

Overview:

Impact craters caused by meteorites are believed to be related to lunar craters and possibly to the extinction of dinosaurs. The transformation of kinetic energy of motion of a meteorite into heat, sound, displacement of materials, and the relationship of crater size to velocity at impact are emphasized in this activity.

Objective:

After completing this activity, students should be able to:

1. Describe the transformation of energy that takes place during the formation of an impact crater.

2. Describe the technique of investigating the relation between crater size and impact velocity at impact of a meteorite.

3. Infer how craters on the moon may have been formed.

Motivator:

Show students photographs of planets and moons and have them observe the range of sizes and similarities among the craters. Also examine space photographs of the Earth and view the NASA film *Exploration of A Planet.*

Materials: (Per small group)

Marbles of various diameters

Meterstick or metertape

Laboratory book for recording data

Graph paper

Shoebox or foil pan or similar container filled with wet sand, or soft plaster of Paris.

Procedure:

Instruct students to drop the marble into the sandbox or wet plaster from increasing heights, beginning at a height of 10 centimeters, and continuing in increments of 10 cm, up to a height of several meters if possible. Each time they should measure the height from which the marble was dropped, as well as the depth and width of the crater. Emphasize that good inquiry habits and laboratory procedures require accurate measuring techniques.

Subject:

Earth and Space Science: Solar System, Galaxies, Universe

Group Size:

Small groups (3–5)

Time:

150 Minutes

Teaching Strategies:

Guided Discovery

Inquiry

Simulation

Discussion

Concept:

A meteorite that rushes toward another celestial body, such as the moon, has potential energy by virtue of gravitational attraction between it and the body, and kinetic energy by virtue of its movement. Craters thus produced result from both the potential and kinetic energy of the meteorite.

Processes:

Observing

Measuring

Predicting

Interpreting Data

Controlling Variables

Formulating Models

Formation of
Impact Craters (continued)

Ask students to identify the sources of energy in the act of raising the
marble, releasing it, and the subsequent impact crater in the sand or
plaster. (This will involve *chemical energy* and *metabolism*
in your body; *gravitational potential energy* (GPE = mgh)
that is directly proportional to the height of the drop disregarding *air
frictional energy; kinetic energy* ($KE = 0.5 \ mv^2$) of the
marble that increases to a maximum as it hits the sand; *mechanical
energy, frictional energy,* and *sound energy* as the
impact crater is formed.) Have students summarize the data in a table.

Suggest that students plot a graph, showing crater diameters versus drop
heights. Ask: *What general relationship did you discover? How does
this activity demonstrate conservation of energy?*

As a group, discuss the variables in this experiment that might affect the
size or shape of the crater. (Examples might include marble
characteristics, sand or plaster characteristics, angle of approach, and
similar factors.

Extension/Follow-up Activities:
You might wish to have students drop stones of various sizes into a
large tray of wet plaster of Paris. After a crater has formed, have them
carefully remove the stones and allow the plaster to harden. Compare
this with pictures of the moon's surface.

Have interested students read and report on the possible event 65 million
years ago in which it is suggested that dinosaurs became extinct.

Applications:
Remind students about those "falling stars" that they used to wish upon.
Explain that they weren't "stars" at all, but were really small bits of rock
left over from the formation of our solar system. Also, fossil records
confirm that many life forms periodically become extinct and the
evolution of life to its present state may be drastically affected by large
impacts from space every few million years. In fact, the government has
now started Operation Sky Watch to monitor asteroids that pass close to
the Earth.

However, an even more immediate concern comes from the possible
results of a limited nuclear war. It has been estimated that crater dust,
debris, and smoke from a nuclear explosion might block enough sunlight
so that a significant cooling of the Earth might take place suddenly. It is
suggested that temperatures might drop to -40° C. Ask: *Could humans
survive?*

Careers:
Astronomer
Aerospace Engineer
Geologist
Satellite Specialist

Resources:
The Solar System and Beyond,
4 color filmstrips, 55 frames
each, and 4 cassettes,
National Geographic Society,
Educational Services,
17th & M Sts., N.W.,
Washington, D.C., 20036, 1985.

Discussion

L. Gouguenheim: *I wish also to comment that a similar lab was successfully given by L. Celnikier during a summer school for teachers. The report was published in the proceedings of the TARBES Summer School (1978, L. Celnikier, Editor, Publication de l'Observatoire de Paris).*

J. Fierro: *Covering the sandbox with colored powder or cornstarch will show ejecta and rays for the crater experiment you described.*

J.V. Feitzinger: *In the book* Astronomie Experimentelle (Experimental Astronomy) *by J. Meurers, about 1955, you will find detailed descriptions of experiments on cratering. The production of central bulges is described.*

W. Bisard: Yes, thank you.

J.-C. Pecker: *All these types of experiments are good inasmuch as they help students to make actual measurements, build diagrams, and understand the physical processes that may happen in the universe. But these is a danger of misconceptions: for example, some young people might think that* large craters on the moon are created by objects that are coming from far away, small *ones by objects that are coming from the* vicinity.... *So teachers should heavily comment about the* difficulty *of generalization, and of extrapolation of classroom experiments to the real astronomical world.*

O. Gingerich: *Can you get elliptical craters? How do you measure the sideways component?*

W. Bisard: Yes, especially with very oblique impacts. The astronomical craters are *circular* and are caused by the explosion of superheated gases at the impact site. No evidence can be found of initial sideways velocities of astronomical impacts.

LABORATORY ACTIVITIES IN THE TEACHING OF ASTRONOMY

Wayne Osborn
Central Michigan University, Mt. Pleasant, Michigan 48859, U.S.A

Astronomy is one of the most popular of the sciences. That this is so can be seen in the frequent news articles about astronomical discoveries and happenings, in the number of questions about the sky fielded by planetarium and observatory staff, and in the large turnouts for public observing sessions at observatories or astronomy club star parties.

Another reflection of astronomy's popularity is the fact that thousands of university students study the subject each year. Only a few of these are training to become professional astronomers. The vast majority of these students seek mainly to obtain a general background. An important subset of the latter group are those students preparing to teach in the schools and who have the need to understand, and be able to explain, common astronomical phenomena clearly. Those of us who teach introductory astronomy courses have an obligation to ensure that these prospective teachers are well trained.

Traditionally, astronomy courses consist of lectures and associated laboratory activities. The purpose of the lectures is to present and explain astronomical concepts and describe the various astronomical objects. Depending on the goals and backgrounds of the students, the level of lecture presentation ranges from purely descriptive to highly technical and quantitative.

It should go without saying that the laboratory exercises associated with an astronomy course also should depend on the type of students being served. Those intending to become professional astronomers must learn the techniques of observing and the methods of analysis of astrophysical data. Most astronomy educators would agree that in this case the activities should resemble true astronomical practice as closely as time, the available equipment, and the sophistication of the students permit. The chief difficulty encountered is having access to modern instrumentation, especially at less-developed institutions. The new "traveling telescope" project of IAU Commission 46 is a step toward solving this problem.

The laboratory exercises designed for professional training are, however, of little importance to the student with only a general interest in astronomy. For these students simple equipment is entirely suitable for the observing and measurement involved in laboratory work. The major concern is to select the most appropriate activities. I would argue that often little thought is given to this matter.

Traditionally the laboratory components of general astronomy courses stress activities that illustrate technical concepts. Examples are exercises on spectral classification, the H-R diagram, and the redshift-distance relation for galaxies (Hubble's Law). These activities reinforce material in the lecture course and provide the student with a feeling for how science is done, but they have little everyday practicality. Demonstrating the process of science is important, but the topics through which this is done need not be those of most interest to professional astronomers. Of more immediate and practical use to general interest students are exercises on common astronomical effects and on simple observing. Examples include activities that demonstrate the causes of day and night, the seasons, and lunar phases, and ones that involve constellation and planet identification and the use of a small telescope. Simple observing projects can be used as the basis for showing scientific methods.

Clear knowledge of the everyday astronomical phenomena is particularly important for school teachers because this is the astronomy that is required to be taught in schools. As an example, my home state of Michigan specifies the following seven topics as the astronomy that must be taught (grade level is in parentheses):

the cause of day and night (4–6), the relationship between the earth and the moon (4–6), the motion of the earth and planets (4–6), major structures in the universe including the solar system (4–6), the phases of the moon (7–9), theoretical origins of the solar system and the universe (7–9), and astronomical equipment including telescopes (7–9).

At Central Michigan University we have developed a laboratory course for general astronomy students that is based almost entirely on practical and elementary concepts. Activities include learning to identify the prominent constellations, bright stars, and naked-eye planets; charting the motions of planets and seeing how this affects their visibility; observing and understanding diurnal and seasonal changes in the sky; knowing the phases of the moon, their times of visibility, and the cause; understanding the different types of eclipses; learning the components and the use of a small astronomical telescope; learning the use of star maps and the celestial globes; and viewing the sun, moon and planets. A copy of the laboratory manual may be obtained by contacting the author.

TEACHING OBSERVATIONAL ASTRONOMY AS A LABORATORY COURSE FOR NON-MAJORS

Yong H. Kim
Department of Astronomy, Geosciences & Physics, Saddleback College,
Mission Viejo, California 92692, U.S.A.

1. Introduction

Since antiquity, doing astronomy means basically stepping outside, looking upward, and considering the widest environment. Thus any undergraduate astronomy program, no matter how diverse its course offering, is incomplete without observational astronomy. For example, some California community colleges offer several courses including such titles as "Man and the Cosmos," "Final Stellar States," "Astronomy Enrichment," and "Astronomical Myths, Mysteries & Fallacies," but do not offer "Observational Astronomy." As a teaching astronomer, I question the wisdom and honesty of such practice of proliferation solely based on sensationalism. An introductory lecture course and an observational lab course must be the core of lower-division undergraduate astronomy education. Anything else, in my opinion, is peripheral.

This paper intends to address the importance of the teaching of observational astronomy as a liberal studies lab course. Instead of delineating specific course materials or instructional methods, I aim to discuss epistemological goals which, I believe, should help to answer clearly (a) what a student should learn and (b) what pedagogic attitudes and approaches should be taken to ensure effective teaching.

2. A Survey

It is generally believed that astronomy is so popular that a college (if we exclude highly specialized schools) cannot afford *not* to offer at least introductory astronomy courses. However, my survey of 200 randomly chosen colleges and universities across the United States gives an unexpected result. I found that 52 schools (26%) offer no introductory astronomy courses at all. Among the 148 colleges and universities that provide at least one introductory-level astronomy course, 98 offer lecture-only courses. The remaining 52 institutions have introductory courses that require laboratory hours. Only 45 schools offer observational astronomy courses that are clearly designated as being taught at night. Interestingly, 35 colleges and universities list multiple lower-division courses beyond the introductory level without offering observational astronomy.

According to my survey, observational astronomy belongs to a minority (only about 30%) of the schools with lower-division astronomy programs. Whatever the reason, this tendency for the majority to ignore observational astronomy is disturbing. Some may argue that introductory courses with lab hours are sufficient. They are definitely much better than those without lab sessions. Any astronomer would agree that "daytime" lab hours simply cannot substitute for a "nighttime" observational course.

3. Some Insight About Undergraduate Teaching

According to some critics, our liberal education, even at the best colleges, has been so cheapened that it is like a carnival with too many competing and contradicting courses and with no substantive vigor and clear direction (Bloom, 1987). Our students are wasting their "charmed" four years of undergraduate learning with many questionable courses and studies, all overemphasizing trendiness and popularization and heading towards a dead-end. It is truly tragic that good courses in natural sciences must compete with these other courses for resources.

College students today generally avoid taking courses in science and mathematics unless these subjects are required for their major or graduation. Those who involuntarily take science, particularly physical science, usually bring with them self-destructive attitudes about their ability to learn. It is almost impossible, unless we change human nature, to make involuntary learning a pleasure. However, if we make sure that the level, quality, and standard of the course are consistent with what we should expect of the student, we can possibly overcome this problem.

Many students find astronomy fascinating in itself. Also, astronomy is a wonderful subject because of its inexhaustible nature (Bondi, 1970). Thus astronomy as a college-level subject has tremendous possibilities for transforming non-majors into scientifically astute citizens. For many students, introductory astronomy is the last (for some probably the only) exposure to science in their lives. Teaching astronomers must not waste this opportunity of winning future pro-astronomy taxpayers. This opportunity is not without a challenge: These students have little or no background in science. Most of them are scared to death of mathematics. How do we engage such students to learn a quantitative science like astronomy? It can only be done by the teacher's creativity, flexibility, and adaptability in dealing with the immediate circumstances.

4. Laboratory Experience: What Should It Mean?

For non-science students, lab courses are unique among undergraduate educational experiences. Lab courses are supposed to provide opportunities "to learn by getting one's fingers dirty." We find a deeper insight into what a lab science course should mean to the student in the following quotation: "An introductory course in physical science done entirely with the printed word, pencil, and paper cannot convey a true picture of the nature of scientific enterprise... The essence of learning in science is participation: doing and asking and making errors and learning from them... The lab work is not primarily to train you in certain dexterous actions. It is only to give the genuine feel of that world which we can at best pallidly describe on paper in words or symbols. It is above all to give the scene of what is meant by abstraction" (Holcomb and Morrison, 1974). Any practicing scientist can intuitively appreciate the power and insight of the above quotation. Yet that is the problem of every teaching scientist: we understand so well what a lab experience means that we have the difficulty of reading the same statement through the eyes of a student, who may not have even the vaguest idea why a lab course is required.

All lab courses are taxing propositions to all parties involved: the student, the teacher, and the department. The student is required to put in more hours than in other courses (for a 3-unit observational course perhaps even 4 hours per week plus 12 hours to be additionally arranged). For the teacher, a lab course requires a lot more work than a normal science course by a factor of 2 or 3. Particularly in the case of observational astronomy, the teacher's enthusiasm, energy, and interest are imperative for success because of the nighttime observing activities. Some argue that actual telescope work by non-majors is of very little use (Booth and McNally, 1980), with which I strongly disagree. The department must be resourceful enough to provide at least a dark-night observing site and related necessities (teaching assistants, portable or transportable telescopes, means of transportation, *etc.*).

The most important epistemological goal of any lab course is to provide the students with an opportunity to *verify* themself some pre-ordained knowledge (or scientific truth) by practicing experimental skills. The student may gain from lab exercises the sense to recognize the difference between what is real and what is based on someone else's judgments, perceptions, and assumptions. Thus the ideal lab course should be one in which we can challenge the student with the existential question of what scientific truth is and how scientists go about discovering it. The goal is specifically not to tell the student what truth is (even if we could know), or even to tell him when and if he has verified the truth. On the contrary, the worst possible lab course is one that emphasizes memorization as the main task, or requires mostly cook-book type experiments, or stresses nothing but the plug-and-chug manipulation of numbers.

5. A Few Maxims Worth (I Hope) Considering

Quality is not accidental, but must be designed. A successful course is based on the basic concept of "design integrity," which means designing a course so it will fulfill its goals well and reliably despite unavoidable imperfections.

Perfection of means cannot correct confusion of goals. Setting a manageable and cohesive set of well-defined goals is the utmost important aspect of effective teaching. The goals must match the backgrounds of the students and resources available.

For better products, use fewer parts. Less detail does not necessarily make the course less proper. Relatively minor subjects should not be brought up just for the sake of completeness. We cannot give knowledge neatly wrapped up with no loose ends. The strength of astronomy, I think, is that it allows us to present science as unfolding, advancing, and never completed.

Popularity alone will not always make a course interesting. Astronomy is very popular. However, students often find that it is interesting up to the first lecture or those first few pages in the text. Astronomy can be made dreadfully dull like any other subject unless the teacher makes an active effort to make his or her lectures and presentations interesting. Also, we should not be afraid to spend more time and effort on those topics that students enjoy most.

Seeing it once is incomparably better than hearing about it one hundred times. This old Chinese proverb sums up very well what observational astronomy is all about. Therefore, weather permitting, every class session must be held out outdoors. The night sky provides inexhaustible things for seeing.

6. Conclusion

Despite all sorts of logistical problems, complex issues, and unexpected glitches, I still believe that observational astronomy is very worthwhile, not only because it is the core curriculum of astronomy education but also because we can create in the minds of the educated public of the future a foundation of public appreciation of what astronomers are doing. Many alumni of my observational course have joined the ranks of amateur astronomers.

References

Bloom Alan. *The Closing of the American Mind.* New York: Simon & Schuster, Inc. 1987.

Booth, R.S., and D. McNally. "Undergraduate Astronomy Teaching." *Quarterly Journal of the Royal Astronomical Society* **21**:32-53; 1980.

Bondi, H. "Astronomy of the Future." *Quarterly Journal of the Royal Astronomical*

Society 11:443-450; 1970.

Holcomb, D.F., and Philip Morrison. *My Father's Watch: Aspects of the Physical World.* (Englewood Cliffs, New Jersey: Prentice Hall), 1974.

Comment

Jay M. Pasachoff: We must not forget solar observations — we have one star that we can observe in the daytime, when it happens to be most convenient for many students. Further, the sun changes from day to day, so there is always something new to be seen.

Chapter 5

Computers

With the widespread availability of personal computers, the role of computers in education is rapidly increasing. The chapter begins with an overview of microcomputers in astronomy teaching and a discussion of some programs either available or under development. We then read about specific personal-computer programs worked on in Italy, the U.S., Bulgaria, India, and Belgium. The next paper discusses the use of powerful microcomputers to generate classroom displays. Finally, we hear about one "paperless classroom" that uses a central computer to manage virtually every aspect of the teaching process.

MICROCOMPUTERS IN THE TEACHING OF ASTRONOMY

Robert J. Dukes, Jr.
Physics Department, The College of Charleston, Charleston, South Carolina 29424, U.S.A.

1. Introduction

There are many ways to use computers in teaching. This paper discusses one of those ways: as an adjunct to the introductory astronomy course as taught in the United States.

The computer allows the demonstration of phenomena in a graphic dynamic manner, so its use can enable students to gain a deeper understanding of the material than they might when they passively listen to a lecture, look at slides, or study a textbook. In a very real sense, material presented in a computer is similar to that presented by film or video, but with one major difference: viewing movies is essentially a passive experience while interacting with a computer is an active one. Using computer simulations, the user can vary initial conditions and observe the effect of changes. This ability is the case whether the person doing the interacting is the instructor during a lecture or the student during a lab or while independently executing a homework exercise.

Over the years, a number of papers have described specific computer programs for use in teaching astronomy. A list of available astronomical software is maintained by the Astronomical Society of the Pacific and published periodically (Mosley and Fraknoi, 1986). In this paper, I am going to take a slightly different tack and describe how to use a variety of programs throughout a one-year introductory course. I am also going to discuss a rather complex project that doesn't fit into the traditional astronomy course and that is currently being mounted at the College of Charleston.

Most of this paper discusses using computer software either for demonstrations or for laboratory exercises, since these are the prime ways I use computers in teaching introductory courses. The characteristics of the software and the preparation on the part of the instructor vary according to the type of use and caliber of students involved. Obviously, if the software is being used as a classroom demo, then it is important for the instructor to be familiar with commands and for the software to function rapidly. On the other hand, if the students are to interact with the software as homework or in a laboratory setting, then the speed of operation is not quite as important while the clarity of the instructions is paramount. When students are interacting with the computer directly, it is important to have worksheets or suggestions for report formats to accompany the programs.

2. A Sample Course

In the succeeding discussion, I am going to consider a sample course presented in the "traditional" manner. That is, I will start with the observable phenomena of the night sky and move on to the historical presentation of the development of explanations for astronomical observations, the solar system, the physics of stars, our galaxy, and external galaxies.

Observable Phenomena of the Night Sky

A very popular use of the computer (judging from the number of different programs available) is as an indoor planetarium. These have been reviewed in a number of papers elsewhere (Mosley, 1986; Mosley, 1987, Kanipe *et al.*, 1988). In this mode, the computer screen represents the night sky with the positions of stars, the moon, the sun, and the planets plotted. These programs demonstrate many of the phenomena that were observable by the ancients and that form the foundation of modern astronomy. While the same phenomena are visible today to anyone who takes the time to look (sometimes over a period measured in years), the computer enables them to be demonstrated in minutes. Early attempts at writing such programs resulted in ones that were not especially useful for educational purposes because of both the low resolution and slow speed of the early microcomputers. More recent computers, such as the IBM AT and Series 2 computers and the Apple Macintosh, have overcome these problems and can present remarkable displays that are useful as substitutes for or adjuncts to planetariums. Mosley (1988a) has suggested that two of the better ones for IBM compatible computers are "Superstar" (PicoSCIENCE, 1988) and "Visible Universe" (Parker, 1988). Superstar seems unique in that one version of it contains the entire SAO catalog. An excellent program of this type for the Macintosh which I have used is "Voyager" (Mathis, 1988). This program has been reviewed by Mosley (1988b). Screen replotting is very rapid, which makes the program useful for demonstrating time-dependent phenomena such as the diurnal motion of the celestial sphere or the motion of the planets with respect to the background stars. The program also allows relocating the observer to a point up to 100 astronomical units from the sun. This relocation allows the solar system as a whole to be observed in motion. This program can be very useful as a demonstration device for homework or as a laboratory exercise. It (and programs like it) will gradually supplant many special-purpose programs that were written for slower computers and showed only a single aspect of motion on the celestial sphere (such as retrograde motion).

Retrograde motion is one of the most important planetary motion phenomena to explain. Texts usually illustrate it with a cryptic diagram that shows the orbits of two planets and a retrograde loop drawn in the sky with numbered lines joining various planetary positions as projected onto the sky. A much more effective way to demonstrate this phenomenon is to show a simulation of a planet executing a retrograde loop on the computer screen using a program like Voyager or one written just for this purpose (Johnson, 1984a). Ptolemy's system of epicycles for explaining retrograde motion can also be demonstrated on the computer. In fact, a very effective demonstration is to plot the actual motion of the planets from a geocentric view (Schwartz, 1983). This plotting allows the idea of epicycles, which were first used to explain the phenomenon, to be tied to the actual phenomenon. As well, the Copernican view may be shown in an demonstration in which one planet overtakes another. The line of sight connecting the two planets can be drawn and will show the back and forth motion of the retrograde loop (Johnson, 1984a).

There are other examples. The apparent motion of the sun as a function of season and latitude can be demonstrated very effectively with the computer. A horizon view that shows the rising or setting sun demonstrates that both the position of sunrise and sunset and the angle of the sun's path with respect to the horizon change with the season (Sumners, 1983a). While a view showing the entire visible hemisphere is impossible to show due to the limited resolution of the computer screen, it is possible to show various sections of this hemisphere and the motion of the sun through them. Thus, the diurnal motion of the sun may be shown in three plots representing the eastern, southern (for observers in the northern hemisphere), and western horizons (MECC, 1985). Lunar phases may be demonstrated by programs that animate the motion of the moon around the Earth and simultaneously show the appearance of the moon as seen from Earth (Sumners, 1983b). There are several ways in which the computer may be useful in discussing eclipses. For solar eclipses, a map of a portion of the Earth may be drawn on the screen with the shadow of the moon moving across it (Parker, 1988). A diagram showing the parts of a shadow can also be drawn. One interesting program shows the positions of the Earth, moon, and sun, and the lines of nodes of the moon's orbit schematically. This program will indicate the possibility of lunar or solar eclipses when the new or full phase is coincident with the moon being at one of the nodes of its orbit. Again, care must be taken in using these programs since the drawings cannot be made to the proper scale (Zink, 1983a).

The computer makes the job of describing Kepler's Laws much easier. To start with, we no longer have to try to sketch ellipses. We can present a plot either as a whole or piece by piece. First the ellipse is drawn, then the semi-major axis is drawn and labelled, and finally the foci are drawn and labelled (Johnson, 1984b; Zink, 1983b). The first law is illustrated by a body moving in an elliptical orbit, its speed varying according to its distance from the focus. But the splendor of the computer emerges with the discussion of Kepler's Second Law. How often have you tried to describe the meaning of equal areas in equal times? Using a computer, we can plot the radius vector and color in sectors of equal areas as we go (Johnson, 1984b; Seeds, 1985). We can use orbits of different semi-major axes to demonstrate the third law. In this case, there may be several objects on the screen moving in orbits of different semi-major axes with periods given by Kepler's Third Law. Finally, bodies with orbits of various semi-major axes, eccentricities, and orientations can also be shown (Zink, 1983b).

A very good simulation allows the student to investigate Newton's Laws of motion. Objects can be moved on the Earth under varying conditions of friction or in space with or without a gravitational field. One exercise lets students move an object around a track under various conditions. The goal is for the student to apply forces in such a way as to have the object go around the track without colliding with one of the sides (Schwartz, 1985).

A simulation that might serve as a programming project for advanced students and one of the earliest applications of computers in physics is to solve for the motion of three or more bodies interacting gravitationally (Cromer, 1981; Bork, 1981). This simulation differs from the Kepler's Laws simulations since the computer calculates motion in real time while in the former case the results were presented based on calculations previously made and stored as data. With the more modern computers, these calculations can be done rapidly enough to allow the results to be displayed as the calculation proceeds. A good example of this type is Gravitation, Ltd. (Rommereide, 1988). This program allows several bodies to be set in motion simultaneously and the effects of their mutual gravitational interaction to be observed.

A number of interesting simulations treat various aspects of space travel in the solar system. These range from one in which the student has to choose proper positions in the orbit of the Earth and Mars to launch a probe in a least-energy ellipse (Simon, 1984a) to a much more complicated one in which the student has to get a space probe to Saturn and achieve an orbit inside the inner ring (Huntress, 1982).

Stellar Astronomy

Much of an introductory astronomy course is concerned with the properties of stars and the means by which we know these properties. These include the absorption and emission of radiation by atoms, the formation of spectral lines, spectral classification, parallax, proper motion, binary-star motion, the H-R diagram, calculation of stellar models, and stellar evolution.

There are several programs that treat the emission and absorption of radiation by atoms. Some work using the Bohr-atom approach (Johnson, 1984c), while others use energy-level diagrams. In either case, a photon is shown approaching the energy levels of an atom. When the photon reaches the atom, it disappears and the electron in the atom moves up one or more levels. A similar representation of photon emission is also possible.

An impressive program is available that teaches some of the basic principles used in classifying spectra (Simon, 1984b). Students compare an unknown spectrum with two standard spectra. They then give a classification for the unknown spectrum. If this classification is incorrect, they choose two new spectra and try again. Though this program has been criticized by some astronomers since the representations of spectrum drawn on the computer screen (due to limited resolution) do not effectively resemble the appearance of a photographic spectrum of the same type, I do not agree. A non-science student should learn how science functions and not specific details from an introductory science course. Therefore, any exercise that teaches some of the methodology of spectral classification is worthwhile.

Parallax is another area in which computer simulations are valuable aids for student visualization. These should be used to supplement more traditional demonstrations involving, for example, outstretched fingers. In one program, the computer screen show the Earth's motion around the sun with a line from the Earth to a nearby star (Zink, 1983c, A'Hearn *et al.*, 1988). At the same time, another view shows the apparent motion of the star with respect to the background star field. Additionally, the distance of the star from the Earth may be varied and the resulting variation in parallactic shift observed.

Stellar proper motion is a fascinating topic for computer simulations (Dukes and Roskoske, 1979). Constellation patterns are plotted on the screen and allowed to evolve with time using extrapolations of the best current values of the proper motion. Students are especially fascinated when the time span is long enough that the pattern we observe today completely disappears. Of course, the warning must be given that the measured proper motions are imprecise enough that these long extrapolations are not valid. However, students can use such plots to discover moving groups such as the Hyades or those members of the Big Dipper that are part of the Ursa Major moving group.

One of the best uses I have found for the computer is in describing the motion of binary-stars systems. One program useful for this purpose plots the motion of two bodies around a common center of mass as seen from above the orbital plane (Seeds, 1985b). The eccentricity of the orbits is varied as well as the mass ratio of the bodies. A program like this is very effective in demonstrating the positions of the two bodies and the center of mass at various points in the orbit. Another class of programs involves plotting the motion of two stars as seen from the orbital plane. These programs demonstrate the properties of eclipsing systems. Two stars are shown in a magnified view. The motion can be referred to the center of mass of the system or to the center of the primary star. The light and velocity curves of the system can be plotted as a function of orbital phase. I use two such programs. The first is definitely a demonstration program (Zink, 1983d). It shows two views of a binary system: one from the orbital plane and the other from the pole. The two stars move as long as is necessary for students to comprehend the plots. Next, the planar view is plotted along with a light and velocity curve. The second program is similar except here the student can exercise control over the properties of these stars and the inclination of the orbit, making it useful as a laboratory or homework exercise (Reitmeyer, 1980).

The H-R diagram is one of the most valuable tools of the astronomer. Certainly, we spend considerable time discussing it. One very good program lets us quickly plot an H-R diagram with various groups of stars (Alexander, Foster, and Unruh, 1983a). I realize that similar plots can be drawn on the board (slow and messy) or projected from slides (not flexible). With the computer, we can vary the plots as class discussion progresses. For example, we can first plot an H-R diagram showing a main-sequence line. Next we can add plots of the nearest stars, the apparently brightest stars, or both, or we can plot a series of cluster isochrons and add stars from star clusters such as the Pleiades or Hyades, demonstrating how the series enables us to date star clusters. Additionally, we can take one star and watch the change in position on the H-R diagram over time. In another program, one of the most effective uses of the computer I have seen, we can take a star cluster and watch how the position of the stars change on the H-R diagram as the cluster ages (Simon, 1984c). With this, we see that massive stars evolve the fastest and move off the main sequence long before the lowest mass stars reach it.

At least two programs enable simple stellar models to be presented to a class (Hall, 1977; Simon, 1984d). Graphically, these programs plot variations in properties such as density and temperature with distance from the center of the star. While neither is correct in all details, at least one of them calculates a crude set of models. The surface properties of these models can be used to infer the presence of a main sequence and a mass-luminosity law.

An area that is peripherally treated in astronomy courses, but which has great fascination for the average student, is the possibility of interstellar travel. Some of the problems associated with the relativistic effects of interstellar travel are treated in one computer simulation (Simon, 1984e). The student controls an interstellar ship on a journey from one stellar system to another by varying the acceleration of the ship. The computer calculates and displays the relativistic time dilation and length contraction effects involved. The student must decide when to start slowing the ship in order to arrive at the destination at zero relative speed. The main result of using this simulation is to illustrate that relativity and common sense have little in common.

Moving from properties of individual stars to groups of stars, relatively few programs are available. One shows the differential rotation of the galaxy (Thomas and Grayzeck, 1983). With an animation, students much more easily see the effects of this type of rotation. Several objects at different distances from the center move around the center at differing speeds. A more recent program (Dykla, 1988) simulates the gravitational interaction of a star cluster. In this simulation, a number of stars move under the influence of their mutual gravitational attraction. Additionally, the effects of a background of dark matter may be included.

The only simulations that have been prepared involving systems outside our own galaxy have dealt with the Hubble law. These can be divided into two classes. The first is similar to the rubber band or rubber balloon analogy. A number of points are plotted on the computer screen and moved such that their separations obey the Hubble law (Johnson, 1984d). The second class plots graphical representations of the Hubble law. A very good program, but one that is too complex for the average introductory astronomy student in this country, plots the Hubble law and shows its evolution with time illustrating that early in the history of the universe the slope was steeper than at the present (Alexander, Foster, and Unruh, 1983b).

An excellent use of the computer allows students to assimilate the knowledge they have gained from a number of different areas to construct varied planets circling different stars. There are several such programs available though they are, of course, based on rather speculative science. Some of them postulate the existence of alien species with certain characteristics, requiring the student to construct a world habitable by these species (Simon, 1984f; O'Brien and Roney, 1987).

3. The Indoor Telescope

An unusual mode of computer use we have had under development at the College of Charleston for the past five years involves what we call an indoor telescope. We recognized a number of years ago that astronomy laboratories are different from chemistry and physics laboratories in that students generally do not gather quantitative data. Rather, they analyze data gathered by others or make qualitative observations of astronomical phenomena. The reasons for this difference include the fact that obtaining quantitative astronomical data requires much more sophisticated equipment than that required for physics and chemistry. We recognized that one way of circumventing this problem was to use the microcomputer as a simulated research telescope to enable students to gather and analyze quantitative data. Each student is provided with a simulated universe of 1000 stars. The properties of these stars are calculated based on general luminosity function with the assumption of a limiting magnitude of 11. This data set contains variable stars of various types as well as high-proper-motion stars and an open star cluster. In addition, interstellar reddening is present. Each student has his or her own diskette with individualized data. The object of the simulation is for the students to find out as much as possible about the 1000 stars in their data set. With their simulated telescope, students can measure positions and magnitudes as well as obtain spectra.

In interacting with the program, the student first requests one or more nights of telescope time. Once this is accomplished, they choose whether they wish to obtain photographs, Cassegrain spectra, objective prism spectra, or photoelectric measurements. When the observing session begins, students are presented with a view of a small portion of their star field. They move the telescope to the position at which they want to take data and obtain it. A simulated clock keeps track of the progress of the night and forces observations to cease at sunrise. Once the student has obtained one or more photographs, the data can be analyzed. To do this he or she uses a simulated blink comparator. Two plates at a time are loaded and just as in a real blink comparator, they are intercompared. Choosing plates obtained with different filters enables detection of either very blue or very red stars. Choosing plates obtained at different times allows detection of variability or proper motion. In order to detect proper motion, the existence of a first generation set of plates taken a number of years earlier is postulated. Loading a current plate with one of these older plates usually reveals one or more stars with a detectable proper motion on each plate. Our simulated comparator includes a built-in measuring engine and iris photometer so that positions and magnitudes of stars may be measured. From the measurements, students can determine proper motions easily and distances with difficulty. In the latter case, they must remove the effects of interstellar reddening to get a true apparent magnitude and obtain the absolute magnitude from either the unreddened colors or the spectral type. When a student finds a variable star, he or she can determine the period of variation and use information such as the spectral type to determine the type of variable.

After students have obtained and analyzed the data, they can draw conclusions about the properties of their stars and report them to the computer, which judges the validity of their work and awards them appropriate points. Types of conclusions that may be reported include the colors and spectral types of stars, their distances, and the periods of variable stars.

4. Equipment Required

Many instructors are hesitant to attempt to incorporate computers into their curriculum because of the difficulty of obtaining funds to equip a computer laboratory. From the above discussion, you have realized that one of the best ways to use the computer in teaching is as a demonstration device. For this purpose, one computer is sufficient for an entire class. Ideally, there should be a means of projecting a large image of the computer screen. However, I have worked for a number of years with 2 small monitors for a class of 60 students. While the students can't see what is written on the computer screen, they can see the graphics. Since these are the essence of most computer simulations, this equipment is sufficient. However, a better means is to connect the computer to a projection device such as a projection television or a liquid crystal display that fits onto an overhead projector. If you want students to interact directly with the computer, then a computer lab is necessary. A minimal lab is one computer available 8 hours per day for every 30 students. This could be used for take-home labs or homework exercises. In a formal lab, the best situation is one computer for every two students. We run with one for every four students. Many times this still requires take-home labs.

5. Problems Involved in Teaching with Computers

I do not want to leave the impression that computers are a cure-all for the problems facing astronomy education or that their integration into the curriculum is easy. A number of difficulties are present. First, a wide variety of computers are available, many of which are incompatible with each other. Obviously, the hardware you have will limit the programs you can use. Since to date in this country, the most popular hardware configuration for educational use has been the Apple II series, the widest variety of software is available for this machine. However, IBM-compatible computers and the Apple Macintosh are becoming more prevalent (especially at the post-secondary level) and so more software is being developed for these machines. Another problem concerns the setup time required to use computers as demonstration devices. Unless permanent projection TV or monitors are installed in the classroom with provisions for easy attachment of a computer, setup could require as much as 20 minutes per class. Even with matching hardware and software and a good projection setup, other problems remain. For example, a number of software programs that are not equivalent treat the some topic. In the discussion of Kepler's Laws above, I referred to several programs by different authors. Since I like to use pieces of each of these programs, time is required to switch from one program to another. It would be much better if I were able to put together my own selection of pieces from each. Unfortunately, this is not currently possible. Student motivation is also a problem. No matter how interesting the software is to the instructor, the student is unlikely to be equally interested. It is not sufficient to tell a student to run a particular program to find out what he or she can from it. Guidance is needed to ensure that students use the program properly and some method of evaluation is required.

It might be asked why many of the programs described are relatively old, dating from the early eighties. Unfortunately, the educational market in astronomy is not a popular one for software developers. The development of the bulk of astronomy software is driven by the market among amateur astronomers. This is the main market for the planetarium-type programs mentioned earlier. Except for these programs, all of the programs discussed have been developed by astronomy educators for use in education. The time required for such a person to develop a program for a new machine seems to be about five years. Thus only now are we seeing programs developed by educators begin to appear for the Macintosh and more recent IBM compatible machines. Another factor is the relatively low level of funding for educational developments available in the United States in the last few years. Recently we have seen an increase in such funding from both Federal and private sources. In addition to the project by A'Hearn *et al.* mentioned earlier, Shapiro and Sadler (1987) have described the preliminary software development being carried out with N.S.F. funding by the staff of Project STAR. I am hopeful, therefore, that there will soon be similar but more sophisticated material than that described above available.

I would like to thank Michael Francisco, Alan Johnson, Keith Johnson, William Kubinec, and William Lindstrom for their assistance with "The Indoor Telescope." Keith Johnson and John Mosley provided valuable information used in the present work. Finally, Katina Strauch and Jay Pasachoff made valuable comments on an earlier version of this manuscript. Work on "The Indoor Telescope" was funded by NSF Grant SPE-8263018.

References

Note: There is no accepted system for providing references to computer software. In some cases authors are uncertain or unknown. I have adopted the convention of using embedded text references giving the author or in a few cases the publisher of the software. If a program has been discussed in the astronomical literature, I have cited that reference. Disks consisting of collections of software have been treated as books with individual programs as journal articles. The availability of many of these programs is uncertain. Where possible I have given an address for information.

A'Hearn, M.F., Bell, R.A., Blitz, L., Freedman, I., Greenblatt, J., McDaniell, C., Ohlmacher, J.T., and Trasco, J.D., 1988, "Parallax Lab" part of *Introductory Astronomy Minilabs for Non-Science Majors* (College Park, MD: Astronomy Program and Instructional Computing Division, University of Maryland).

Alexander, D., Foster, D., Unruh, H., 1983a, "H-R Diagrams" on *Modern Astronomy Demonstrations* (New York: John Wiley and Sons).

Bork, A., 1981, *Introductory Mechanics* (Iowa City, IA Conduit).

Cromer, A., 1981, *Three Body Orbits* (Newton Centre, MA: EduTech, 634 Commonwealth Avenue, Newton Centre, MA 02159).

Dukes, R., Jr., and Roskoske, T., "An Interactive Computer Exercise for Elementary Astronomy," *AAPT Announcer* **9(2)**, 79, 1979.

Dykla, John J., 1988, "Star Cluster Dynamics Simulation with Dark Matter," presented at The Conference on Computers in Physics Instruction, Raleigh, NC.

Hall, D.E., 1976, "A Stellar-Structure Model for the Classroom," *B.A.A.S.*, **8**, 546.

Huntress, W., 1982, *Saturn Navigator* (Sub-Logic, 713 Edgebrook Drive, Champaign, IL 61820).

Johnson, K.H., 1984c, "The Bohr Atom" on *Astro-Demos*, (Reno, NV: Fleischmann Atmospherium/Planetarium; privately distributed).

Johnson, K.H., 1984d, "Hubble Law" on *Astro-Demos*.

Johnson, K.H., 1984b, "Kepler's Laws" on *Astro-Demos*.

Johnson, K.H., 1984a, "Retrograde Motion" on *Astro-Demos*.

Kanipe, J., Korchmal, M., Paul, L., Berry, R., Bailey, J., 1988, "Personal Planetariums: The Night Sky in Your Computer," *Astronomy*, **16(8)**, pp. 36–41.

Mathis, T., 1988, *Voyager* (Carina Software, 830 Williams Street, San Leandro, CA 94577).

MECC, 1985, *Sky Lab* (St. Paul, MN: Minnesota Educational Computing Consortium).

Mosley, J.E., "Software for Classrooms," *Journal of Computers in Mathematics and Science Teaching* **VI**(**1**), pp. 69–70.

Mosley, J.E., "Astronomy on an Apple Macintosh," *Journal of Computers in Mathematics and Science Teaching* **VI**(**3**), pp. 55–57.

Mosley, J.E., 1988a, private communication.

Mosley, J.E., 1988b, "Review Corner," *Sky and Telescope*, **76**, 676.

Mosley, J.E., and Fraknoi, A., 1986, "Computer Software for Astronomy," *Mercury*, **XV**, 152.

Reitmeyer, L., 1979, "Demonstration of a Microprocessor Program of Binary Stars for General Science Instruction," *AAPT Announcer*, **9**(**2**), 81.

O'Brien, T.C., III, and Roney, M.L., *Planetary Construction Set* (Sunburst Communications, 39 Washington Avenue, Pleasantville, NY 10570-9971).

Parker, R., 1988, *Visible Universe* (Parsec Software, 1949 Blair Loop Road, Danville, VA 24541).

PicoSCIENCE, 1988, *Superstar* (41512 Chandbourne Dr., Fremont, CA 94539).

Schwartz, J.L., 1983, "The Four Inner Planets" on *Microcomputer Software for Undergraduate Physics* (New York: Addison Wesley).

Schwartz, J.L., 1985, *Sir Isaac Newton's Games*, (Sunburst Communications, 39 Washington Avenue, Pleasantville, NY 10570-9971).

Seeds, M., 1985, "Orbital Motion" on *Horizons: Exploring the Universe* (Belmont, CA: Wadsworth).

Shapiro, Irwin I., and Sadler, Philip, 1987, "Project STAR Annual Report to the National Science Foundation Washington, D.C.," (Cambridge, MA: Harvard College Observatory).

Simon, S., 1984a, "Expedition to Mars" on *The Astronomy Disk* (Englewood Cliffs, NJ: Prentice-Hall).

Simon, S., 1984b, "Spectral Classification" on *The Astronomy Disk.*

Simon, S., 1984c, "Stellar Evolution" on *The Astronomy Disk.*

Simon, S., 1984d, "Stellar Model" on *The Astronomy Disk.*

Simon, S., 1984e, "Starship" on *The Astronomy Disk.*

Simon, S., 1984f, "Build-a-World" on *The Astronomy Disk.*

Sumners, C., 1983a, "Sunrise/Sunset/Sun Locator" on *Astrografix* (Houston, TX: The Astronomy Data Center, Houston Museum of Science; privately distributed).

Sumners, C., 1983b, "Moon Plotter for Phases/Locations" on *Astrografix.*

Thomas, P., and E.J. Grayzeck, 1983, *B.A.A.S.*, **15**, 996, 1983.

Zink, J., 1983a, "Lunar Phases and Eclipses" on *Astronomy Demonstrations* (University Park, PA: Department of Astronomy, Pennsylvania State University; privately distributed).

Zink, J., 1983b, "Kepler's Three Laws of Planetary Motion" on *Astronomy Demonstrations.*

Zink, J., 1983c, "Parallax" on *Astronomy Demonstrations.*

Zink, J., 1983d, "Binary Stars" on *Astronomy Demonstrations.*

Discussion

T.J. Balonek: *In addition to using microcomputers as demonstration devices, it is desirable to expose introductory college-course students to how scientists use computers in their research. Since astronomy is the only physical science and/or laboratory course that many college students take, these exercises should also be representative of how computers are used in other sciences. One such example is to develop laboratory exercises that use inexpensive microcomputers to perform image processing of astronomical images, in which students learn the problem-solving techniques and thought processes used in image analysis.*

B.W. Jones: *Microcomputers can also be used in an adaptive mode with frequent branching to meet individual students needs at a variety of points in the program. This applies as much to astronomy programs as to programs in other subject areas.*

G.S. Mumford: *What is the availability of the software you have discussed?*

R.J. Dukes, Jr.: Demonstration programs are available from many sources. Prentice-Hall, Wiley, and Wadsworth have packages available for sale or to adopters of their textbooks. A list of available astronomy software is available from the Astronomical Society of the Pacific. The American Association of Physics Teachers publishes descriptions of public domain software in their quarterly *Announcer*. The Physics Coursework Project at North Carolina (Raleigh, North Carolina, U.S.A.) State University maintains a collection of public-domain physics software as well as a list and review of commercial software. Some of these programs are suitable for astronomy. The *Physics Teacher* publishes software reviews. The *Planetarian* likewise. *Sky & Telescope* and *Astronomy* publish monthly columns that consist of program listings and/or reviews. I will be happy to send a list of my personal recommendations to anyone writing me. I can also provide some Apple programs that I have written if a floppy disk is enclosed. I am maintaining a list of people interested in using *Indoor Telescope*. I will notify these people of its availability when it is in a form suitable for distribution.

THE PERSONAL COMPUTER FOR TEACHING ASTRONOMY

R. Andreoni[1], G. Forti[1], P. Ranfagni[2]

[1]Osservatorio Astrofisico di Arcetri, 50125 Firenze, Italy
[2]Istituto di Astronomia dell'Universitá di Firenze

1. Introduction

The two interactive software packages, which we called *ASTRONOMIA 1* and *ASTRONOMIA 2* (in the following A1 and A2) and developed for a personal computer under MS-DOS, are primarily intended to be used in junior-high and high schools as tools for teaching astronomy. With the help of these programs and using the sky as a laboratory, the teacher can explain the most difficult parts of the astronomical geography: time and coordinates. Very few schools in Italy own the necessary educational tools such as a small planetarium, a sidereal clock, an oriented celestial sphere, *etc.*

We tried to exploit the capability of a personal computer to visualize and to present the results of a calculation as diagrams and tables. A limited interaction permitted by the programs allows the user to compare the results with other computations.

All the programs are written in such a way that it is difficult to make a mistake when entering data. The programs were first assembled and tested by us, then presented to the students and teachers to be evaluated and criticized. We obtained from them suggestions, advice, and, generally good reviews. Obviously, this special kind of reviewers had good backgrounds in astronomy in order to evaluate the programs completely.

A small booklet containing a description of the programs and a few worked-out exercises is available for both packages.

2. ASTRONOMIA 1: Time and Coordinates

The A1 package is devoted mainly to the understanding of the main astronomical phenomena, such as the rotation and revolution of the Earth starting from direct observation of the sky (Ranfagni, 1985). The programs allow the user to answer the question which, unconsciously or not, a beginner observer wonders about: "Where do I have to look to see a certain object?" The long and tedious computations necessary to answer this question usually discourage the beginner. In our case, the computer takes care of the computations and the user only has the task (undoubtedly more pleasant) of trying to understand the meaning and to verify the results directly from the sky.

The A1 software is divided into four sections: the programs in the first three (OROL4, EUROPA, EQTEMPO) and those dealing with the systems of coordinates in the last one (APPLICAZIONI) aim at explaining the measure of time and the angular quantities used in astronomy. The results are visualized in the most suitable form for the quantity considered: time is represented by analog and digital clocks, the geographical coordinates by a cursor moving over a map, and the astronomical coordinates by a drawing of the celestial sphere.

In the section OROL4, four analog and digital clocks are shown: a couple of them tick the civil time, and the others the local sidereal time, the mean local solar time, and the true local solar time. The equation of time and other parameters are shown in a small window in order to understand the various kind of times.

In the section EUROPA, the longitude and latitude of a place in Europe or in Italy can be obtained and propagated to the other programs with the help of a cursor moving on the screen and a small menu.

The section EQTEMPO deals with the equation of time and its variation.

The remaining programs of the last section, called APPLICAZIONI, are utility tools: it is presumed that the user is more interested in the value of the quantities rather than in their meanings. They include the change of the coordinates, the calculations necessary for the pointing of a telescope, a method to find the observer's geographic coordinates by taking only two height measurements of a known celestial object. These programs may be useful to the student as well as to the amateur astronomer since the results are immediately applicable.

It is possible to pass from a section to another one with the help of a small menu. Appropriate explanations can be obtained in each program.

3. ASTRONOMIA 2: Movements on the Celestial Sphere

This second package, A2, is mainly dedicated to obtain on the screen and on the printer (if available) the sky for every time and place on the Earth (Andreoni, et. al., 1987). The projection of the sky on the plane of the screen is chosen such that all the angular distances of the objects remain almost unchanged and the representation of the sky is rather similar to what we could really see by looking at the sky in the same direction.

The programs of A2 are divided in four sections, as those of A1.

The first section, TERRA, is the natural evolution of Europe in A1: the geographic coordinates of the observer can be found with a cursor on the celestial globe. The coordinates found can be passed to the next section, CIELO. This set of programs deals with the representation of the sky. The input data are: the date, the hour, the geographic coordinates, the limiting stellar magnitude and a flag to compute or not the precession (necessary if the date is very far from 2000.0). The time zone is automatically computed as soon as the location coordinates are entered (time must be specified in U.T.). When the data are ready to be plotted on the screen there might be two choices according to the time of the day: a) the daily movement of the sun across the sky and the sun among the stars; b) the night sky. Other features can be obtained when the sky is drawn: a cursor to identify the objects, height-azimuth and equatorial grids, the possibility of bringing an object to transit, *etc.*

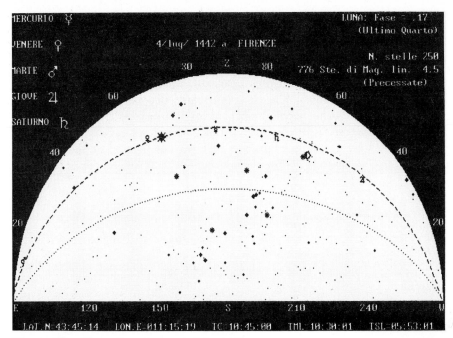

Fig. 1. *The Florentine sky on 4 July 1442.*

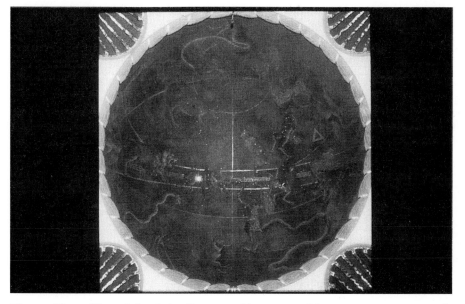

Fig. 2. *The ceiling of the Saint Lorenzo old sacresty, Florence.*

The section, COSTELLAZIONI, shows each constellation (with or without objects other than stars) with the possibility of going back and forth in time to see their distortions due to proper motion.

The last section, CATALOGO, is a small catalog of about a thousand stars (limiting magnitude about 6 and equinox 2000.0), where a few physical data are stored for each individual object; since it is possible to select one star at a time, we provided the possibility of precessing its equatorial coordinates.

In order to illustrate a different use of this package, we have reproduced the Florentine sky of 4 July 1442 to compare it with the one painted on the ceiling of the Saint Lorenzo old sacresty in Florence, Italy. For more details see Forti *et al.*, 1987.

Acknowledgments

One of the authors wishes to thank the "Comitato per la Divulgazione dell'Astronomia" for providing him travel funds.

References

P. Ranfagni, "Presentazione di alcuni programmi per il personal computer utili all'insegnamento dell'Astronomia," *Giornale di Astronomia*, vol. 11, n. 4, 1985.

R. Andreoni, G. Forti, P. Ranfagni, "Astronomia con l'ausilio del personal computer," *Giornale di Astronomia*, vol. 13, n. 3, 1987.

G. Forti, I.L. Ballerine, P. Ranfagni, B. Monsignori Fossi, "Un Planetario del XV secolo," *L'Astronomia*, n. 62, 1987.

MICROCOMPUTERS IN AN INTRODUCTORY COLLEGE ASTRONOMY LABORATORY: A SOFTWARE DEVELOPMENT PROJECT

David D. Meisel[1], Kenneth F. Kinsey, and Charles H. Recchia
Department of Physics and Astronomy, State University College at Geneseo,
Geneseo, New York 14454 U.S.A.

We have developed software for the Apple IIe series of microcomputers for use in labs in an introductory astronomy course. This software emphasizes a toolkit approach to data analysis; it has been class tested with over 170 students and was a resounding success as a replacement for previously used graphical approximations. A unique feature of this software is the incorporation of image-processing techniques into a course designed for non-science majors.

The five software packages are:

(a) **Datasheet** - A six-column spreadsheet with columnwise operations, statistical functions, and double-high-resolution graphics.

(b) **Image-Processor Program** - Allows 37 × 27 pixel × 8 bit video captured images to be manipulated using standard image-processing techniques such as low pass/high pass filtering and histogram equalization.

(c) **Picture-Processor Program** - Allows 256 × 192 bilevel pictures to be manipulated and measured with functions that include calipers, odometer, planimeter, and protractor.

(d) **Orrery Program** - Simulates planet configurations along the ecliptic. A movable cursor allows selection of specific configurations. Since both relative times and angular positions are given, students can deduce the scale of the solar system using simple trigonometry.

(e) **Plot Program** - Allows orbital positions as observed from above the pole to be plotted on the screen. By entering trial values of elliptical orbit parameters, students obtain and the program plots the best fitting ellipse to the data. The sum of the squares of the residuals in the radial coordinate is given after each trial so that students can discover convergence more easily than by simple visual examination of a plot comparing the trial theoretical points with the raw data points.

The video-captured materials used with the Image Processing and Picture Processing programs were obtained first on VHS video tape using the 0.6-m telescope at the C.E.K. Mees Observatory (near Bristol Springs, New York State) or the 0.35-m telescope at SUNY/Geneseo. Selected video frames were subsequently captured into

[1] Also Associate Director of the C.E.K. Mees Observatory, University of Rochester

256 x 192 pixels x 8 bit video images using the Imageworks^{TM} system (Redshift Ltd., P.O. Box 4335, Mountain View, California, 94040) on an Apple IIe and recorded to floppy disk. Using our own extensively modified version of the Imageworks^{TM} software, we were able to create images and bilevel pictures that can be loaded directly into *any* Apple IIe for processing. Images presently available include selected lunar features, Jupiter, Saturn, Jovian satellites, Venus, double stars, solar activity, and low dispersion stellar spectra. We are striving to expand this list to other objects. Astronomers having public-domain images they might be willing to share should write to us giving details.

Our approach to computer usage is (1) to provide menu-driven programs that furnish tools that are analogous to those used in classical laboratory tasks; (2) to give students as much control as possible over the order in which the tasks are done; (3) to restrict computer use to those tasks where it is essential or has greatest impact; (4) to provide a consistent interface to more advanced features that will be used in later courses; and (5) to provide students with their own copies of software and images, at the price of a diskette, to allow work outside the laboratory.

In summary, we have successfully brought low-cost image processing techniques into elementary astronomy laboratories through an extension of a toolkit programming approach. Our materials are all in the public-domain and therefore available at minimum cost. The software runs under the PRODOS system, and has been designed for the 128k Apple IIe (enhanced) series, and is being extended to the Apple IIgs model.

Distribution copies including ASCII readable notes are now available for: DATASHEET 6.0 (one 5 1/4" disk); PICTURE/IMAGE (one 5 1/4" disk); ORRERY/PLOT (one 5 1/4" disk); Imageworks^{TM} Enhanced Software (one 5 1/4" disk)*; Imageworks^{TM} Compatible Images (one 5 1/4" disk)*.

Requests for these should be accompanied by the appropriate type and number of blank diskettes. Requestors will be put on a mailing list for notices of updates. Address all inquiries to: Dr. David D. Meisel, Department of Physics and Astronomy, State University College, Geneseo, New York 14454, U.S.A.. Other inquiries can be made by mail or BITNET (MEISEL@GENESEO).

*Requires Imageworks^{TM} board to function properly. Xerox copy of user manual should accompany request.

———————————

AN APPLICATION OF PERSONAL COMPUTERS IN ASTRONOMY EDUCATION

A.S. Nikolov
University of Sofia, A. Ivanov Str. 5, 1126 Sofia, Bulgaria

Observations play an important role in the process of teaching astronomical knowledge. Practical observations of astronomical phenomena lead to analysis and explanation based on natural laws and so form the basis of cognitive processes in the education. Evidently the observations are an integral part of acquiring astronomical knowledge. Giving up observations, no matter what the reasons, is equivalent to losing quality in the educational process. It decreases the possible influence over the personal development of pupils and students. At the same time, observation and observational results are important for success in education.

Carefully planned observational time has a substantial influence on cognitive and educational processes. It leads to considerably more active participation in astronomy lectures.

Implementing observations with pupils beyond doubt has objective problems: a suitable time and conditions, *etc.* The training and qualification of the lecturers for this specific kind of activity also play an important part here.

Pupils in Bulgaria receive their initial teaching about astronomy as early as in the primary schools. Afterwards, this knowledge is expanded, arranged systematically, and deepened in high schools. This fact also impedes the training of tutors and teachers for observational work in astronomy. At the same time it does not abolish the need to teach in timely fashion information for the students — future teachers — about the appearance of the stellar sky and the basics of planning observations, including the visibility of the moon, the planets, and other interesting sky objects from a site at a given moment of time.

Students' interests are so varied that the training of future teachers in their primary and secondary school education gives Bulgarian specialists in education the task of effectively and rapidly solving the problem of how to include the most necessary astronomical observations.

Since 8-bit personal computers are widely found in Bulgarian schools, their usefulness for modeling different processes and their easy operation made possible the implementation of experimental programs. Such programs included acquainting students with the appearance of the stellar sky (the constellations) and with programs for planning nighttime observations with preliminary introductions to the location of the planets among the stars, recognizing the periods of visibility, *etc.*

The first of the suggested programs offers several possibilities: a) the computer displays a constellation with its name chosen from a list on the monitor; b) the computer shows in successive series the constellations with their names, and c) it displays on the monitor an arbitrary constellation and asks about its name. If the answer is correct, another constellation is displayed on the monitor. When the student has proposed two incorrect answers, the name of the displayed constellation is shown on the screen after which the next constellation is displayed. When this program mode is stopped, the computer show how many correct answers a student gave.

The second of the programs calculates the coordinates of a planet for an arbitrary moment of time and displays them on the screen along with the surrounding constellation. This program allows a preliminary detailed preparation for night observations in "laboratory conditions."

Our experience shows that the natural interest of pupils and students in the sky combined with their desire to handle computer techniques increases their motivation and their respective active participation in observations. Their preliminary acquaintance with the constellations on the screen of the monitor lessens considerably the moderate possibility that they would not recognize them in their natural appearance in the sky. This fact is of utmost importance for future teachers, who could swiftly train themselves for each separate observation and analyze easily both the constellations visible for the planned moment of observation and the visibility and location of the planets among the stars in laboratory conditions.

At the same time, a rapidly increasing wish to compare the materials learned with the aid of the computer with the natural appearance of the constellations on the celestial sphere under suitable conditions finally leads to the natural individual learning of the constellations.

The described positive features of the programs and the absence of suitable conditions in heavily populated cities for carrying out night observations (an obstacle for the quality training of teachers for such observations) make the suggested programs very effective and useful.

INTERACTIVE COMPUTER PROGRAMS FOR TEACHING ASTRONOMY

S.R. Prabhakaran Nayar & Rebecca T. Thomas
University Observatory, Kerala University, Observatory Hill, Trivandrum, India

1. Introduction

Astronomy is the most popular and oldest of all sciences and it has had profound influence on human thought. Unfortunately, astronomy does not find an appropriate place in our school-college syllabi. This may be due to the fact that teaching of astronomy encounters a large number of problems with regard to visualization and practical experiments. Popularization of astronomy depends heavily on a large variety of astronomical events, such as the arrival of comets, eclipses, supernovae, *etc.* Visualization of dynamics in several directions, wide variance of time scales, concepts of space, *etc.*, create problems in teaching/learning processes in astronomy. Our world of human experience is limited to within a narrow frame, whereas in astronomy we speak of size, time, and temperature in gigantic scales. To bring all these parameters onto the human level, one has to think of effective teaching aids and the right type of techniques. We have been using a large number of tools in teaching astronomy, including star charts, globes, models, photographs, slides, *etc.* Microcomputers act as an effective medium in teaching astronomy. They can even replace most of the above mentioned teaching aids (Hunt, 1986; Marx and Szucs, 1985; Sparkes, 1986). A microcomputer can also act as a textbook, a black-board, or even a planetarium. The computer acts also as a mediator between the student and the model of some real-life situation. The process of building and using models, called *simulation*, helps us to investigate systems that would otherwise be inaccessible. We have developed a set of software to teach basic concepts in astronomy, such as the solar system, constellations, and the physics of stars. We have explored the possibilities of making the software interactive, using the observed data so that a real life situation can be experienced by the users, just as for practical experiments.

2. Microcomputers in Astronomy Teaching

The available teaching software for microcomputers can be grouped into a few major categories: 1) Planetarium-type software — which uses the monitor screen to show the position and motion of stars, sun, moon, planets, and other objects in the sky for any place or time, according to the user's choice. These programs are technically of high quality, and this type of software can be well used in education to present the basics of the clockwork universe. 2) Celestial-mechanics software — useful in revealing the finer details of Keplerian motion. The concepts of gravity and the central gravitation field, the solar system and the revolution of planets, and elliptical orbits and their comparison with realistic models can be shown using this category of software. 3) Astrophysical software — compared to the other two types, it has a wider scope in teaching astrophysics, showing stellar/solar models, evolution of stars, *etc.* Instead of restricting the software to techniques like page turning or routine-type description, one can think of a genuine interactive system where the models built could be used to simulate real life situations and comparison can be made with the sets of observations. Interactive video systems are a relatively new technique. Video recorders guided by microcomputers promise to make interactive picture-and-movie presentations possible.

3. Interactive Software for Teaching Astronomy

In interactive software, the user has an active role in the software operation. The interaction can be carried out in a large variety of ways. In this paper, we have restricted the interaction to the use of keyboards. Software for teaching astronomy at an introductory level can be generally grouped into three categories, as given in Table 1: 1) Solar system; 2) constellations; and 3) physics of stars. In this course, in addition to routine software techniques, we have explored the possibility of making some of the software segments interactive so as to make them efficient teaching material. The solar-system group consists of two segments. The first is a graphic representation of the orbits of planets. For convenience and for clarity, the inner and outer planets are represented in separate frames. From the details of the orbit, one can visualize the relative orbital motion of planets and other phenomena like planetary conjunctions, oppositions, *etc.* By selecting the proper initial epoch and accurate orbital parameters, one can visualize and learn many of the planetary phenomena of the past or future. The second part of the solar system program is a quiz to identify a given planet for which the properties are given. The parameters selected for the program are planetary mass, size, density, orbital period, orbital radius, number of satellites, orbital velocity, inclination of the orbital plane with the ecliptic, major constituents of the atmosphere, surface temperature, *etc.* The user can ask questions regarding one of the parameters of the selected planet and from the answer given by the computer make the guess. The software selects one of the planets at random and continues selection until all the planets are selected. This program was found to be very interesting for introducing details of the planets among children.

The part that introduces constellations consists of a few frames describing important constellations of the night sky and starts with some of the stories associated with them. The learner is introduced to important constellations, prominent stars, directions in the night sky, and stories associated with them in the epics. The program can be compared with planetarium software.

The segment on physics of stars gives an introduction to the concepts of apparent magnitude, distance to stars, absolute magnitude of a star, color of stars, meaning of color, surface temperature, *etc.* With the examples of prominent stars and with this background, the H-R diagram is introduced, explaining the hierarchy of stellar evolution. An attempt is also made in this part to develop interactive software to find the position of a star in the H-R diagram (Kaufmann, 1978) as it evolves, taking the sun as an example. As an initial step, one can simulate the evolutionary track of a star in the H-R diagram by giving a set of data points corresponding to certain initial conditions. This method will be modified by using equations describing evolutionary parameters or simulating the evolution of a star using available computer codes.

4. Conclusion

Astronomy is the fastest developing science and it requires support at the basic level. The teaching of astronomy encounters a large number of difficulties and teaching materials and tools may help to a large extent to overcome these difficulties. The use of microcomputers is one of these tools. We show that the microcomputer can be made more effective by preparing interactive rather than non-interactive software. The interaction can be either in the form of a question/answer type, which give a familiarization or an introductory course in a specified topic, or in the form of an interactive real-life situation, as in the case of planetary motion or stellar evolution. The interaction can be effected either by using the specified data points collected from the observations or calculations or by simulating the real-life situation with the help of computer codes. This type of software opens up a new field of experimental work for astronomy students, showing stellar evolution, orbital motions, and other astronomical phenomena.

References

1. Hunt, J.J. (1986) *Cosmos — An Educational Challenge*, GIREP Conference Proceedings (ESA SP-253).

2. Kaufmann, W.J. (1978) *Stars and Nebulas*, W.H. Freeman and Co. (then San Francisco, now New York).

3. Marx, G., and Szucs, P. (1985) *Microcomputers in Science Education*. Microscience International Workshop, Hungary 20-25 May.

4. Sparkes, R.A. (1986) *Microcomputers in Science Teaching*, Hutchinson Co. (U.K.).

Table 1. List of interactive and non-interactive astronomy software.

Category	Name of the Program	Description
Solar System	Solar 1	Introduces the concepts of orbits.
	Solar 2	Relative sizes, relative velocities, *etc.*, for inner and outer planets separately.
	Planet X	Interactive quiz to expose the user to elementary ideas about planets.
Constellations	Rasi	Familiarization of some of the important stars and constellations.
Stellar	Star 1	Introduces the concept of color, spectra, and surface temperature.
	Star 2	Introduction to H-R diagrams.
	Star 3	Evolution of sun in H-R diagrams.
	Star 4	Stellar magnitude.
	Star 5	Interactive quiz to familiarize with parameters of important stars.

COMPUTER SIMULATIONS OF ASTRONOMICAL PHENOMENA

A. Noels and R. Papy
Institut d'Astrophysique, Université de Liège, B-4200 Ougrée-Liège, Belgium

Introduction

The basic idea of this project is to introduce a senior student to an astronomical phenomenon that he analyses carefully and simulates on the graphic screen of a personal computer. In a second step, this simulation serves as a demonstration for junior students.

Three examples are presented here: the retrograde motion of the planets, the Earth's rotation, and the precession and the equilibrium figure of the Earth due to tides. Programs have been written on an IBM PC/AT.

It would be highly desirable to start an exchange program of material of didatic interest among the international astronomical community[1]

1. The Retrograde Motion of the Planets

In a first option, a part of the celestial sphere centered on the ecliptic circle is developed along a cylinder; the ecliptic latitudes are magnified. As many as five planets can be visualized simultaneously with different colors. Observations can be made for any planet chosen in the menu. For a terrestrial observer, the annual motion of the sun is a horizontal mid-screen track. For a given planet, the time interval between two successive retrograde loops is its synodic period. Observed from the Earth, the exterior planets present a loop nearly every year while the interior planets show a loop nearly each time they make a revolution around the sun.

The second option shows the chosen planets and the Earth orbiting the sun, as seen from the ecliptic pole, while their apparent paths are represented in the lower part of the screen as in option 1 but with a smaller magnification in latitude.

A third option gives the apparent orbits of the planets in the ecliptic plane, viewed from a distant point on an axis perpendicular to the ecliptic plane. The sun orbits the Earth and the planets show their retrograde loops in the ecliptic plane. The origin of the deferents and epicycles in the Ptolemaic system is clear. The lower portion of the screen still shows the same tracks as in option 1.

---

[1] Anyone interested should contact A. Noels, Institut d'Astrophysique, 5 Avenue de Cointe, B-4200 Ougrée-Liège, Belgium.

2. The Earth's Rotation — Precession

The rotation of the Earth is viewed in an inertial frame centered on the Earth. In that frame, stars would occupy fixed points on the screen and the horizontal (x,y) plane is the ecliptic plane. A red meridian serves as a reference, and a red spot is the place on Earth where the mean sun is at the zenith at local noon time. A red γ marks the vernal equinox. The origin of time (year 0, day 0, hour 12) is arbitrarily taken when the reference meridian, γ, and the mean sun coincide.

In the first mode, the time interval is 12 solar minutes and the diurnal rotation of the Earth is seen. A solar day is equivalent to 24 h. Starting from the mean sun, marked by the red spot, the red meridian makes more than a complete revolution in 24 h, to coincide again exactly with the mean sun; γ remains fixed on the screen. The reference meridian makes a complete revolution in 24 h.

The second mode has a time interval of 2 solar days. It shows the annual revolution of the Earth around the sun. The angular distance between the reference meridian and the mean sun is constant and γ is nearly fixed on the screen. The red spot makes a whole revolution in a tropical year, as does the reference meridian, which demonstrates the difference between the sidereal and the solar day.

The time interval is set to 72 years in the third mode to show the precession of the axis of rotation. The γ point makes a complete revolution in the retrograde sense in about 26000 years.

3. The Equilibrium Figure of the Earth due to Tides

The tidal effect of the sun and the moon is simulated, assuming the Earth to be a homogeneous spheroid. The Earth's center occupies the center of the screen and the positions of the moon and the sun are computed with great accuracy as a function of time. The distance ratio is not to scale and the deformation of the Earth is magnified by a constant factor. The sun is at a fixed point at the right of the screen while the moon moves around the Earth. The equilibrium figures of the Earth due to the tidal force of a single body are drafted in white and the resulting figure is fully painted. Near the sun, a red line scales the Earth-moon distance and a white line scales the flattening of the Earth due to the presence of the moon. The equilibrium figure of the Earth is the same whenever the Earth, the moon, and the sun are on the same line, whether in opposition or in conjunction, the only possible difference coming from a change in the Earth-moon distance.

TEACHING UNIVERSITY ASTRONOMY IN A MEGALOPOLIS: THE "ASTROTINO" APPROACH

Charles D. Keyes and Mirek J. Plavec
Dept. of Astronomy, University of California, Los Angeles, CA 90024-1562, U.S.A.

1. Observing the Students, the Textbooks, and the Skies

About every other undergraduate student eventually takes the UCLA elementary astronomy class *Astronomy 3*, and the two of us have been teaching about 1,000 of them every year. We will not repeat complaints about their poor and patchy high-school preparation in science, but we must mention it as an important phenomenon in U.S. education. The European system appears to be preferable, in which mathematics, chemistry, physics, and biology are mandatory and taught every year, although only 3 hours per week. It prevents large memory gaps and the spiral-wise parallel development in all four subjects makes it possible to teach on a higher level every subsequent year.

In our astronomy teaching, we attempt both (1) to explain the structure and evolution of the universe by systematically using fundamental physical concepts (related, if possible, to analogous terrestrial phenomena), and (2) to encourage the students to observe the sky and watch the phenomena. In order to realize this plan in just 10 weeks, we avoid voluminous textbooks, in particular those that devote too much space to the solar system. Regrettably, this has recently been a dominating trend. Another unfortunate tendency in the textbook business is to suppress actual color photographs of important objects, and to replace them by X-ray pictures in false colors. Thus, in an entire textbook, there may not be one direct optical picture of a planetary nebula or bright spiral galaxy. We are afraid that this trend will spread into popular magazines and calendars, and that in the next *Swimsuit* issue we will see only false-color maps of skin-surface temperature distribution for pretty girls, or perhaps X-ray images of their kidneys.

We also encourage students not only to observe the sky with our telescope on regular Open Nights, but also to watch the stars without a telescope, recognize the constellations, and follow the motions of the moon and planets. The limiting magnitude is seldom better than 3 from the UCLA campus, but we have many clear and reasonably warm nights on which the main constellations and bright planets can be seen even from the campus. And we rely on the students' interest in back-country trips. However, only about 15 per cent of the students listen to our exhortations; it seems to us that inertia plays a significant role.

2. ASTROTINO

Our Way of Emulating and Explaining the Phenomena

Astronomy not only enables but also badly needs visual demonstrations, and today it is possible to go beyond slides (which are beautiful but static), and beyond films (of which few are available and some are becoming rather obsolete; and which are frequently cumbersome as teaching aids). Many of the great strides in our knowledge of the Universe have not yet been adequately covered by films.

We are attempting to strike a middle road between static slide displays and immutable film strips, and are developing a project called ASTROTINO (Astronomy Teaching Innovation), which utilizes multi-color, very high resolution *interactive* computer graphics. The visual programs are generated by a MicroVAX or Sun-class computer graphics workstation, and then displayed on a large wall-screen directly as an integral part of the lecture presentation. Our system is capable of animation of images and graphics that simultaneously display up to 1024 different colors.

The instructor runs a program on a terminal in the classroom which tells the MicroVAX to select previously stored images (or sequences of images) or to run a code that generates in real-time one of many available displays. Previously established parameters or interactively-selected set-ups can be used to vary the demonstration. Thus, when a student asks about a specific item, the program can be re-entered at the critical spot; and if it is necessary to show how the initial parameters affect the outcome, a new run can be started.

We have developed a library of graphics functions, all written in FORTRAN, that are used to produce the plots and animation sequences. New display packages are continuously being developed in this framework.

Among the demonstrations currently available are: The appearance of the night sky at any chosen time or place, including daily motion animation with user-selected time-steps; the phases of the moon (an extremely confusing topic for city students!); a realistic emulation of twinkling; an extensive demonstration of the emission and absorption line spectra of a variety of common elements; a very popular animation of the excitation and subsequent de-excitation of an electron via the processes of absorption and emission; an observational H-R diagram; an animation that allows the students to watch a star move across the H-R diagram and change size in time steps that are proportional to actual evolutionary time scales; depiction of constellations and their secular changes in shape due to proper motion; animations illustrating the geocentric and annual parallax; an orrery and an animated solar system model that explains retrograde motion and the most important aspects of planetary configurations; and Roche models of selected interacting binary systems with animated trajectories of gas streams.

Discussion

D. Hurst: *You make extensive use of visual aids in your course. Do you include exam questions that cover these visual aids (e.g., identifying a type of object from a slide)?*

C.D. Keyes: Yes, we occasionally do show slides, asking the students to recognize the object, its type, its role in the universe, *etc.* It can be done on the multiple choice system, even in large classes, but it does require special preparations (starting the slide show at the beginning of the exam; having different questions on different versions of the test to prevent cheating.)

THE PAPERLESS ASTRONOMY CLASSROOM

James C. LoPresto
Observatory, Edinboro University, Edinboro, Pennsylvania 16444, U.S.A.

The "Paperless Astronomy Classroom" was instituted at Edinboro University of Pennsylvania in the fall semester of the 1985–1986 academic year.

The goal of this endeavor was to teach an elementary and some advanced astronomy courses "on line" with the use of multiple terminals supported by a Digital VAX 11/785. Currently the system used is a clustered VAX system using two VAX 11/785's and one VAX 8550.

The procedure of the "paperless classroom" eliminates the need for the transfer of paper between professor and students (homework, exams, supplementary hand-outs) in the first phase and eliminates the need for a textbook in the second phase.

An equally important goal is to provide additional enhancement for the student that may not be convenient or easily available from a textbook and the traditional classroom. Such an enhancement is series of large data bases, some of which can be the small data bases that traditionally appear in the appendices of textbooks. Another example might be a data base of graphs and charts (*e.g.*, H-R diagrams, periodic table, abundance tables and graphs, binding energy curves, *etc.*).

The advantages to such an approach are many: The system

- is less expensive for the student.
- is less expensive for the college or university (paper costs).
- gives students access to enormous data bases.
- allows students to take instant tutoring lessons on a chapter-by-chapter basis or on a topical or subject basis.
- gives students access to powerful calculation capabilities.
- allows the student to interact on-line by changing parameters of a calculation and seeing the instantaneous changes in the results, *e.g.*, change the sum of

the masses or individual masses in a binary star system and watch the effect on the relative orbital sizes or change the relative positions, the masses and watch the orbits change shape.

- gives the students access to powerful graphics and allows the student to inspect, change, and make graphs.

- to varying degrees depending on the presentation, makes the student computer conversant (a valuable asset in today's society).

The paperless astronomy classroom is available in four formats, Digital VAX VMS, IBM PC MS-DOS, Macintosh Hypercard, and the traditional hard copy. In the material that follows, "VAX/VMS files" means VAX/VMS, IBM PC MS-DOS, and Macintosh Hypercard cards.

Material that can be considered either optional or supplementary or, most importantly, of an enrichment nature will be presented with the use of "VAX/VMS files," tutoring lessons, menu-driven programs, and graphics (both interactive and non-interactive). The tutoring lessons will emphasize the subject material in the hard-copy presentation. Additional tutoring lessons will be arranged in a topical fashion that will not follow the order of presentation in this "hard copy" or any textbook.

Another option is to have the "hard copy" available on magnetic tape, for subsequent placement on a system disk for electronic access along with the rest of the paperless classroom. Furthermore, as indicated above, this can be done in both the IBM PC version and the Macintosh Hypercard version.

In my opinion, students respond very positively to such a method of presentation. The reasons for this are perhaps many, but include the novelty compared to most courses they take. Further, the idea that instantaneously calling up data or information has an element of fun, power, and adventure associated with it.

The Great Comet of 1881, by Etienne Léopold Trouvelot. Sawyer Library, Williams College. Further information on page 432.

Chapter 6

Textbooks

*Two papers on astronomical textbooks through history begin the textbook
section. Following, we reproduce here two of the three papers on personal
experiences of authors of American astronomy textbooks, who discuss the
factors that motivate and influence the process of writing and publication.
A discussion of textbooks for developing countries follows; a summary of
an additional discussion is included in the chapter on developing countries.*

FIVE CENTURIES OF ASTRONOMICAL TEXTBOOKS
AND THEIR ROLE IN TEACHING

Owen Gingerich
Harvard-Smithsonian Center for Astrophysics, Cambridge, Massachusetts, U.S.A.

At the turn of the century, astronomy was required of every senior in what
is now the American University of Beirut, for the reason that "the heavens declare
the glory of God"; this was especially appropriate for what was then the Syrian
Protestant College. And in the late Middle Ages, astronomy was one of the seven
liberal arts, a required part of every basic university education — for much the same
reason. Thus, for many centuries, astronomy was considered essential for what every
educated person should know. The organizers of this colloquium thought it would
be informative to learn more about the historical background to our colloquium
and asked if I would speak on the history of astronomy teaching over the ages. I
demurred at such an overwhelming topic, which would require a major research program, and have offered instead something about the history of astronomy textbooks,
because this subset provides answers to at least some of the broader questions of
how astronomy has been taught.

I begin with a venerable textbook whose popularity remained unabated for several centuries. It is the *Sphere* of John of Hollywood, or Sacrobosco in his Latinized
form, a thin manual of spherical astronomy composed in Paris around 1220. The
Sphere was among the earliest science books printed, an incunable (that is, pre-1501)
edition appearing in 1472. It is sometimes said that Sacrobosco was the all-time
best seller among the authors of astronomy texts. In its heyday, in the sixteenth
century, something over 100 different editions were printed, slightly more than one

a year. Typical press runs for popular books in those days were about a thousand copies. So, including a few editions from the fifteenth century and a few more in the seventeenth, I would guess that perhaps as many as 200,000 were printed. There are among us authors of popular texts who have surpassed Sacrobosco's success.

By our present standards the *Sphere* of Sacrobosco is painfully elementary and narrow. Its four short chapters begin with a topic found in every astronomy textbook since then, namely, the spherical nature of the Earth. In the sixteenth century, a key center for astronomy teaching was the Lutheran University of Wittenberg, and many of the Sacrobosco editions were produced there. It is interesting to examine how the arguments for the spherical Earth were actually displayed in these editions. That the Earth was round in a north-south direction depended on the aspects of the stars such as the observed height of the pole. That the Earth was round in an east-west direction could be demonstrated by comparative times of a lunar eclipse as viewed from different longitudes. At first, these points were made with small and rather busy diagrams, but beginning in 1538 the books incorporated pages with moving parts, the so-called volvelles, to aid the student's comprehension. From then on, virtually every edition produced anywhere in Europe carried the same three volvelles. The second volvelle actually did double duty, for it helped to introduce the material in Sacrobosco's second chapter, on the equator, ecliptic, the zodiac and so on (see Fig. 1).

Sacrobosco's third chapter continued with the rising and setting of stars throughout the seasons. Like modern texts that attempt to attract non-scientists with enrichment from cultural material, Sacrobosco's chapter is illustrated with numerous poetic quotations, from Virgil, Ovid, and Lucan. For example, he quotes a line from Virgil, "And the dog, yielding to the adverse star, sets" to illustrate the heliacal setting of Sirius as the seasons advance.

In contrast to the cheap, small, octavo-format texts that were probably required at major universities, some editions of the *Sphere* were on a larger scale and often very attractively printed. There were also translations into the vernacular — German, French, Italian, Spanish, English — generally substantially edited to include fresh material.

Sacrobosco's treatment of planetary motions came in a very short final chapter. It was so perfunctory that a demand quickly arose for a more advanced text to explain, qualitatively, the nature of Ptolemy's epicyclic models. An anonymous text called the *Theorica Planetarum* soon described Ptolemy's geocentric planetary system. By the middle of the fifteenth century, Georg Peurbach undertook a revision, which he called the *New Theory of the Planets* — not a new theory, just a new book. Thus, alongside the Sacrobosco there were also many editions of the *Theoricae Novae*. As was the case with Sacrobosco, Peurbach's brief work attracted commentators who explained and expanded the text. One of the best was the edition produced by Erasmus Reinhold, the senior professor of astronomy at Wittenberg and the leading astronomical pedagogue in the generation following Copernicus. In fine Wittenberg tradition, his version included two volvelles to exhibit the Ptolemaic epicycle on a movable deferent disk.

Fig. 1. Sacrobosco's Sphaera *(Wittenberg, 1538), the first edition with moving parts. The motto "No day without its lines" is an ancient proverb cited by Erasmus; traditionally it meant no day without its special work, but here as a pun it refers to the daily linear path of the sun shown below the movable part.*

With the mention of Copernicus comes the question of how new material was introduced into the curriculum. Copernicus' 1543 *De Revolutionibus* was far too technical and much too expensive to serve as an undergraduate textbook — it cost about $150 by modern standards — but it went into the hands of most university astronomy professors, who undoubtedly alluded to Copernicus in their lectures. In Italy, the Jesuit astronomer and professor at the Gregorian University in Rome, Christopher Clavius, enlarged the Sacrobosco by an order of magnitude, so much so that it has to be considered his own textbook and no longer a mere edition of Sacrobosco. In the revised edition of 1581, he mentioned the Copernican astronomy at least indirectly, saying that Copernicus had made astronomers realize that the Ptolemaic arrangement might not be the only one. Clavius revised his textbook half a dozen times, and in the edition placed into his collected works (which was much too monumental to be carried around as a textbook) he even mentioned Galileo's new telescopic observations.

Michael Maestlin, the astronomer at Tubingen University and best known as Kepler's teacher, produced his own new, independent textbook in 1570. It contained scattered references to Copernicus, but nothing about the heliocentric cosmology.

Nevertheless, Maestlin regularly discussed Copernicus in his classes, and there the young Johannes Kepler first learned of Copernican astronomy. Kepler himself ultimately produced a squat, fat textbook containing more words than any of his other writings. Called *The Epitome of Copernican Astronomy*, it is generally believed by historians of science to have been one of the most widely used texts of the early seventeenth century, but they are wrong. Just because a historian of astronomy once said so does not make it true. I think it unlikely that a text that went through only two printings in that century had a widespread use as an assigned university text. Maestlin's pocket-sized volume went through seven editions, for example.

The really common text from the middle of the seventeenth century seems to be Pierre Gassendi's *Institutio Astronomica*. The French cleric's book was reprinted both on the Continent and in England. It naturally began with the standard material on the sphere, but it also mentioned the phases of Venus (newly discovered by Galileo) and had an extensive discussion of the Copernican and Tychonic systems. Gassendi's book served as an early astronomy text at New England's Cambridge, which meant that Harvard men not only learned about Tycho as well as Copernicus, but they also could read Galileo's *Starry Messenger* and Kepler's *Dioptrice*, which came in the same volume.

By the time Newton's *Principia* was published in 1687, the Tychonic system was a dead issue, although this would not prevent Lalande, nearly a century later, from including several pages of refutation even in the abridged version of the text. More of that presently, but first we must note that Newtonian physics placed a new and heavy burden on the astronomy teacher. Never before had there been so much difficult mathematics that really spoke to a physical understanding of the heavens. The early English textbook authors did not hesitate to treat their students to ample doses of geometry. David Gregory, John Keill, and William Whiston all included the standard topics of spherical astronomy, but added plenty of mathematics as well as timely new topics such as planetary satellites and comets. The English version of Whiston's Cambridge lectures, published in 1728, began, as did Sacrobosco, with the size and shape of the Earth. He next took up diurnal motion and precession, followed by Huygens' photometric method for deducing the distances of the fixed stars. The unsolved problem of trigonometric parallaxes loomed large, with a discussion of Robert Hooke's attempts to observe them. For the fourth lecture, there was apparently not enough material on nebulae, so Whiston immediately turned to the ellipse and Newton. With that, he had plunged deeply into planetary motions, which fill the heart of his text. By the seventeenth lecture he discussed planetary satellites, and in the next, the rings of Saturn. What amounts to Kepler's third law was shown here in numerical form, but completely anonymously, with no mention of Kepler. There was also the method for finding the velocity of light from Jupiter's satellites, all of this showing the strong influence of Huygens.

If we skip to later in the seventeenth century, the most remarkable textbook is Joseph-Jérôme de Lalande's multi-volume compendium, literally a mini-encyclopedia of astronomy. His abridgment, the *Abrégé d'astronomie* of 1774, provided the information on a more tractable scale. As usual, it started with the spherical form of the Earth and the height of the pole. But early on it mentioned stars, including variable ones. The planetary system and theory of motions occupied a significant part of the volume. At the outset he stated Kepler's three laws in the order we use them today. Now Kepler never considered them in one place nor did he number them, so it is something of a mystery as to when this ordering took place. Lalande implied that the threefold numbering was already in use, so I wonder if it was introduced by LaCaille. Since, of the two, Lalande was the more original, perhaps he invented the 1-2-3 order without claiming credit.

Lalande, a vehement atheist, took several pages to refute the objections against the heliocentric cosmology, saying "I respond with the utmost pleasure to the objections Tycho Brahe had against the Copernican system." He also took a crack at Riccioli, the Jesuit who, a century earlier, had defended the Catholic opinion on the Copernican doctrine as it related to Galileo's trial, and he mentioned that certain Copernican books were still included on the latest edition of the *Index of Prohibited Books* — in fact, they didn't disappear until the edition of 1838. To us all this seems anachronistic, but perhaps it is like including a section against astrology or against a 6000-year-old Earth in a modern astronomy textbook. Elsewhere, Lalande included a short section on aberration, which had been found by Bradley in 1728, and he briefly mentioned the problem of parallax but nothing about Huygens' photometric method. At the end he included what is today a familiar parting shot: a section on the plurality of worlds.

Space does not permit a review of all the significant textbooks of the past two centuries, but let me mention what I believe to be the most important astronomy text of the nineteenth century: John Herschel's *Outline of Astronomy*, first published in 1849 as a considerably expanded version of his *A Treatise on Astronomy*, which had appeared in 1831 in the Cabinet Cyclopaedia. Herschel's *Outlines* appeared in fourteen revised London editions between 1849 and 1883. Herschel began with the shape of the Earth, a distant echo of Sacrobosco, but he quickly turned in other directions: refraction, stellar parallax, and then instruments, terrestrial map making, celestial cartography and so on. As for Kepler's laws, he first mentioned the third "as it is called" and then gave Kepler's original order for one and two, that is, calling the law of areas Kepler's first law and the ellipse Kepler's second law, contrary to our present convention. With commendable modesty he described the discovery of Uranus by his father, and he recounted the advantage of the Leverrier-Galle team in having a relevant star chart to aid in discovering Neptune, an advantage that cost the Englishmen Adams and Challis the priority. There was a substantial section on planetary perturbations and a discussion of binary star orbits. Of 600 pages, 100 were devoted to sidereal astronomy, including star clusters and the nebulae in Andromeda, Orion, and Argo (that is, the Eta Carina nebula).

The most important American textbook of the last century was Charles Young's *General Astronomy*, first issued in 1888. Explicitly targeted for the general "liberally educated" student, it nevertheless had a level of mathematics almost never found in our contemporary American astronomy texts (see Fig. 2). A cursory glance at Young's 1900 edition shows about 100 out of 600 pages devoted to stars, nebulae, and cosmogony — cosmogony for Young meant Norman Lockyer's meteoritic hypothesis of solar heat — just about the same ratio as in Herschel's 1849 *Outlines*. But that initial page count is slightly misleading, for Young devoted 90 pages to his own specialty, the sun, which included a large section on spectroscopy, topics that were then beginning to be called *astrophysics*.

288   CENTRAL FORCES.

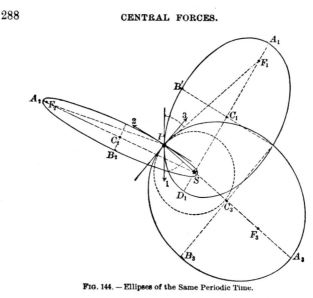

FIG. 144. — Ellipses of the Same Periodic Time.

Fig. 2. "Ellipses of the Same Periodic Time" illustrating the mathematical level of Charles Young's General Astronomy *(Boston, 1900).*

The direct successor to Young's college-level text was another from Princeton: Russell, Dugan, and Stewart's two-volume *Astronomy.* Although it was considered too long and detailed for many American colleges, it was nevertheless extremely influential. For the first time the universe of stars and astrophysics (Volume II, 1927) got as much space as the solar system (Volume I, 1926). Secondly, because of Henry Norris Russell's involvement, it included many ideas on the cutting edge of astronomical research. As an example, the second volume included the temperatures of stars from Cecilia Payne's 1925 Harvard thesis, and the statement that "Miss Payne concludes that 'the uniformity of the composition of stellar atmospheres appears to be an established fact.'" In fine print the text goes on to consider the puzzling behavior of hydrogen lines, which indicated how very abundant hydrogen must be. Payne, like Russell, Shapley, and the other astronomers of that time, firmly believed in the homogeneity of the universe and therefore assumed that iron, the most abundant element on the Earth, should also dominate in the cosmos at large, despite the baffling results of her analysis. Not until two years later did Russell conclude that hydrogen really was immensely abundant in the universe, and not just a thin layer of light nuclei blanketing the outside of the sun. Surprisingly, this section of the book passed unchanged when the authors undertook a somewhat cursory revision of the second volume in 1938.

Not until 1945 was the first volume of "RDS" brought up to date. Because an index to the entire work appeared in the second volume, the publishers, Ginn and Company, allowed Russell and Stewart to revise the first volume as much as they wanted as long as they did not change the pagination of any of the index entries! This head-in-the-sand attitude so discouraged the two surviving authors that it effectively killed the most important astronomy textbook of our century. No full revision of the astrophysics section was ever undertaken.

The second half of the twentieth century has seen a vast proliferation in textbooks, especially in North America, where there continues to be a huge demand for non-mathematical and engagingly illustrated introductions to astronomy. While volumes with a level of mathematics as high as Charles Young's *General Astronomy* are still scarce, at least there has been an unparalleled opportunity to explore fresh ways of presenting both old and new material. I still find two of the American texts from the 1950's particularly memorable in their attempts to replace a somewhat "cookbook" approach with a physically motivated presentation: Wasley Krogdahl's *The Astronomical Universe* (1952) and Otto Struve's *Elementary Astronomy* (1959). Their viewpoints, out of the mainstream defined by the principal texts of that time, have insinuated themselves at least indirectly into several of today's leading guides for beginning astronomy students.

Even if astronomy is no longer required of every university student, probably more students have studied astronomy in the last fifty years than in all the previous centuries combined, more different astronomy textbooks have been printed, and probably in larger quantities than the combined total of all previous eras.

Discussion

C.R. Chambliss: *How was the delicate question of the Copernican versus Ptolemaic hypothesis handled in astronomy texts in Catholic nations such as France in the seventeenth and early eighteenth centuries? I believe that Cassini never publicly accepted the heliocentric system, for instance.*

O. Gingerich: Gassendi at mid-seventeenth century presented both the Copernican and Tychonic cosmologies, I think without a strong commitment. A century later, Lalande, a militant atheist, waged war on those who had defended a fixed Earth. At the end of the eighteenth century, a professor at the Gregorian University in Rome, Settele, had trouble getting an imprimatur (Catholic license to publish) for his heliocentric astronomy textbook, which created a scandal resolved only by a papal decree that finally had the effect of removing the Copernican books from the *Index of Prohibited Books.*

J.-C. Pecker: *It is interesting to note that the abridged versions of textbooks were often named "Astronomie pour les dames": this was the title of one of Lalande's books, an excellent introduction to astronomy for the beginner — of all sexes! It is true, I believe, in all languages in the eighteenth and nineteenth centuries.*

O. Gingerich: Yes, there was a whole genre of this form of book. In England there was even *The Ladies Almanac*; in it William Herschel first appeared in print.

INTRODUCTORY ASTRONOMY TEXTBOOKS IN 19TH AND 20TH CENTURY AMERICA

Norman Sperling
Chabot Observatory, 5248 Lawton Avenue, Oakland, California 94618, U.S.A.

1. Introduction

A survey of 138 introductory-astronomy textbooks spanning 152 years reveals growing consensus regarding each topic's proportion, with some clearly gaining space at the expense of others. The tables in the texts cite curious numbers, claim too many significant digits, neglect to note uncertainties, and are frequently inconsistent with, or badly behind, the research of the times. This study investigates apportionment of topics, planet and star data tables, and categorization of nebulae.

To probe the student/textbook interaction, I used one copy of each of 40 recent introductory textbooks when teaching astronomy in Fall 1986. Students swapped books each session. Texts' treatments were surveyed in daily recitation as well as term papers comparing and contrasting them on specific topics. Most books sound much more positive than current data justify. There was lots of confusing phrasing, shoddy proofreading, and careless assembly of data tables. A few books are shamefully erroneous. There are numerous impressive examples of the imperfection and transience of "textbook learning." The biggest and most pervasive sin was writing in the passive voice "Official Style" of interminable sentences laden with prepositional phrases. Illustrations, while important, are secondary to phrasing. Students need and use chapter summaries, glossaries, and indices. About half prefer paperbacks and half hardbacks. Students rated the books for appropriateness to their needs. The experience may stimulate others to develop controlled experiments to probe the student/textbook interaction.

2. Topics in Textbooks

A text is a worthy, good-faith reference, but in 20 years it will be hopelessly inadequate, and in 100 years, laughable. As knowledge of astronomy has grown, so has the number of topics and the amount of data that could be presented. Thus, the proportions of various topics have changed over the centuries. The accompanying graph demonstrates some of the changes.

Indices can also be derived by dividing one portion by another. *Stellar divided by positional* shows a dramatic growth. Thus, modern stellar astronomy horned into textbooks by shoving aside older topics such as positional astronomy. Conversely, *(asteroids + comets + meteors) divided by (nebulae + clusters + galaxies + cosmic)* plummets strikingly. As the space devoted to the universe beyond the stars grew, it did so largely at the expense of older, less-spectacular topics like solar system debris.

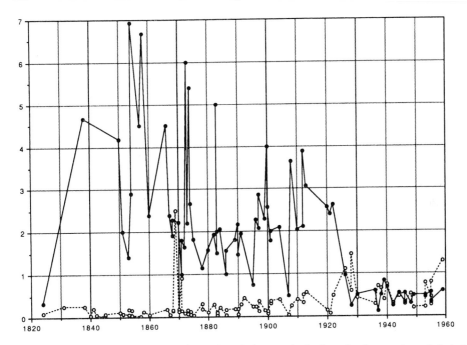

A graph of the changing stellar/positional *(open circles) and* solar system debris/ beyond stars *(filled circles) in introductory astronomy textbooks in America.*

Most topics show a narrowing of the scatter, demonstrating a strong regimentation about the relative proportions — though their arrangement has always been various, and remains so today.

Most books include data tables for planetary orbits and physical characteristics, and recent ones include tables of the nearest and brightest stars. The data are strikingly varied! Since only a few independent determinations of diameters and rotations had been made, it is very curious to find such a spread in quotations. Some may result from poor proof-reading and negligent copying. Too few authors told their readers that the data weren't as definite as the numbers implied — and too few still do.

Classification is one of science's most fundamental techniques for coping with nature's diversity. We owe it to our students to show them how science categorizes celestial phenomena. A problem gaining prominence when this run of texts begins, and retaining it through the present, has been how to categorize the "nebulae." Categories of nebulae that became resolved as clusters were the first to go, but didn't leave textbooks until the 1880s — many decades after relegation by the astronomers at the forefront! Galaxies were recognized as such by about 1928, but didn't lose the term "extragalactic nebulae" until 1940, when they were carved off and given their own sections as galaxies.

The only category of nebula that has survived from beginning to present is "planetary." However, three other categories have been merged into it: stellar nebulae into nebulous stars before 1880, nebulous stars into planetaries about 1910, and ring or annular nebulae into planetaries in the mid 1920s.

As technology has progressively brought giant telescopes, spectroscopy, and photography into play, understanding has undergone several paradigm shifts. Categories, such as those most popular at present, did not all spring up together: dark nebulae entered the texts after 1920, the interstellar medium in the late 1940s, reflection and emission nebulae about 5 years later, and supernova remnants in the mid-1960s. However, category sets may fairly often disappear sharply.

Though most nebular categories sound pretty familiar, specific objects we all use to exemplify those types have not always been considered thus. Messier 31 in Andromeda exemplified elliptical or oval nebulae from the 1830s to the turn of the century. Greater light-gathering power in the mid-1800s portrayed it as irregular instead, as reported in several texts from 1870 to 1912, and a couple in the mid-1800s even called it annular. Photography recast it as a spiral before the end of the 1800s, with most books calling it so after the turn of the century.

3. Students' Views of Existing Textbooks

A 1986 experiment which I conducted sought to learn what students think of textbooks, and what differences among texts are important. Every student came to understand that textbooks are fallible, because in the course of the semester every student had books with obvious faults. Several students applauded personal notes, and narratives of science as process instead of "pat" answers.

I was definitely confirmed in my long-standing objection to books that sound too "positive." Those become the most hilarious at age 100, but they're also misleading right now. In the drive to tell the student what astronomy has learned, most authors declare "truths" that aren't all that firmly pinned down. One student logged all the data her assorted books quoted about Pluto, and found not only the spread in data but also mysterious variations, even in long-established quantities like orbital inclination and apparent magnitude. She also spotted typographical errors in decimal places — which, of course, make the data wrong by factors of 10. Who verifies all those data tables? The students concluded:

1. Paramount is the ease of reading, particularly for weak students. "Official passive" voice is deadly! Interminable sentences laden with prepositional phrases definitely inhibit reading, and turn off marginal students.

2. Chapter-end summaries are very useful. Chapter-end questions challenge some

students, frustrate others, and are skipped by most except when assigned. Weak students depend heavily on glossaries, and all need a thorough, competent index. The rest of the chapter-end and appendix material was less important. (Personally, I recommend keeping the "further readings" listings if only to demonstrate that such exist.)

3. Students view being "up-to-date" more as the teacher's responsibility than the book's.

4. Illustrations are very useful (the less-well-illustrated books were less-well-remembered) but didn't actually sell them. Students will be hard-pressed to identify anything in the sky from the star charts in most books.

Two books placed head-and-shoulders above all others: Zeilik and Pasachoff. But what I'd really like is a textbook with Zeilik's personal flair, lively language, and thorough history; Pasachoff's passionate accuracy and marginal notes; Snow's 3rd edition's color; Kaufmann's thoroughness; Friedlander's cautions; Shu's feel for the physics; Robbins' and Jeffreys' observing; Hartmann's art — and my outline.

Table 1: How Appropriate My students Thought Each Book Was[a]

3.375–3.5:	Pasachoff; Zeilik
2.78–2.9:	Abell; Friedlander; Goldsmith; Hartmann; Kaufmann; Parker
2.44–2.57:	Bash; Dixon; Field, Verschuur, and Ponnamperuma; Long; Protheroe, Capriotti, and Newsom; Seeds
2.17–2.3:	Berman and Evans; Chapman; Jastrow and Thompson
1.5–2.0:	Apfel and Hynek; Branley, Chartrand, and Wimmer; Cole; Devinney, Smith, and Sofia; Hodge; Hoyle; Jefferys and Robbins; King; Pananides and Arny; Shu; Snow[b] Wyatt
0.625–1.25:	Brandt and Maran; Clotfelter; Ebbighausen; Oriti and Starbird

[a]Scored as conventional grades: excellent = 4; good = 3; adequate = 2; barely OK = 1; unacceptable = 0.
[b]Probably influenced by instructor's negative comments.

I hope these experiences will provoke other instructors to probe student/textbook interactions. Can someone design rigorous, controlled experiments with assorted books, formats, and types of students?

Discussion

T. Dennis: *When do you see problem sets appear, and can you tell if they are meant to be solved by the student, or are they just there?*

N. Sperling: If you count predecessors of textbooks, the discourses on the use of globes: my oldest (Moxon, 1674) has copious problems for each student. Problems clearly predate Moxon. As discourses on globes incorporate the rest of astronomy, they keep the problems about the globes, and during the 1700s become more about the cosmos and its workings, and less about globes. My (very few) 1700s British texts have many "examples" but no "problems" beyond globes. My earliest American texts (1820s) have lots of problems all through. Problems abound in them ever since.

B.W. Jones: *This might seem a silly question, but how did you decide, particularly for the older books, which were* textbooks, i.e., *how did you prepare your sample?*

N. Sperling: On the basis of the stated goals of the author of the book.

TEXTBOOKS: A PANEL DISCUSSION

Textbooks for Developing Countries

Mazlan Othman
Physics Department, Universiti Kebangsaan Malaysia
43600 UKM, Bangi, Selangor D.E., Malaysia

I would like to address the issue of textbooks from a developing country's point of view.

Firstly, I believe that in the developing countries a textbook should be written in the native language if it is to reach the very people whose awareness of astronomy needs to be raised. And when we talk about books in the vernacular, the problem is one of dearth: a dearth of writers and a dearth of resources.

The lack of writers comes about simply because in the developing nations there are inevitably very few astronomers and of these only one or two will be inclined to write books. If there exist writers, they are faced with several choices and problems. There are three types of books which can be written: a university text, a school text, and I shall include a book for the public. Each type of book targets a different kind of audience, so the writer must be attuned to the needs of the nation to know what the priority should be. If, for example, astronomy is about to be introduced into a school curriculum, then a school textbook should take the highest priority because it tackles astronomy education at the grass-roots level. Naturally, the final choice will depend on other factors as well, including promotion, fame, and pecuniary considerations.

A more intractable problem for the developing countries is one of getting materials for the book. Photographs are properties of big observatories or individuals and special permission must be sought to use them. Personal contact is the only easy way to overcome this problem but it is precisely these personal contacts that are hard to cultivate because of isolation of the Third World astronomers from their counterparts in the more developed world. If there are observatories that allow photographs to be used for the price of an acknowledgment, then these invaluable resources should be listed and made known to all.

I should also mention that I believe that direct translations of foreign books into the local language may not necessarily be the right thing to do because differences in culture are significant. Geographical location is an obvious disparity and it is also true that different countries have disparate needs; therefore, the emphasis of the book should subsequently be dissimilar. A jointly written text may be more desirable, as this will allow the local author to inject the local bias of the book.

Astronomy in American Textbooks

Jay M. Pasachoff
Williams College, Williamstown, Massachusetts 01267, U.S.A.

Over the last dozen years, I have written textbooks on a variety of levels, starting with books for university students, proceeding to work with Naomi Pasachoff on books on the junior-high level, and, most recently, working with her and with others on an elementary-school series. I can testify that, in the United States, at least, the world of college and university texts is as different from the world of "el-hi" (elementary-high) texts as night is from day.

1. College and University Texts

When I began teaching astronomy, in 1972, I looked at the available texts to find one that emphasized the things that I thought were most important to teach students. Since I think students are interested by contemporary astronomy in general, and I am sure they are excited by black holes and quasars in particular, I looked for books with extensive treatments of the latter topics. To my surprise, I found none.

I nonetheless made out my course with topics distributed according to what I thought was appropriate emphasis, including whole lectures on pulsars, black holes, quasars, Mars (with Viking approaching), and so on. Eventually I had the basis of a book from my lecture notes; I had even recorded some of the lectures, though it turned out to be more difficult to transcribe them than it was to write from scratch (given that one of my major skills is typing). Eventually, I showed some of my materials to a publisher, who expressed interest — and who led me on for over a year as one editor replaced another over and over, an interesting introduction to the world of publishing. Eventually, a traveling field representative from another publisher came to my office and I showed him my material. His editor was forceful and active, and within a week or so I had good progress toward a contract to write a book. This book turned out to be *Contemporary Astronomy*, whose fourth edition has just appeared (1989). Obviously, many professors found the distribution of emphasis from topic to topic to their liking.

Contemporary Astronomy is still unique among texts in having separate chapters for the most important contemporary material, now including supernovae, neutron stars/pulsars, black holes, the Milky Way Galaxy, quasars, and the individual planets that have been visited by spacecraft. Further, I thought a major advantage was that by starting with stars, students could enter real astronomy immediately, without having to wade through many chapters of historical and positional material. Still, even with special attention to modern topics both in separate chapters and throughout, I paid and continue to pay great attention to having a fair and balanced coverage of all astronomy.

The book has appeared in many versions, including *Astronomy: From the Earth to the Universe*, with an increased historical emphasis and with the solar system first. Though I had originally thought that teachers could use the chapters in various orders, one professor pointed out that many students feel that difficulty correlates with page number. So now there is the "stars first" order, the "planets first" order, and a short version.

In our course at Williams, the students not only learn about general astronomy and contemporary astronomy but also do both daytime observations of the sun and nighttime observations of farther celestial objects. In addition, they do indoor laboratories with astronomical photographs and with spectroscopes. I feel strongly that a balance is necessary, and the material in my textbooks is supplemented by laboratory exercises and references in a *Teachers' Guide*. Students who spend too much time observing with small telescopes may not appreciate the advances of modern astronomy or comprehend that the new ideas and technology sparking those advances need support not only from individuals but also from governments and foundations. Students who spend too much time redoing methods of the past, even if they wind up with good comprehension of certain basic topics, may be lost to astronomy in particular and physical science in general when their interest is sparked by the discoveries about DNA they learn about in biology class or other contemporary aspects of various sciences.

One of the things I learned from being a textbook author is about the need to check everything. It is amazing how much I didn't know, or how much I knew that turned out to be wrong! Publishers often send drafts of textbooks to half a dozen readers, and I always sent each section to additional experts. One can learn a lot about the accuracy of a text by looking at the number and professional expertise of those acknowledged in the preface.

I also learned about economies of scale. The more copies a textbook sells, the less it can be sold for, since the publisher's receipts are, obviously, number times price. The expenses of photos and other materials, and of setting type, have to be amortized over the number of copies sold. So anything that sells more copies — including, to the surprise of many, color plates — means that the book can be sold more cheaply. A major problem now is that the used-book business has become very well organized, with whole classloads of books shipped around the country. These books are bought from students very cheaply, but by the time profits are taken by the middlemen and the bookstores, the price to students is not too much lower than that of new books. And not only the authors' royalty but also money to the publishers is not paid. Sales for the second year of a book plummet, and publishers are less willing to invest the effort needed for quality books. Allowing bookstores to sell used books in this way seems shortsighted. And any professor who allows a used-book salesman to buy the sample copies that publishers send out is causing harm.

I am continually reminded how nice it is that astronomy is my field, for so much new material is constantly available. It is interesting to see how many new discoveries have been made, often requiring new chapters, from one edition to the next.

2. Elementary, Junior-High, and High-School Texts

Whereas for college and university students, publishers print what the author writes (after suitable editing and review), "el-hi" texts are committee efforts. The author certainly plays a role in defining the book, but the in-house editors at the el-hi publishers have final say about all aspects of the book.

In 1980, my wife (Naomi Pasachoff) and I signed on with Scott, Foresman and Co., one of the largest el-hi publishers, known to Americans of my generation as the publishers of the "Dick and Jane" juggernaut. The number of editors involved with the junior-high texts in *Earth Science* and *Physical Science* that we wrote together (and my wife does the lion's share) with a third author they assigned, T. Cooney, approached 10. But the number was justified by the eventual sales; perhaps 2,000,000 students a year take junior-high science, and our books have for each of the last five years sold about 200,000 copies.

Still, it is a continual struggle to see that the material that appears is correct and well put. Our original manuscript passes through so many hands and revisions by non-astronomers that errors often creep in. Every time the manuscript or proof goes by us, we make it correct, but sometimes we wind up correcting the same point several times. Apparently the authors of competing junior-high texts aren't as compulsive as we are about corrections, or, because they aren't astronomers, weren't as knowledgeable in the first place, because I often find egregious errors in these other texts. I would like to see the accuracy and up-to-date nature of our texts be major selling points, but the el-hi publishers seem to advertise only on the basis of "readability" and seem to minimize the fact that a "college professor" in a discipline (that is, not a professor of education) is associated with the project.

We have heard several times at this Colloquium that students don't understand such basic matters as tides and seasons. Perhaps part of the problem is that their textbooks may have explained them unclearly or incorrectly. Because of all the hands the material passes through, it is difficult even for my wife and me to get correct explanations into our own books. If only we scientists can think of a way — perhaps by convincing statewide adoption committees, who often must approve textbooks on this level — to require accuracy, the texts students use would be much improved. It would be good if scientists in each discipline were asked by such adoption committees to comment on the accuracy of the texts being considered. Perhaps *you* can volunteer to advise your local school board.

Whatever projects are being developed for the future, there are millions of students taking school science courses now. Thus one can exert a tremendous influence by participating in ongoing textbook projects. We have, in our books, succeeded in enlarging the quota of astronomy taught to hundreds of thousands of junior-high students each year; since each book is often used for five years in a given school, our books are equivalent to 1 million student-users each year, or a 5-million-user total so far. Our junior-high effort was successful enough that not only have second editions of *Earth Science* and *Physical Science* appeared (1990) but also we have been included in an "author team" for *Discover Science* (1990), a series of 7 books for grades Kindergarten and first through sixth. All these books are lavishly laid out and illustrated in full color, and represent major investments for the publisher.

We hope that others join us in trying to spread astronomy and accurate science through the schools. More scientists participating as authors would be great. And more scientists advising their local, county, or state adoption committees that accurate and up-to-date science texts are necessary would be very helpful to today's students.

3. Conclusions

It is important to be clear about our goals when teaching. For example, is it more important to teach about the advance of knowledge in astronomy during the current era or is it more important to teach about basics like phases and seasons? Or what balance should be struck? The question of what we should teach goes, of course, with the question of what students should learn. My own evaluation is that it is important for students of all ages to know, and therefore for us to teach, about both modern advances in astronomy and about the general basics. If I had to choose, though, I would choose contemporary astronomy, for some students might be attracted to astronomy as majors as a result of exciting advances. Further, even non-major students become citizens and voters, and it is important for them to understand that astronomy continually progresses. If their understanding were limited to basics, they might feel that astronomy had been worked out long ago, that new investments for astronomical research are not necessary, and that there would be no point for them to consider going into astronomy or voting for politicians who support research. On the other hand, though a gap in their understanding of basics should be plugged, students with such gaps who understood the vibrancy of contemporary astronomy would nonetheless feel that astronomy was exciting and that it was worth supporting.

In essence, the astronomy non-major survey course can be an "astronomy appreciation" course, similar to "music appreciation" courses. It should not dwell excessively on conceptual basics, just as it would be a pity for a music appreciation course to limit itself to understanding chords without ever hearing a symphony. Similarly, elementary-school and high-school courses should also include the exciting results of modern astronomy in addition to providing astronomical basics.

Another analogy might be with teaching English. University professors of English (or, in non-English-speaking countries, of the respective dominant languages) are teachers of literature. They assume that grammar is known; they are sometimes willing to correct students' grammatical deficiencies, but those corrections are not their main job. These professors are to teach Shakespeare (or, in France, Victor Hugo) and the contemporary novel, rather than to teach punctuation. Even in schools during the lower grades, good English teachers teach about poetry and prose, and do not limit themselves to grammar. Analogously, we astronomers must bring our students to see the wonders of the universe in the view of our latest understandings as demonstrated by the glories of contemporary research.

Textbook Panel Presentation — IAU Colloquium 105

R. Robert Robbins,
Department of Astronomy, University of Texas, Austin, Texas 78712, U.S.A.

You might be somewhat surprised by the presentations [as delivered] so far, in that there has been relatively little discussion of pedagogical matters (*e.g.*, what should be the content of a text and what should be its order and manner of presentation). It seems that rather personal factors are in the foreground of everyone's thinking. But I think that is probably inevitable, because the process of writing a textbook is not only a huge undertaking that comes to dominate the life of the author, but it is also an endeavor that places the author directly at the interface between the idealistic world of education and the very practical world of economics as dictated by the realities of the publishing markets. In addition, a textbook has a strong effect on an astronomy class. It has been suggested that an important reason why so many educational research studies fail to come to a definite conclusion is that the textbook is such a dominant influence that it swamps all the other factors!

We have already seen that much of the astronomy teaching in the United States is in courses for *non-science majors*. These courses generate huge enrollments; at my institution, we teach some 7000 students per year (about 20,000 credit hours). It is these enrollments that make it possible for commercial publishers to make a profit on a textbook, because every student is expected to buy a copy of the book. And of course we see the evidence of this — a great proliferation of non-mathematical texts for non-science majors, *in English*.

Where it is not profitable to publish a book, the books do not appear. Note the shortage of books for astronomy *majors*, even in English. The market is just too small. Technical symposia do make their way into print, but the prices of the volumes are often incredible (in some cases up to $1000!). And in the European educational system (and in countries following a similar pattern), enrollments are smaller and students are not expected to buy a text. In these circumstances, publishing is often not economically feasible.

One solution is perhaps to create larger enrollments (for example, by teaching classes for non-science students, or requiring the books). But this requires large changes in educational systems, which by their very size are conservative and hard to change. However, if *some* markets were to be created for commercial publishers, they would make some profits and then perhaps be more willing to produce some of the advanced volumes that are needed, perhaps even at a loss. Or be more willing to permit translations into local languages, perhaps at no profit to themselves.

If a way to produce some commercial profits is not found, then books will only appear if they are subsidized in some way, and this is traditionally a slower and less-dependable route. I have proposed that Commission 46 of the IAU attempt to locate some modest funding to support prospective textbook writers in developing countries to produce texts in their local languages. I feel that sponsoring a semester or a year's sabbatical for a local author would be only a modest expense and would bring a substantial educational return.

From a personal point of view: if you are considering writing a text book, you should realize that (1) it takes a very long time — much, much longer than any estimate you might project and (2) the market is actually very competitive and finite, with a number of good books already on the market. As an indicator, I have just finished compiling a comprehensive listing of new astronomy books in English for IAU Commission 46. Between January 1, 1985, and December 31, 1987, new astronomy books of all kinds in English appeared at the rate of one every 28 hours!

As a further practical consideration, there is a serious war going on between the publishers and the book stores in the U.S. market over used textbooks. Bookstores can make more money reselling used books than they can selling the new texts. Such sales generate no income for authors or publishers. The publishers retaliate by having their authors put out a new edition every year and a half, which wastes the author's time and often has no sound pedagogical justification.

The idealistic side of you probably wants to rebel against having to worry about all of these economic considerations, but you cannot. Because if you are going to actually put in all of that work to write a textbook, then in order to have the continuing benefit of the text you have written — *it must survive in the marketplace.* And this means that there is no way to avoid interfacing with the real world of economics. So while you will hear various horror stories of author-publisher interactions, realize that things can go wrong in any endeavor and prepare your mind for all possibilities. Your best protection is to know your editors well, and, as best you can, evaluate them as people and as professionals and hope for the best.

You will certainly not get rich by writing a textbook, so I would advise you to undertake this endeavor *only if you have a very important academic or personal reason for doing so.* For example, I wrote my book because of my personal dissatisfaction with the existing texts on the market: 95 per cent of the students across the country were taking astronomy without having any contact with astronomical observations, the sky, or the measurement-inference process that is at the heart of science. I had no success in trying to graft one of the existing "lab" manuals onto my lectures, so I ended up writing a book in which some observational activities were integrated into the text itself where appropriate. Now my students can acquire some of the knowledge in the syllabus by astronomical investigations — the active discovery process advocated by Piaget and other cognition researchers, as opposed to passive lecture experiences. Not all of the material can be taught by discovery methods, but I am very happy with that portion that can be structured in this way and the student response has been very favorable.

Textbooks: A Survey

After the panel and open discussions on textbooks, John Stull and Scott Weaver of Alfred University, Alfred, New York 14802, kindly prepared a questionnaire on textbooks, and collected and analyzed the results. [The Organizing Committees are grateful to them for undertaking this survey.] The following is a summary of their report.

Responses were received from 23 people from 14 countries. Although this limited response precludes meaningful generalization, it generally supports the views expressed in the panel discussion. U.S. texts, with their frequent editions, up-to-the-minute information and many illustrations (frequently in color) are not only inconvenient for non-English speakers, but also out of reach in countries where dollars are scarce. They may also have an annoying amount of cultural and geographical bias. The lack of locally-written texts adds to this difficulty. Many non-U.S. respondents indicated that they write and distribute lecture notes; these notes may eventually be the basis for locally-written texts. The main problems of textbooks may well have been identified at this Colloquium, but were by no means solved. Further thought and effort seem justified.

Discussion

H.F. Haupt: *Comment on "Textbooks" from a non-American in a non-English speaking country:*

- *at the university in Austria, we teach only science-majors*
- *we mention a few (a dozen) textbooks (both in German and in English) that students can look into at our libraries; but normally students will not buy these*

- *we (as teachers) extract from those textbooks and from our own experience and put out lecture notes*
- *these lecture notes are more or less obligatory for the students; they can be obtained at very low cost.*

L. Gouguenheim: *The situation in many countries outside North America is quite different. The students are not expected to buy a textbook, or to follow a specific one. In France, generally the students are given typewritten lecture notes from the teacher together with a rather large bibliography (including generally French and English language books). They also have access to a library. N.B. Astronomy courses at various university levels concern only science students.*

J.-C. Pecker: *We often (in France, in graduate school) do not use any particular book, but like better to send the students to bibliographical sources (documentation of all kinds, Annual Reviews, even original papers, be it in English, or French, or else. . .).*

L. Houziaux: *Along the some line as the previous comment, I would like to stress that the situation about textbooks is very different in U.S. or in Canada from other parts of the world and namely in Continental Europe, where astronomy is not taught to non-science major students. And these students represent a huge part (90 per cent) of the American market.*

Therefore the textbooks for astronomy teaching in universities in Europe rely on quite a different knowledge of physics and mathematics. Therefore the structure of the textbooks is very different. Production costs are very high and the market is very restricted, of the same size about as books published in the U.S. for graduate students in astrophysics.

P. Legault: *Laurentian University is a bilingual university, both French and English being used. Many English textbooks exist but only a few French ones, mainly for science students. Most of my students are non-science students.*

J.V. Narlikar: *My experience of writing an astronomy text in the Marathi language illustrates some of the difficulties of producing books in native languages in under-developed countries. Publishers (private) do not wish to undertake such books as they are not sure of making any profit. My book was brought out by a Government literary agency. It was a reasonably produced, moderately priced book, but the agency had no mechanism of making it available to the general public through retail outlets.*

Such agencies therefore do not do the "publisher's" job. A private publisher is anxious to publicize through reviews, advertisements, etc., which an official agency does not do. Can we think of IAU sponsorship for such books? Some mechanism for giving incentives to publication of astronomy texts needs to be evolved.

B.G. Sidharth: *In India, there are a number of textbooks geared to astronomy cur-*

ricula that are deplorably bad, but make good economic sense from the publishers' point of view. On the other hand, there is a definite market for good astronomy textbooks outside the stream of a formal curriculum, and there are enough authors, but no publishers because they are not confident of returns.

H.S. Gurm:

1. *Much of the discussion has been about high school courses or courses for non-science major students of astronomy. Presently, we have either books for introductory and popular astronomy or symposia. There is a dearth of textbooks for those majoring in astronomy or for courses like that offered by University College London.*

2. *The book business is in a crisis in the developing countries. Most of the booksellers collect remainders from the west and dump them in the third world. For example, Pasachoff's paperback (an older edition) was sold in my town by a bookseller for 70 Rupees (about $5.00). How about having cheap editions for the third world? It could be initiated through U.S.-held rupee funds in India.*

S.E. Okoye: *I would also like to draw attention to the cultural dimension of textbook writing. It is our experience in Africa and other developing parts of the world that the cultural background and orientation of authors affect the way they present their materials and this can be limiting for readers of a different cultural background. The tendency now is for the local readership to prefer textbooks written by local scientists.*

The second point relates to the worsening economies of the developing countries, which has made imported books too expensive, and almost unaffordable. This further provides an additional motivation for local textbook writers. But alas, the incidence of book piracy in some countries, arising from the same poor national economies, tends to work against the emergence of local textbook writers.

G.A. Carlson: *I appreciate the number of excellent astronomy textbooks on the market. It allows me the option of changing texts every three years or so. By doing this, I keep current on most aspects of general astronomy. I feel that each author is strong in certain areas. By changing texts, I feel I do a better job of teaching.*

J.C. LoPresto: *What about having a text "on line" for computer subscription?*

J.M. Pasachoff: I suspect that not enough students are "on line" to make it worthwhile for mainline publishers to make their texts available in this way. Further, current computer systems do not have adequate picture quality. I await the availability of CD-ROM.

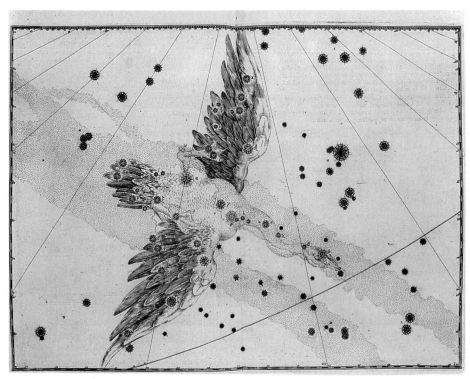

Cygnus, by Alexander Mair; Plate in *Uranometria*, by Johann Bayer. Collection of Jay M. Pasachoff. Further information on page 435.

Chapter 7

Teaching Aids and Resources

Astronomy lectures benefit from the availability of many beautiful and interesting slides, and the first paper discusses how these are made more valuable by providing background information to teachers. Several papers then discuss innovative demonstrations. One set for elementary-school students uses, memorably, the students themselves as participants. Another uses special materials to show concretely how the seasons result from the tilt of the Earth's axis, a concept that a paper in Chapter 9 shows is widely misunderstood. Finally, we read of a valuable book to aid observing the sky, and about an International Astronomical Union project that provides lists of useful astronomy educational material.

INFORMATIVE SLIDE SETS IN ASTRONOMY

Andrew Fraknoi and Sherwood Harrington
Astronomical Society of the Pacific, 390 Ashton Ave,
San Francisco, California 94112, U.S.A.

In astronomy, perhaps more than any other science, the cliché is true: A picture *can* be worth a thousand words. Unfortunately, with the illustrations in some astronomical textbooks, a picture can sometimes *require* a thousand words. And for beginning or inexperienced instructors, the worst thing can be that a picture may *deserve* a thousand words, but the instructor has only 50 words at his or her disposal.

Even for experienced instructors, unless the image happens to be in one's own area of expertise, it can sometimes be difficult to appreciate or convey the full significance of a modern deep-sky photo, radiograph, or satellite multicolor composite image.

In most cases, the image suppliers are of little help. Most audio-visual companies or astronomers will provide no more than a few words or at most a brief paragraph of caption materials. Yet, when we hear a colleague give a wonderful lecture with slides, using just a few key images to weave a marvelous story of discovery, we see that slides can be wonderful educational aids, even though we ourselves may not know or remember enough about each slide to put them to such effective use.

At the Astronomical Society of the Pacific, we have been working for the last ten years to change this situation for astronomy instructors. (This work has been supported in part by the A.S.P.'s Bart J. Bok Memorial Fund, designed to underwrite some of our Society's programs in astronomy education. Those of you who knew Bart Bok will remember how much he loved showing and explaining good images in his lectures.)

We have been producing a wide variety of astronomical slide sets assembled by noted experts in each area of astronomy or by the A.S.P. staff in consultation with such experts. Each set contains 6 to 50 excellent images, plus a booklet up to 40 pages in length, with detailed captions, background information, a thorough reading list, and, in some cases, classroom activities and projects.

The slides are carefully reproduced from early generation masters whenever possible, and the cost of the sets is below the price of commercial slide sets.

Among the sets currently offered by our nonprofit Society are:

1. *Telescopes of the World*: Assembled with the assistance of observatory staff members around the world, this set of 50 slides shows some of the most important astronomical instruments on Earth and in space. It comes with a 32-page book featuring a detailed introduction to telescopes and descriptions of each specific instrument. It can be used by a middle-school general-science teacher, a high-school physics teacher, or an astronomer teaching a university course.

2. *Astronomers of the Past*: These 50 portraits of noted astronomers were selected by the A.S.P. History Committee — chaired by Donald Osterbrock (Lick Observatory) — and range over time from Copernicus to Rudolph Minkowski. The set includes a 24-page book of capsule biographies and a thorough introductory reading list about the history of astronomy.

3. *The Radio Universe*: This set features 50 radiographs constructed from data gathered with the VLA, including the sun, planets, supernova remnants, planetary nebulae, the galactic center, active galaxies, and quasars. It is accompanied by a 24-page booklet written by Gerrit Verschuur, with a clear introduction to radio astronomy, detailed captions, and a bibliography. It won an award for "one of the outstanding non-print materials of the year" from *Choice*, the magazine of the Association of College and Research Libraries.

4. *Worlds In Comparison*: In these 20 "visual analogy" slides, NASA's Stephen Meszaros compares the sizes of objects and features in the Solar System. For example, Earth is seen projected to scale near Jupiter's Red Spot; radar maps of Earth, Mars, and Venus are compared; and a map of the U.S. is drawn to scale on a detailed photo mosaic of Valles Marineris on Mars. A 20-page book includes captions, classroom activities, a solar system reading list, and a table of the characteristics of all known planets and satellites.

Other slide sets available from the A.S.P. include:

1. *The Moon Kit* (18 slides)
2. *Mars Kit* (6 slides)
3. *Venus Kit* (6 slides)
4. *The Solar System Close-Up* (2 sets of 50 slides, 1 set of 30 slides, with captions by David Morrison)
5. *Voyager at Uranus* (15 slides)
6. *Halley's Comet Revealed* (17 slides, with a booklet by John C. Brandt)
7. *Splendors of the Universe* (15 slides with captions and background information by David Malin; shows color views of Southern Hemisphere objects)
8. *The Sky at Many Wavelengths* (11 slides, with an extensive booklet by Christine Jones and William Forman)
9. *The Infrared Universe: An IRAS Gallery* (25 slides with captions and background information by Charles Beichman)
10. *The Search for Extraterrestrial Intelligence* (20 slides with captions and a basic introduction by Frank Drake)
11. *Portraits of the Solar System* (20 slides of paintings by William Hartmann)
12. *Supernova 1987A* (6 slides)
13. *Science from the Space Shuttle* (30 slides with captions by Michael Lampton)

We welcome suggestions for other sets we should do and also offers to put together a set for us on a subject of special interest to you.

For a complete catalog of A.S.P. slide sets, please write to: Catalog Requests Dept., Astronomical Society of the Pacific, 390 Ashton Ave., San Francisco, CA 94112, U.S.A.

DYNAMIC HUMAN (ASTRONOMICAL) MODELS

Jeanne E. Bishop
Director, Westlake Schools Planetarium, 24525 Hilliard Road,
Westlake, Ohio 44145, U.S.A.

Many educators have found that models can add a lot to the understanding of astronomical concepts. Attractive commercial models of the solar system and celestial globes are readily available. Many models can be constructed of common materials — I found in my doctoral dissertation work that models made and manipulated by students increased retention of concepts of the celestial sphere and rotation, the seasons, lunar phases, and planet motions.[1]

But there is another type of model which I have found very successful. I call it the "dynamic human model." Actually, the concept is not new. I know that many good teachers have a favorite demonstration in which one or more students represent the Earth, sun, or moon. Also, the Elementary-School Science Project of the University of Illinois contained activities for student models to distinguish between Ptolemaic and Copernican systems.[2] However, I do not think the idea has been extended as far as it should be. I would like to suggest ideas for dynamic human models beyond what I have seen elsewhere.

Before describing examples, let me indicate good reasons for utilizing the technique. First, students *enjoy* a model in which they both participate and cooperate with others, and this appears to be true for elementary school through university. A recent review of research studies involving student-student interaction concludes that a group project requiring interdependence is likely to help everyone in the group master the concepts studied. Also it appears that retention is enhanced in the cooperative setting.[3]

There are other reasons. The spatial visualization required in many astronomical concepts requires time.... The more complex the spatial array, the more time is needed. Dynamic human models take time to explain, arrange, and then enact. Spatial visualization is sometimes difficult — frequently girls have more trouble than boys, and the more detailed concrete three-dimensional models incorporated in astronomy studies, the better learning will be. Models that show motion or changes are superior to those that do not, and mechanical models are usually expensive, difficult, or impossible to construct.

As I developed ideas for dynamic human models, I learned that several conditions (at least) are necessary for success. First, students must be serious and cooperative so that they become quiet at appropriate moments and assume the required roles. It is important to sit down prior to the activity and discuss what is to be done, the value of the activity, and the roles to be taken. Not only do students become sympathetic to the purpose of the activity, but they frequently make suggestions or ask questions that lead to modifications that will improve the model. With everyone united in purpose, the model probably will be a success.

Second, the concept and degree of detail should match the ability and previous knowledge of the group of students. It would not be effective to have a group of primary students (6–8 years old) model intricate details of star evolution or the electromagnetic spectrum. I have found, however, that very young capable children in gifted and talented programs can learn advanced concepts via this method of dynamic models better than via any other procedure.

Third, a fairly large area is needed for the dynamic human models. A gym, a lecture room without chairs, and outdoor areas work well. Particularly if you would like to provide some semblance of scale, the large spaces are required.

Today I would like to demonstrate two of my ideas on dynamic human models appropriate for upper elementary students, secondary students, and university students. The first model is for dynamic equilibrium in a stable star and the evolution of that star. The second model is for different wavelengths of light.

Model: Dynamic Equilibrium and Star Evolution

Two groups of students (two halves of a class) are identified. One group becomes inner gas parts of a giant star, who hold up hands toward the outside, exerting radiation pressure. The second group becomes outer gas parts of the giant star, who hold up hands toward the inside. The inner group pushes out against the hands of the outer group, who push inward. The star is in equilibrium, and it neither expands or shrinks. On cue, the dynamic equilibrium is upset, and inner forces push the outer back quickly and dramatically. This models a supernova explosion. One inner person remains spinning rapidly (conserving angular momentum), assuming the role of a black hole/neutron star. The model can be greatly enriched with a) a flashlight for the neutron star, making a pulsar, b) different "winds" or envelopes leaving the outside in stages, as shown happened in Supernova 1987A, and c) some exploded material initiating new star formation and/or becoming incorporated into a next-generation star.

Model: Wavelengths of Light

Mark parallel lines on the floor or ground. Different numbers can be used, although three works well. The three lines represent paths of red, yellow, and blue light. Place sticky dots of different colors on the lines. The red light line has red dots, the yellow light line has yellow dots, and the blue light line has blue dots. Place the red dots about 1.0 m apart, the yellow dots about 0.8 m apart, and the blue dots about 0.6 m apart. One student walks along each line simulating a traveling transverse light wave. All students begin by crouching on the first dot of a line. On reaching the next dot in a line, the student is to be fully upright. On each subsequent dot the student is crouched or upright, alternately. All students should progress down the light lines at the same rates, illustrating that all light waves travel at the same velocity in the same medium. To accomplish this, students moving on the lines with closer-spaced dots will have to go up and down faster. This demonstrates that yellow and blue light have more energy than red light, and have higher frequencies. Waves in non-visible sections of the electromagnetic spectrum also can be modeled. the scattering of short waves and the ability to see red light on the other side of clouds can be shown.

Other Ideas for Dynamic Human Models

— Eclipses, in which students assume roles of the Earth, the sun, and the moon.
— Binary-star system, in which two students revolve about a common center of gravity. The "stars" can form an eclipsing binary if the head of one student partly or totally blocks the head of the other from a distant student's perspective.

- Different types of galaxies.

- Interacting galaxies, in which groups of students pass with little or no bumping. Some students in roles as "gas clouds" are pulled away in the direction of the galaxy, as in M51. (Galaxies do not need to be the same size.)

- The sun and planets. "Planets" closer to the sun have greater velocities. To show how Neptune can be closer to the sun than Pluto, mark (with chalk, string, or tape) a long eccentric orbit for Pluto which is closer to the "sun" than Neptune in one section. "Earth" can note sidereal and synodic periods of the planets.

- The Oort Cloud, developing comets. A comet "tail" can be a prop of a long piece of aluminum foil, unrolled as the "comet" approaches the "sun." The "comet" walks swiftly about the sun at perihelion, trying to keep its tail pointed away from the sun. "Comets" can be periodic or not.

- Formation and evolution of the sun.

- Comparison of evolution of a sun-sized star and a supergiant, showing different late-life stages and the much longer life of the smaller star.

- The big bang. This can be simple or quite complex. In a complex model, students can model the formation of different types of matter and the separation of forces.

- Adiabatic and isothermal scenarios of galaxy formation.

- Structure of the sun.

- Structure of the center of the Milky Way Galaxy.

- Proton-proton and the CNO cycles, in which students assume roles as different nuclides, protons, neutrinos, and energy.

- Chemical species and structures: giant molecular clouds, nucleus of Comet Halley, including the clathrate hydrate structure, composition of Earth's early atmosphere and its evolution.

- Lunar Phases.

- Latitudinal and longitudinal lunar librations.

- Seasons.

- Sidereal and synodic days compared.

- Sidereal and synodic months compared.

Two final comments about using this teaching procedure. . . . First, I have found that it is helpful to have labels students can wear, when the model is complex. With many different roles, the labels help everyone comprehend what is happening. Labels can be signs with string or yarn worn about the neck, or pin-on cards. Sometimes props in the form of spheres held by the students or colors they are wearing will enhance the model.

Second, I believe it is important to explain what is wrong with any model, as well as what is right about it. Student misconceptions can be minimized in this way. For example, in our dynamic equilibrium model, the process is not present only at one radius from the center but at every point within the star. In our wavelenghts model, the waves actually travel much more rapidly than students can move, and unless polarized, vibration is in many planes centered along a line.

I would appreciate hearing from others who have tried dynamic models and who may now create come new ones.

References

[1] Bishop, Jeanne E. *The Developing and Testing of a Participatory Planetarium Unit Emphasizing Projective Astronomy Concepts and Utilizing the Karplus Learning Cycle, Student Model Manipulation, and Student Drawing with Eighth Grade Students.* Unpublished doctoral dissertation, University of Akron, Ohio, U.S.A., 1980.

[2] *Astronomy: The Universe in Motion* (Book 2). Elementary-School Science Project, University of Illinois, Urbana, U.S.A., 1963.

[3] Johnson, Roger T. and Johnson, David W. "Encouraging Student/Student Interaction." *National Association for Research in Science Teaching News.* U.S.A. December, 1986.

Discussion

R.W. Clark: *You have stated that females have more difficulty with spatial concepts than males. Is this statement based on any factual reserach, or on some assumptions?*

J.E. Bishop: There are *many* studies that have found that there is both a difference in spatial ability between males and females and a difference in learning material involving spatial ability between males and females. Some of these, including my own doctoral research, have found this for astronomy concepts. the effect is not strong at young ages, but develops through adolesence. *Why* this happens is not adequately explained. Many feel it is related to environmental treatment.

TWO INNOVATIVE DEVICES FOR THE TEACHING OF ASTRONOMY

Roland Szostak
Institut für Didaktik der Physik, Universität Münster, Fachbereich Physik,
Wilhelm-Klemm-Str. 10, D-4000 Münster, Federal Republic of Germany

The teaching of astronomy in schools should start by explaining the most elementary phenomena, which are a part of everyone's daily experience. Before proceeding to advanced topics, teachers of astronomy in school have to make sure that these very basic phenomena are fully understood. Since many programs have a gap at the elementary level, many people have foggy notions or disconcertingly wrong ideas of elementary matters. The seasons are a well-known case of a badly-understood phenomenon even for highly educated people. Still these people are honest and should not be laughed at. They do not understand the basic concepts of our contemporary scientific world, simply because these topics have not been taught properly. This problem raises a challenge for developing better ways of teaching astronomy. In order to show that there are specific new solutions to this challenge, I am presenting two innovative devices, which have been developed at the Universität Münster . The first example shows how the seasons may be successfully explained with a physical model that changes color when heated. The other example refers to the visibility of the stars and their disappearance in the daytime.

1. Thermal Method for Describing Seasons

The standard method for teaching these basic phenomena uses simple geometric optics. By depicting the rays of light from the sun, it is easy to recognize the illuminated parts of the Earth and moon. So, for the phenomena of night and day, and the phases of the moon, we do not feel that it is necessary to visualize the related rotational and orbital movements. But the situation is different when explaining the seasons. There, the angle of incidence and the duration of irradiation have to be considered for each place on the rotating Earth. To keep it simple, we focus our analysis on the solstice positions. For these two positions, a cut through the meridional plane allows us to see that the rays of the sun are steeper at the summer solstice than at the winter solstice. Although this analysis is only static, it affords some skill in geometry. Trying to include motions as well would involve superimposed rotational transformations, whose details become very sophisticated. Geometric optics would then no longer be sufficient for the model.

But we may ask whether the seasons should be explained in this way. Children identify winter with low temperatures, with snow and ice, and summer with high temperatures, but they are not so much aware of the changing position of the sun in the sky. So it would be much better to use a temperature effect for explaining the seasons. I report such an approach here.

Our technique uses a thermally sensitive material (Ag_2HgI_4) that looks yellow at room temperature and turns reversibly red at about 40° C. This material can be obtained from suppliers such as the Merck Company in the form of a salt of tiny crystals in a glass container. These crystals have to be suspended in a colorless transparent paint to fix them to a surface. I have coated a sphere of styrofoam with this material, and exposed this model globe to a heater or a spotlight. The most intensively irradiated parts of the sphere turn red. A rotating sphere irradiated from its equatorial plane shows a red ring on its equatorial zone. Children in a classroom accept this as being obvious, and are fascinated by this play of colors. Assigning the color red to the hot areas seems intuitively natural.

How does this method work with respect to our seasons? Where we live, the sphere should be yellow in winter and red in summer. In a first naive attempt, this effect would be achieved by putting the rotating sphere somewhat closer to or further from the heater. Then the red ring would become broader or narrower. But as the ring would remain symmetric with respect to the equator, and the some seasons would occur simultaneously in the northern and southern hemisphere, this turns out to be a wrong idea. In order to get *opposite* seasons in the two hemispheres, the red ring has to shift somewhat towards the north or south, oscillating around its symmetric equatorial position. But this cannot be achieved by changing the distance to the sun. Children in a classroom find out, simply by playing, that the red ring is shifted by tilting the axis. This visual experience promotes a more effective understanding.

Now one is ready to carry the sphere around the heater to observe how the hot zone is shifted by the orbital path of the Earth. It becomes evident that, without the inclination of the axis, there would be no seasons. The orbit leads to a year with seasons because the axis is tilted. In my experience, this procedure works very satisfactorily in classes with children down to the age of nine. The children even make the statement that the orientation of the axis in space must remain constant. The idea of checking this orientation with the poler star seems to suggest itself here.

A crucial difference between the common optical method and this new thermal method is the irradiated heat is stored, whereas the incident light is available instantaneously only. By virtue of this storage, the sphere displays the temperature distribution on the whole globe, independently of the day-night position. This averaging over the short day-night period leaves just the smoothed behavior of the seasonal temperature variation, which is not accessible by the optical method.

Let me add some technical remarks: The model sphere needs to be heated up for some time (a minute or so) before it changes its color. Similarly, it holds the stored heat for some time, so that the temperature distribution can be seen by everyone in the classroom at leisure. This thermal inertia offers an additional interesting feature: If the rotating sphere moves around the heater slowly, like the Earth in its orbital path, then the displayed temperature distribution shows a characteristic time lag, just as lowest and highest temperature occur not at the solstices but several weeks later. For proper display, the orbital speed has to be adjusted to the thermal reaction time of the model sphere.

Figure 1. The chemical on the ball turns from yellow to red (lighter to darker on this black-and-white photograph) when warmed. a) The appearance of the ball if the axis were upright and so with the sun shining directly down on the equator. b) The appearance of the ball for the northern summer solstice, which the Earth's axis is tilted toward the sun. (The sun is off to the right in both photographs.)

This reaction time is determined by the specific heat capacity and the thermal conductance of the material. Both of these values should be as low as possible for practical purposes, in order to reduce the power of the heater and to get a reaction quickly. Styrofoam turned out to be a suitable material. To lower these values further, we coated the sphere in some cases with a velvet layer about 0.5–1.0 mm thickness before applying the temperature-sensitive paint. This velvet consists of short pieces of organic fibers which are fixed to the surface electrostatically by a standard industrial procedure. Generally, the styrofoam has to be very homogeneous in order to keep the areas where the color is changing from looking somewhat stained.

2. Star Charts

Let me now deal with the starfinders that are used by amateurs. In order to familiarize students with this tool, we made a planisphere for overhead projection in the classroom. Instead of using a transparent foil, we took an opaque material with little holes for the stars. In order to make this stellar sky rotate around the pole star, the star disk is fixed by a snap onto an underlying transparent sheet that has an opaque cover for the parts outside of the horizon, so that the stars can be seen entering and leaving this horizon area.

Compared to other ways of visualization, this projected stellar sky is surprisingly impressive in a darkened classroom. It may initiate students into how attractive it must be to observe the stars in nature. Although such an experience can be better supplied by a planetarium, the advantage of this cheap overhead projection is the individual interaction between students and the teacher that is not available in the planetarium during a show. As a result, the teacher can secure the knowledge of constellations like the Big Dipper, Orion, *etc.* The teacher can select these stars very simply with a sheet of paper having a suitable aperture to prevent all other stars from being projected. Thus the constellation can be seen and memorized very effectively. Such a step is not trivial, because most constellations are barely known. Further, it is worthwhile to show how the Big Dipper changes its orientation hour by hour, thus making students aware of the rotation of the earth.

When using these planispheres, one very soon arrives at the question of the visibility of the stars in the daytime. At this point, one can make clear that the stars still there, and that they are not visible because the sun is flooding the atmosphere with light. It helps to remind students of the fact that some bright stars or planets can be seen in the early morning or when dusk is falling. Children become aware of this effect when they are shown how the stars emerge from twilight. This process is much better visualized on a quicker time scale. We simulate sunrise and sunset on an overhead projector in the following way: we take two polarizing sheets about 30 cm by 30 cm covering the projector's flat surface. One of these two foils has a pattern of small holes drilled at the positions of the stars. When the polarization axes of these foils are parallel, the projector's light passes through and the stars are almost invisible. But when the second foil, which contains no holes, is turned into the crossed position, the "sky" slowly darkens. Meanwhile, the stars emerge from twilight as if dusk is falling; they finally appear as brilliant spots on the dark background. This procedure can be very impressive, bring the children in the classroom into contact with what can be observed in nature with some patience.

Some additional optical effects can add further intellectual content. One such technique allows us to introduce extra motions of celestial bodies into this projected stellar sky. We use the fact that light passes again through the crossed filters when a quarter-wave plate is inserted between them, transforming the linearly polarized light from the first polarizing sheet into circularly polarized light. By shaping this quarter-wave plate like the silhouette of the moon, one can have a bright moon projected together with the other stars on the dark background. But this moon can now be moved by hand through the pattern of the other projected stars. In the same way, of course, a planet instead of the moon can be introduced into the projected stellar sky. So the apparent motions of the planets through the stellar sky, including their retrograde loops, can be shown and discussed very specifically and individually in the classroom.

A second technique allow us to introduce colors into scenery by using the above optical effect's wavelength dependence. Not all the light being turned into the circularly polarized mode can pass through the second filter; only light within a certain window of wavelengths does so. As a result, this projected light looks colored. It is fortunate that foils with a suitable dichroic behavior are extremely cheap and readily available. Clear plastic used for wrapping bunches of flowers or candy boxes and even clear stickytape works perfectly. By simply inserting such colorless foils between the two polarizing filters, the projected dark sky turns a color; meanwhile, the stars stay bright and colorless. By turning the foil and the second polarizing filter into adequate orientations, the color itself and its saturation can be chosen at will. In this way, the night-black sky can be turned into a deeply saturated blue and continuously into a light blue as the stars stay unchanged and so fade away in the rising light. In this simulation of the morning hours, the blue can also be changed into red or other colors. Further, the moon, when rising above horizon, may be projected in orange and can be gradually turned into white.

All the above can be done in a classroom with a few foils on an overhead projector. The process is very effective and allows individual discussion of astronomical phenomena between children and teacher. Although the process allows some effects of a planetarium into the classroom, it is not meant to be a substitute for a planetarium. But it can be an effective preparation for a planetarium visit, because several phenomena that are shown in the planetarium will be better understood by the children after such prior discussion.

Discussion

G.L. Verschuur: *This contribution highlights what I believe is an important point regarding astronomy courses in the U.S.A. The great majority of students are in the class to fulfill their science requirement. The goal is to give them a sense of what science is about, how scientists think and how they approach problems. The fact that we use astronomy as an example is almost secondary (although it may be heretical to suggest that here). Thus we do not have to teach all of astronomy. We do want to show how the scientific method is used to answer questions regarding the nature of phenomena such as the seasons, as you have so elegantly demonstrated.*

J.-C. Pecker: *One should realize that, at the university level, there are two ways to teach astronomy; one is to do it as a "natural science," using physics of course, but as a tool; another one is to do it as a series of applications of physical concepts. The latter is becoming quite a fashion in some places, and I would object to it. The remark by Verschuur might lead to similar difficulties, i.e., losing the unity of astronomy seen as a "natural science."*

A PORTABLE DEMONSTRATION OF ORBITS IN CURVED SPACE-TIME

David B. Friend and Kevin Forkey
Department of Astronomy and Department of Physics, Williams College,
Williamstown, Massachusetts 01267, U.S.A.

Many introductory astronomy students, when confronted with the idea of curved space-time in discussions of relativity and cosmology, don't have a very good grasp of what this really means. They are told that the presence of mass "curves" space-time, but it is often not clear to them what is meant by that statement. We have developed a simple portable demonstration of what is meant by curved space-time, by using a two-dimensional analog. We have stretched a thin rubber sheet over a circular metal frame which is supported a few inches above a table top. By placing a heavy weight (about a kilogram) in the center of the sheet, we can mimic the effect of the curvature of space-time around a massive object on an orbiting body (such as a planet orbiting a star). We roll a small ball around the sheet, and the ball describes approximately elliptical orbits around the weight (see the photograph). The orbits aren't exactly elliptical for two reasons: energy is dissipated by friction between the ball and the sheet, and the sheet doesn't quite have the correct shape (it is shaped roughly like ln r instead of $1/r$: see this by solving Laplace's equation in two dimensions with circular symmetry). We like to project this demonstration with a video camera placed directly above the sheet, so that the students can see the "orbit" without actually seeing the curvature of the sheet. This curvature is analogous to the actual 4-dimensional structure of space-time: we can see the effect of the curvature without noticing the curvature itself. By letting the central mass

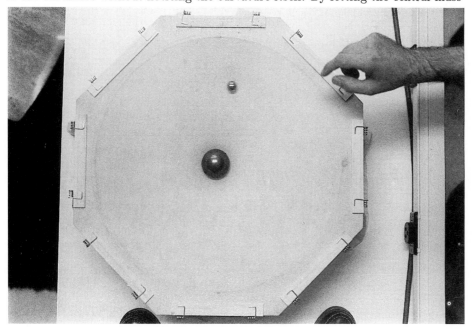

Figure 1. The demonstration of curved space-time

be free to roll, we can also show how the more massive member of a binary system also moves in response to the gravitational field of (or the warping of space-time by) the smaller mass.

DEMONSTRATIONS

Participants were invited to present demonstrations that they found helpful in the teaching of astronomy. Two participants did so.

Roy L. Bishop (Acadia University, Canada) demonstrated the use of a small mirror for projecting an image of the sun on a screen. He recommends a piece of mirror, a few mm in diameter, mounted on a small stand. This combination can be placed on a window sill, or other sunny place, to catch the sunlight. Alternatively, a larger mirror can be used, if it is covered with cardboard or tape, with only a small area, a few mm in diameter, exposed. The screen should be located in a shadowed place. Note that the shape of the sun's image is not affected by the shape of the mirror!

Harry L. Shipman (University of Delaware, U.S.A.) described a new method for getting feedback from students. He attaches an answering machine to his telephone, and encourages students to "dial-a-prof" with comments or questions about the course work. Since the students can phone anonymously, they are less hesitant or embarrassed to do so. Questions left on the answering machine are dealt with at the next class. Harry finds this an effective way to determine which parts of the course are giving students most difficulty.

THE R.A.S.C. *OBSERVER'S HANDBOOK*

Roy L. Bishop
Department of Physics, Acadia University, Wolfville, Nova Scotia B0P 1X0, Canada

Except for two years, 1909 and 1910, the *Observer's Handbook* of The Royal Astronomical Society of Canada has been published every year since 1907. It was founded by Clarence Augustus Chant, who also was responsible for the founding at the University of Toronto, of Canada's first and largest department of astronomy, and who established the David Dunlap Observatory, which contains the largest optical telescope in Canada. In addition, Chant served as Editor of the *Observer's Handbook* for 50 years, which is the main reason I am only the fourth editor since 1907. I should mention that John Percy, Chairman of the Scientific Organizing Committee of this Colloquium, was the third editor.

The sales of the *Observer's Handbook* provide a major source of financial support for The Royal Astronomical Society of Canada, a non-profit, educational organization that is devoted to the advancement of astronomy and allied sciences. The RASC is similar to the Astronomical Society of the Pacific or the British Astronomical Association in that its membership is composed of people from all walks of life, from children to professional scientists.

All of the twenty-nine contributors to the *Observer's Handbook*, including the Editor, are volunteers. Their time, effort, and expertise represents an enormous contribution by these individuals and their institutions to The Royal Astronomical Society of Canada, and to astronomy in general. Although the Handbook exists primarily for the benefit of the 3500 members of the Society, about 11000 copies are sold to non-members in Canada and in many other countries. The use made of the *Observer's Handbook* is as diverse as the membership of the RASC. The Handbook is found on the bookshelves of schools and major libraries; in the eyepiece cases of amateur astronomers; and beside the control consoles at many professional observatories. It is an invaluable reference for astronomy teachers at all levels, from elementary schools to universities.

The 1907 edition contained 108 pages. The 1989 edition, the eighty-first, is the largest ever, with 224 pages. Each edition is published for the following calendar year and contains predictions of astronomical phenomena, plus much other information that is relatively static from year-to-year, although each year significant revisions are made to approximately three-quarters of the pages. The Handbook is organized into eleven major sections, with symbols in the left-hand margins to facilitate rapid locating of each section. The material presented is in various forms: maps, star charts, text, diagrams, tables, lists, and graphs. Many of these maps, charts, *et cetera*, have been expressly created for the Handbook and make it a unique resource for anyone interested in astronomy.

The *Observer's Handbook* represents a labor of love by many volunteers for over three-quarters of a century. For thousands of astronomers and astronomy teachers in many countries, this modest publication from Canada is a valued companion.

Discussion

H.S. Gurm: *I wonder if it is the right time and place to raise an issue. When it comes to the prices of the slides, posters, the* Observer's Handbook *of the RASC, and shipping these in small units to the third world even in the case of India, the local currency leads to a situation where it becomes difficult to make purchases. So there is a problem.*

R. Bishop: Regarding the difficulty teachers in underdeveloped countries have in obtaining educational materials (due to problems in sending money out of the country, or even finding the money to purchase materials), in the case of the RASC's *Observer's Handbook* I offer three comments:

1. Although the single-copy price of *Observer's Handbook* is $10.95 U.S. (ppd) ($13.50 overseas airmail), bulk rates are available involving savings of about 35 per cent (contact the RASC for details).

2. The RASC *will* send single free copies to educators in underdeveloped countries in cases where it is not possible, due to currency restrictions, to send the usual payment. Requests of this sort must be well documented.

3. If it is not essential to have the latest edition, after September 1 (and if copies are available) the RASC will consider requests for bulk orders of the expiring edition (possibly at the cost of shipping in the case of underdeveloped countries).

To facilitate processing of special requests, please write directly to the Editor (Roy Bishop, Avonport, N.S. B0P 1B0, Canada).

———————————

IAU COMMISSION 46: ASTRONOMY EDUCATIONAL MATERIAL

M. Gerbaldi
Institut d'Astrophysique, 98 bis Bd Arago, 75014 Paris, France and
Université de Paris XI, Centre d'Orsay, Lab. d'Astronomie Bât 470, 91405 Orsay,
France

During the IAU Assembly in Prague (1967), Commission 46 on the Teaching of Astronomy decided to prepare a world-wide list of available material suitable for astronomy education at various levels and as in many languages as possible.

The first list was published in 1970 by E. Müller, under the title: "Astronomy Educational Material." Since then, in order to keep the list up to date, an "Addendum" has been edited every three years.

Each Addendum is subdivided into three parts:

part A: material in the English language

part B: material in Slavic languages

part C: material in all languages with the exception of English and Slavic.

These Addenda are sent to everyone on the mailing list of the Commission 46 Newsletter. To receive the Newsletter, write to Prof. Leo Houziaux, Institut d'Astrophysique, 5 Ave. de Cointe, B-4200 Cointe-Ougrée, Liege, Belgium.

Table 1 presents the number of books listed in Addendum C according to language. The very high figures in Chinese (period 1982-1985) and in Romanian (1985-1988) resulted from the fact that these lists were then published for the first time and contained materials since 1949.

From a quick look at this table, we can surely conclude that there are very strong differences among the different countries. The main question is: are the lists exhaustive or a selection of the best items? In any case, is such an Addendum C useful in all languages?

From personal experience, I know that editing and publishing such Astronomy Educational Material consumes more and more manpower and becomes more and more costly. It became necessary to rediscuss the goal of such a document.

A decision about the future of the AEM was made during the IAU General Assembly held at Baltimore (U.S.A.) in August 1988. It is obvious that the real need is for lists of selected items at different levels in a few languages; English, French, and Spanish were selected. Such lists published by Commission 46 should be effective for astronomically developing countries.

Table 1. Number of books listed in each Addendum C according to their language.

	1973-76	1976-79	1979-1982	1982-85	1985-88
Dutch (Holland)	3	6	1		
French (France)	8	95	88	211	117
German (F.R.G.)	9	12	24	16	112
Greek	33	21	26		17
Hungarian	6	8	8	11	19
Indonesian	5	4	6		
Italian	15	48	20	15	8
Japanese	48	38	45	32	20
Norwegian	2	2		4	–
Spanish (Spain)	1	31	59	19	22
Swedish	14	30	25		6
Danish		8			
Finnish		7	14	12	9
Afrikaner			1		3
Portuguese (Portugal)			3	6	1
Chinese (Rep. Populaire)				424	73
French (Belgium)				23	2
French (Canada)				15	1
French (Switzerland)				4	
German (Austria)				6	2
German (G.D.R.)				9	5
Indi (India)				3	
Spanish (Mexico)				4	1
Arabic (Egypt)					1
Dutch (Belgium)					4
Hebrew					2
Malaysian					1
Nigerian					–
Portuguese (Brazil)					10
Romanian					322
Spanish (Columbia)					2
Spanish (Venezuela)					1
Thai					83

Chapter 8

Conceptions/Misconceptions

Astronomers' views of astronomy in particular and of science in general often differ from the views of the general public. Papers in this chapter discuss this dichotomy. The first paper is a personal view of public misconceptions of topics relevant to astronomy. Next, we see that conflicts between science and the local culture can lead to misunderstandings. The American struggle between "creationists" and scientists about what can be taught in schools is discussed next. Finally, we hear of other conflicts between astronomers' thinking and the understanding of the listeners.

PUBLIC MISCONCEPTIONS ABOUT ASTRONOMY

A. Acker[1] and J.-C. Pecker[2]
[1]Université de Strasbourg, 11 rue de l'Université, 6700 Strasbourg, France
[2]Collège de France, Annexe, 3 rue d'Ulm, 75007, Paris, France

> *"If I give you a fish*
> *you will eat only one fish*
> *but if I teach you how to fish*
> *you will eat fish all your life...."*

The Chinese proverb quoted above means more or less the same thing that our Rabelais did in the middle of the Renaissance, when his famous Gargantua made the statement: "well built brains are to be preferred to overfed brains."

Indeed, when facing the fallacies that invade our modern life, and when facing the very quickly changing world in which we evolve, it seems that the only weapon we can give is a critical approach to science and life. We feel that anyone can assess at least the likelihood of any alleged fact, or, if he cannot, that he should be able to recognize openly his inability to do so.

Now let us consider the common fallacies of modern times. And you will excuse me if I come back several times to the same motto, from my old Rabelaisian education....

Bye bye Ptolemaeus! bye bye Paracelsus!

A public poll given in France some years ago showed that about one-third of the people still believed that the Sun revolves around the Earth, in the best of the Ptolemaic tradition! About the same percentage — perhaps the same people — give some credit to astrological phenomenology.

This survey shows how important and urgent it is to work at avoiding such a situation. Should I remind you that most of the great dictators on Earth and even *bona fide* statesmen had or still have their astrologers? In a quickly changing world, lucidity and awareness are necessary. The public must be prepared to understand what is going on and to evaluate critically information that is broadcast and that is often turned into sensational events. The public should not be tempted by irrational behaviors that would bring us back to the dark ages.

The aim of teaching is not only to give undigested knowledge to children or students, but also to help them to understand what they are learning, to understand how we have proceeded to our current state of knowledge, and how one may proceed to know more. Of course, we must not be severe. One must excite the young, if one wants them to be willing to understand. And therefore, one must show them beautiful things; fortunately, astronomy is a wonderful field in this respect. And we astronomers are somewhat lucky. But this exceptionally attractive power of celestial objects is also linked with their mystery. We must face the fact that a look at the sky often inspires some metaphysical meditation, and often inspires in contemporaries — as in the shepherds of the past — some strange feelings of fear or hope, some idea of transcendence, and at the same time, the feeling that man is so small in the immensity that he must totally submit himself to the will of the universe, whether it is either God or celestial bodies. Indeed, this idea is often terrifying. Just remember Blaise Pascal: "Le silence éternal des espaces infinis m'effraie...." Astrological temptation is thus natural enough.

When faced with the sky, the first move was to pray; the rain dances of early Indians and the sacrifices of the pagan gods were ways to obtain favors from the wild sky. Olympian gods were located quite naturally above the clouds. But in spite of prayers, volcanic catastrophes, floods, or storms still remained the sad rule of our lives. From magic, whose aim was to obtain from the elements whatever was good for man, people turned to the other extreme. Man was no longer in the center of a very small sphere, limited by a tangible Empyrean (the sphere of light or, among Christian poets, the abode of God). The astral universe is eternal and infinite. This Pascalian attitude leads to the fact that, if the machinery of the sky is to determine the destiny of mankind, the lives of men and women are entirely submitted to this strict determinism. In this case, the rules of this determinism are to be established: they are the essence of all astrological thinking.

For a long time, the churches were opposed to astrology, because astrology tended to destroy the idea of human free will. But this opposition was a matter of faith, not a scientific need.

Let us therefore come back to the teaching of science.

The modern development of astronomy, with the fabulous and spectacular conquests of space research, offers young people a world of beauty. And the temptation is great, especially at the elementary or secondary level, for the teacher to show beautiful pictures and to comment on them; he or she often gives the results without explaining how they were obtained. . . .

My own tendency would be to try to explain better the historical evolution and the logic behind astronomical concepts. And, if I limit myself to those concepts the ignorance of which leads people to believe in astrology or in flying saucers, I would probably insert some well known facts in the course of the explanations.

First, let us be observers of the apparent motions of planets and the Sun in the sky. Planetariums are excellent tools for such an approach. One can easily show that there are indeed 13, not 12, zodiacal constellations. One can describe the sky by showing also how the constellations were first described mainly as a guide to show the travels of planets, *i.e.*, only for practical purposes. This motivation explains how subjective the shaping and naming of the constellations was, as demonstrated, for example, by the clear differences between western Assyro-Babylonian cosmography and Chinese sky maps.

It seems sensible to remind students that our ancestors made markers in the sky because they had neither calendars nor compasses; this fact explains the origin of the zodiacal constellations, which indicated the seasons to the farmers, while constellation rings (stars with the same declination) were helpful to sailors. Constellations were memorized by mythological legends; celestial figures were made without reference to the actual arrangements of the stars (*e.g.*, Orion, Andromeda). Thus, when you present a photograph of a part of the sky, say Orion, to some pupils, and ask them to link up the stars in order to form a figure (flower, person, or building), one will obtain almost as many figures as pupils. Doubtless, the brave warrior Orion will not appear among them. This demonstration may show students the total nonsense of allegations like "he's strong and stubborn, because he's born under Taurus."

A look at the motion of the Sun and planets in the different zodiacal constellations shows clearly that they do not stay for an equal time in each of them. The Sun stays 44 days 11 hours in Virgo; it stays only 6 days 8 hours in Scorpius. A planet (such as Venus, for example) does not stay the same time in any given constellation during two consecutive years.

Hipparchus knew about the precession of the equinoxes. The precession can be explained, to the youngest children, as an observed fact, defining the vernal point when you describe the seasons. Then you can distinguish between astrological signs and astronomical constellations.

One can note the loops described by planets in their apparent motion. Though these loops can be explained better in the Copernican system than in the Ptolemaic system, they by no means prove the reality of the former. Still, the demonstration leads to the idea that parallactic ellipses may be a way to determine Sun-planet distances. This idea may be generalized, with modern accuracy, to determine the distances of stars to the Sun.

This determination leads to the fact that stars and the Sun are of the same order of magnitude in absolute brightness. The discovery gives a feeling of the very large distances involved; it also leads to a concept about the unity of the universe.

When speaking about stellar distances, it is easy to show — with animations or by building 3-D models — that the appearance of constellations results from perspective effects. It is easy to show that Vega and Sirius are closer to each other, although in two different directions of the sky, than, say, Sirius and Betelgeuse, which are in almost the same direction in the sky.

But let us come to one essential idea, that of the unity of the Universe. This concept is strictly opposed to Aristotelianism. Let us remember that Aristotle suggested a sublunar world and an astral world in opposition to each other. The sublunar world was subject to violent events, to emotional instability, to the strange and the unexpected; it is essentially made of the basic elements air, earth, water, and fire, whose combinations give diversity. The sublunar world also contains the ephemeral, the putrescent, and the passionate. In contrast, the astral world, all of purity and eternity, is made out of a single element, a fifth one: aether. Between the two worlds, correspondences exist, linking macrocosm and microcosm. Such a link was the basis of medieval medicine, of astrology, and of alchemy.

The Aristotelian construction was based on the regularity of celestial motions, as opposed to terrestrial meteorology and to human weakness.

Aristotelianism was destroyed forever a certain evening of 1572, the 11th of November, to be precise. Tycho Brahe, then a young man 26 years old, was living and working with his uncle, Steen Bille. His uncle was an open-minded person, interested in astrology and alchemy. That very day, Tycho went out of Steen's laboratory, and saw, very near the constellation of Cassiopeia, a very bright star. It was at a place where no star was previously visible. "Nova Stella"! The interesting thing was that all subsequent observations made by Tycho himself, and by many other European astronomers, showed that the star, although weakening in brightness, stayed at the same place in the sky. If one realized that the Moon itself moves by about 13° from night to night, this observation shows that the new star was located much farther than the Moon, in the astral world rather than the sublunar world.

This discovery immediately removed the timelessness of the astral world. An explosive event, the birth of a new star — whatever could be the cause of Tycho's discovery, it implied that the Universe is evolving. The Universe is no longer eternal; it has lost its privileged stature!

The real influence of the Cosmos

Inside the immense spatial structure of the universe, forces and well-known actions "play" with the material bodies.

It is not a matter of obscure and evanescent forces but of well-known actions, particularly the universal gravitation that favors the nearest and heaviest bodies. For example, Venus and Jupiter are able to perturb somewhat the motion of Earth but only slightly, because the "alignment" (which, in fact, never occurs) of all the planets would produce a deviation in the orbit of the Earth by only a few centimeters.

Let us talk about the "star of the night," the Moon, to which so many myths and assumed statistics have been attributed (influences on births, deaths, homicides, humors, *etc.*); the tides (which vary as $1/D^3$) clearly explain the large motions of our oceans, but are not at all in touch with the circulation of physiological fluids. It is definitely by chance that the female menstrual cycle coincides approximately with the lunar month; the cycle varies from one woman to another and females of other species than human have totally different cycles!

As to the influences of the faraway Cosmos upon the "destiny of mankind," one may pragmatically notice that the life of a child born in Greenland, where the daytime is six months long, is different from the life of a Senegalese child, whatever their "sky of birth" might have been.

The Sun acts upon Earth and its inhabitants by its "magnetism." There is nothing strange in this action; it is only the consequence of the circulation of electrons through a magnetic field. These effects are well known and can be accurately measured with appropriate instruments. One can notice the existence of a variable magnetic field in the Sun, which leads through a complicated process to the formation of flares and gaseous protuberances. In addition, the Sun ejects into space the "solar wind" and other particles, often violently. Their impact here on Earth has very spectacular effects: the aurora borealis, distortion of the magnetosphere, ionospherical perturbations, periodical climatic variations in the terrestrial polar regions, and even the formation of induced currents inside the ground.

Finally, one may note the vast motion of the solar system in the Galaxy (once round in 250 million years at a speed of 900,000 km/hr), where the crossing of spiral arms can probably explain effects that last for long periods (geological eras, glaciations) but certainly not playing any part in the destiny of individuals of the same generation.

Why should we look first for "hidden" forces in the universe, whereas quantifiable influences of the cosmos onto Earth are so numerous, varied, and fascinating to study?

So Long, Mister Spock!

But of course, the above ideas do not prove, again, that the astral universe is of the same nature as the physical universe. This idea came very slowly, through spectrography during the 19th century, and through modern astrophysics. Still, it was convincing enough to lead philosophers of the 16th and 17th centuries to believe in the plurality of inhabited worlds, in spite of Aristotle and in spite of the Church. It was first Giordano Bruno, and later Fontenelle, who spread these ideas at the dawn of the Century of Enlightenment.

The plurality of inhabited worlds, an inductive idea emerging from the progress of knowledge, has led us towards another type of misconception, those ideas concerning extraterrestrial life.

In the earlier works mentioning such ideas, they appear as substrata for philosophical thinking. Kepler's *Somnium*, and later, for example, Cyrano's *Histoire comique des Etats et Empires de la Lune* or *Histoire comique des Etats et Empires du Soleil*, or Swift's *Gulliver's Travels*, or again Voltaire's *Micromegas* are nothing but satirical novels, intended at mocking the author's contemporaries, without enabling the authorities to put the author in jail.

More recently, Jules Verne, Arthur Conan Doyle, and H.G. Wells wrote fantasies that, at the time of their publication, appeared for what they were: science-fiction, philosophical reflections.

Eventually, people started really to ask the right questions. When observing the changes of color in the Martian landscape, it was immediately proposed to associate the green with local spring and the red with local autumn. Later on, it was realized that minerals could also change color, depending upon the state of oxidation; it was also realized that winds could blow up yellow sands all over the planet. But, at least, life could be searched for in planets — not in the form dreamed of by H.G. Wells, but in some very elementary form.

Modern science-fiction, with Lovelace or Bradbury, went further. Even the radio broadcasts, such as the famous Orson Welles show, inspired by H.G. Wells, started to put extraterrestrial beings if not on Earth at least in Earthians' minds. After the Second World War, atomic fear and space conquests gave more credibility to the dreams, and to the fears.

I do not need here to speak more about the development, since that time, of "ufology." The Blue Book of the US Air Force, the Condon report, and many other studies, have clearly led to meaningful statistics. Out of 100 cases, 97 could be assigned to known phenomena: meteors, meteorological balloons, aircraft, or simply bright stars. The remaining 3 cannot be identified, but the description seems almost objective; one is so described that it represents such twisted emotional information as to make it unrecognizable.

No more than making a course on astrology should we consider making a course on ufology! But again, we should give pupils a clear basis to enable them to look with a critical mind at the alleged phenomena.

Of course, astronomy comes first. A discussion can be made of what minimum complexity of molecules is needed for the very existence of life. Therefore, some elements of molecular astronomy — astronomy of cold bodies — could be given; discussion could include planets and comets and also the interstellar medium where cold grains of dust are associated with large molecular clouds — such as in Orion.

The next step is the discussion of the Drake formula, complemented by an introduction of the notion of probability. For other planetary systems, we can have a full discussion dealing with detectability either by the study of radial or proper motions, and the subsequent eventual discovery of invisible companions, or by infrared studies such as the ones performed by IRAS. Detectability, of course, is more likely around close-by stars than in distant objects.

Then comes the question of life on the planets. With what physico-chemical conditions could life, if brought there, survive? This question is almost simple! But more difficult is the estimation of the probabilities for life to appear given a suitable planet, for life developing into civilization and then into technological civilization, and for alien civilizations to enter into relations with ours.

At least, let us give reasonable estimates, and let us not hide the strong uncertainties; let us remember that science progresses, hence does not have all the answers.

One should also, in cooperation with the psychologists, explain that observations are not always reliable. Astronomers never observe one thing alone; any given observation needs to be checked and rechecked by others before being announced — as did Tycho with his discovery of 1572.

It is easy to say that eyewitnesses may be strongly deceived, be they looking at a crime, or at a celestial and mysterious phenomenon. But can we evaluate the probability of deception? If we believe the Blue Book or the Condon report, we reach a percentage of 3 per cent or so of incorrect testimony! Can we check this on one single object that has been observed by several observers? Without planning it, one of us (J.-C.P., or "I", in the present paragraph) once did such a thing. On the evening of August 16th, 1966, at Nice Observatory, I received several telephone calls from various people in the area: "What is that bright colored object that we saw falling in the West, behind the town?" As my colleague in Marseilles received similar queries, we decided to publish an appeal to eyewitnesses the following day in all daily newspapers of the French Riviera. More than 400 replies came; about 100 could be assigned to something else, either the time or the direction not being right. But 300 dealt with one single phenomenon. Our opinion was clear: it must have been an explosion, at large altitude, say 80 km, of a large meteorite. An astronomer from Marseilles, Mr. Guigay, made the effort of visiting every one of our 300 witnesses. He went onto their balconies or windows, to measure exactly the apparent direction of the motion. Then he computed by a least-square method the trajectory and the location of the point of explosion. We later found some pieces of the meteorite, a very banal chondrite, in the vicinity of the computed place. It was indeed a well-conducted experiment. Still ten persons (the 3 per cent!) said "Thank you, sir, for having confirmed [!] to us the fact that we have seen a flying saucer." And three (1 per cent) made drawings similar to the classical images of a flat disk surrounded by regularly spaced circular windows. They indeed saw that! I do not question their honesty. But one can claim that 1 to 3 per cent of witnesses add to an objective phenomenon so much subjectivity that the truth can be completely masked and fully unrecognizable!

What about the anthropic principle? And the Big Bang?

At the opposite of the wild look for alien visitors, in spite of all likelihood, another attitude has spread very quickly within the realm of *bona fide* scientific communities. This attitude is based on the remark that, were the constants of physics only slightly different from what they actually are, life would never have been possible, as evolution would have been much too quick, or much too slow.

In other words, the very existence of life is linked with the precise value of these universal constants: the velocity of light, the gravitational constant, the charge and mass of the electron, Planck's constant. On this basis, one went so far as to claim that it more or less means that the universe is made for mankind. And if one follows that line, one may conclude that man is alone in the universe, in the midst of the lost paradise, of animals, of plants, on Earth. Should we consider this so-called "anthropic principle" as a "misconception?" Possibly not. No doubt, the concept, just like that of a "Big Bang," has very strong metaphysical connotations. If the Big Bang is now, in some modified forms, rather widely accepted, the anthropic principle is not.

I feel that, although I personally have strong reservations about both, the Big Bang and the anthropic principle cannot be either eliminated or accepted as fully established facts.

The expression "Big Bang" is too often open to confusion, wrongly recalling a "creation," a "*fiat lux*." I believe that it is absolutely necessary to fix as clearly as possible the limits of our knowledge on that subject. In this way, the basis of the "Big Bang" model is the expansion of the universe, deduced from the redshift observed for the galaxies. But other interpretations are possible. Indeed, quasars could well be nearby objects, with an intrinsic redshift, and this would be accounted for easily by an absolute brightness otherwise badly understood. Another point: in the "Big Bang" theory the universe expands similarly to a perfect fluid. But at present, the strong anisotropy of the extragalactic universe appears more and more clearly, thanks to more and more impressive observations. Consequently, one cannot assert that the "Big Bang" model is definitively the only one possible.

In any case, even this model of a "primordial explosion" cannot, in fact, be understood as describing the very beginning of the universe. The model imagines a "primordial universe" with extremely large density and temperature, in which the matter could appear only in the form of the most elementary particles. But the description of this universe is thus totally dependent on the state of knowledge in theoretical physics, and is facing the same limits. In this way, the universe at a theoretical instant "t = 0" corresponding to an "infinite" temperature cannot be described, as we ignore the behavior of matter for $T > 10^{32}$K; the deficiency leads to an interval of 10^{-43} second after the theoretical time zero; the temporal area between $10^{-\infty}$ (=0) and 10^{-43} second is indefinite and cannot be comprehended with our present knowledge, neither at a spatial level nor at a temporal level. In particular, we are unable to know if the universe has ever had any "beginning."

This scientific attitude has no correlation with a religious belief. Indeed, as God can be thought of as infinite and inexpressible, the man who believes in God cannot put his religious beliefs in the form of equations except by reducing them considerably; and that would properly be... diabolic!

In any case, the study of the limits of Science, or of the doubts that the scientists are facing, is clearly needed to complete the teaching of scientific results.

But I feel that we cannot speak about these limits in the same way as we do about astrology or ufology, which we know to be fallacious. We must insert, here and there, into our teaching, the question marks. We must make clear also that an astronomer is also a person with his or her own ideas, philosophy, and religion. This general statement is justified by the study of the history of science and the history of scientists. And when speaking about the evolutions of concepts in astronomy, we should perhaps insist upon the provisional character of the progressing concepts. The last fringe of knowledge will stay somewhat fuzzy. One should realize (and make our pupils and students realize!) that, although science is never "finished," it is built on firm ground. Even Copernicus broadly used Ptolemaic mechanisms; Kepler worked on Copernicus' ideas; and Newton justified Kepler's laws by universal gravitation. Einstein perfected Newton's work, not replaced it, and his torso will not replace Newton's on its pedestal. Similarly, we cannot say any longer that the proponents and opponents of modern cosmologies are really in contradistinction with the previous trends of research: they are still searching, piling up arguments, some bad, some good. Which are which, we do not know for sure! But the bad ones will fall, some time or another, in the dumping grounds of scientific history.

Concluding

Dear colleagues,

My talk has ranged amongst concepts as different as astrology, ufology, and modern cosmology. Although there is no ultimate and final truth, there are at least some beliefs that are obviously fallacies, but very popular ones. Some of them may be a danger for mankind, and I feel that the best remedy is an always awakened lucidity.

To bring the students, the school children, to this lucidity, it is not necessary to teach in any systematic way the known misconceptions I have quoted. But one must give the students the elements of a critical look, by showing how the human mind has evolved throughout history. This history of ideas and the history of men and women of science, are, in this respect, illuminating; but one has not only to follow the course of events. One has also to explain the logic of the proofs: the proof of heliocentrism, the proof of the universality of physical laws, the proof of *etc.* . . .

And if I must conclude with one sentence, I should say again and again that teaching astronomy does not mean teaching only the digested facts, but also essentially means teaching the methodology of astronomy, helping people of all ages to understand what they know. Only in this way shall we teach lucidity.

References

Some (highly!) selected references of books or book-chapters dealing with similar questions:

Acker, A. 1977, *Initiation à l'astronomie*, Strasbourg, Université Louis Pasteur.
Acker, A. and Jaschek, C. 1985, *Astronomie: méthodes et calculs*, 112 exercises avec solutions, Masson, Paris. (Translated into English: Astronomical methods and calculations. J. Wiley, 1986).

Golay, M. 1985, in: *L'Astronomie Flammarion*, ouvrage collectif (J.-C. Pecker, ed.)
    2 volumes, Flammarion, Paris.
Nottale, L. 1985, in: *L'Astronomie Flammarion*, ouvrage collectif (J.-C. Pecker, ed.)
    2 volumes, Flammarion, Paris.
Pecker, J.-C. 1959, *Le Ciel*, Delpire, Paris (translated into English: Orion Press,
    New York; and in Italian, in German, in Japanese).
Pecker, J.-C. 1974, "Astrologie 1974," *l'Astronomie*, **88**, 241.
Pecker, J.-C. 1975, "On Astrology and Modern Science," *Leonardo*, **8**, 181.
Pecker, J.-C. 1983, "L'Astrologie et la Science," *La Recherche*, **140**, 118 et *les
    Cahiers Rationalistes*, 127.
Pecker, J.-C. 1985, "La Nature de L'Univers" in *Encyclopedia Universalis*, Sympo-
    sium, 279–285.
Schneider, J., ed. 1988, *Aux Confins de l'Univers* (collective book) La Nouvelle
    Encyclopédie, Fayard, Paris.

Discussion

G. Mumford: *Astronomers have established the boundaries of the constellations.
Since they do not believe in astrology they added another constellation to the zodiac?
Was this done just to confound astrology? (Tongue-in-cheek question.)*

J.-C. Pecker: Who knows?

S.R. Prabhakaran Nayar: *While discussing the movement of the Sun across the
background of constellations, it is worth discussing calendars followed in many places
where the month (the time spent by the Sun in one constellation) length varies
systematically to bring out the ideas of variation in the speed of the Earth on its
elliptic orbit.*

INFLUENCE OF CULTURE ON UNDERSTANDING ASTRONOMICAL CONCEPTS

Mazlan Othman
Physics Department, Universiti Kebangsaan Malaysia,
43600 UKM, Bangi, Selangor D.E., Malaysia

1. Introduction

Culture is defined as a society's system of values, ideology and social codes of behavior; its productive technologies and modes of consumption; its religious dogmas, myths, and taboos; and its social structure, political system and decision-making process (Coombes, 1985). The cultural environment in which an individual is raised is thus fundamental to the life view with which he perceives his world. Since his world view encompasses ideas of space and time, it is therefore not surprising that his understanding of astronomical concepts is inextricably bound to his cultural envelope.

This paper sets out to examine some of the ways cultural complexities affect the understanding of astronomical concepts and thus to make the case for a more realistic approach to the teaching of astronomy in a multicultural society, recognizing the importance of culture and the way students learn (Burger, 1973; Teynolds and Skilbeck, 1976).

2. Malaysia's Cultural Heritage

Malaysia is a multiracial country with each group practicing its own culture and religion. The main groups are Malays, Chinese, and Indians, each of which has its own very rich and long established culture. Each of these cultures is, in turn, very different from the others (Ryan, 1971). These cultures also have interesting myths, dogmas, and taboos (see for example Knappert, 1980).

3. The Study

In order to see how cultural belief may affect the student's concepts, the introductory astronomy class at Universiti Kebangsaan Malaysia was given a Malay legend with which they were all familiar and was then required to answer some questions. The legend given is as follows.

The legend of the Owl and the Moon (Tan, 1984): *A long time ago, the Moon fell in love with the Owl and married her. She went to live with him in the sky and they were very happy. Now, in those times the sky was very close to the Earth and there were seven suns instead of only one. One particularly hot day, a woman who was heavy with child came down from her house and she was striken* *by the heat so that she fell very ill. When her husband came back from the fields and saw what had happened, he was very angry. Taking his blowpipe and darts, he shot at the suns. One by one they fell, until finally there was only one sun left in the whole sky. The seventh sun, alarmed to see what became of his brothers, quickly soared up higher to escape a similar fate. In doing so, he pulled the sky up with him to the great height that it is today.*

Now, while all this was happening, the Owl had been combing her hair and she accidentally dropped her gold comb. Spreading her wings, she immediately flew down to Earth to recover it, but alas, when she rose to fly back to the sky again, she found that it had moved up too high for her to reach. Poor Owl, how she wept!

She cried out to the Moon, "Oh, my husband! Take me back into the sky to be with you!"

But the Moon only looked down at her helplessly and replied, "Alas, I cannot, for the rift between us is too wide."

This is why, people say, the Owl comes out at night and looks longingly at the Moon, for she had been pining for her husband ever since.

The students were asked: How does the story teller perceive i) the moon ii) the sun iii) the sky? Does his perception agree with yours? Explain.

4. The Results

The results of the study are summarized in Table 1. As the number of Indian students in the group was not significant, only the Malay and Chinese responses have been tabulated.

Table 1. Results of Students' Response (%).

| | | Cultural Group | |
		Chinese	Malay
Living Moon:	Not Contradicting	0	0
	Contradicting	100	100
Multiple Suns:	Not Contradicting	4	10
	Contradicting	96	90
Sky:	Not Contradicting	24	80
	Contradicting	76	20

The results show that the perception of the moon as a living object by the story teller was in contradiction with all the students' own perceptions. A small number of students in each cultural group thought that it may have been possible that there were more than one sun in Earth's history. However, a dichotomy appears in each group's concept of the sky. Interviews of some of the students who had found no contradiction were subsequently carried out to ascertain the reasons for the trends. The interviews revealed that the Chinese students who had stated no contradiction in the concept of the sky actually had not understood the story. Of the Malays in this group, a small percentage did not understand what the perception of the story teller was, while slightly fewer than 20% thought that the sun carried the moon with it because of gravitational attraction. The majority thought that the sky was organized into layers and therefore, in order to preserve the layer structure, the moon had to move up accordingly with the sun, which was on the same layer as the moon.

This result came as a surprise because it did not seem to matter that they knew that the moon circles the Earth, that the Earth circles the sun, *etc.* When they were required to look at the sky from the Earth, in their minds they still arranged the celestial objects in layers. This finding is similar to that reported by Cole and Scribner (1974). They found in their study of perceptions of people from various cultures that formal education was not the principal determining factor. Informal instruction in the home was found to have played a much larger role. In the teaching of astronomy, this is a very important point to appreciate because unless the misconception about space is cleared up right at the beginning of the course, the students will not be able to follow some of the logical deductions they have to make from observations.

5. Conclusion

This study has shown that although the students are taught a lot of facts, how they subsequently structure and reconstruct these facts in their minds depend very much on some deep-rooted beliefs that may or may not be logical. In order to teach astronomy well, it appears that one must first break some cultural bonds.

References

Burger, H.G., 1973. "Cultural Pluralism and the School," in Brombeck, C.S, and Hill, W.H. (editors), *Cultural Challenges to Education*, Oxford University Press, London, pp. 5-22.

Cole, M. and Scribner, S., 1974. *Culture and Thought*, New York, John Wiley and Sons.

Coombes, P.H. 1985. *The World Crisis in Education*, New York, Oxford University Press, p. 244.

Knappert, J., 1980. *Malay Myths and Legends*, Kuala Lumpur, Heinemann Education Books (Asia) Ltd.

Reynolds, J. and Skilbeck, M., 1976, *Culture and the Classroom*, Open Books, London.

Ryan, N.J., 1971. *The Cultural Heritage of Malaya*, Kuala Lumpur, Longman (Malaysia).

Tan, Lillian, 1984. "Malaysian Tales in The Beginning," in *Wings of Gold*, Malaysian Airlines System Publication, pp 52-53.

Taking his blow pipe and darts, he shot at the suns. (Extracted from Tan, 1984.)

Discussion

S. Torres-Peimbert: *What is the level of the students and how large is the sample?*

M. Othman: Students were from the *Introductory Astronomy* course given at the 2nd year level university and it's open to all students, science and non-science alike. Sample size was 102 students.

M. Zeilik: *Your information and your comments yesterday point out a possible strategy for teaching introductory astronomy: the use of ethnoastronomy, the astronomy of a particular culture. This approach has the advantage of starting with the familiar astronomy, comparing other cultures, and focusing on naked-eye astronomy and students' misconceptions about the sky. P. Sadler and Project Star's interviews are ethnographic interviews revealing the ethnoastronomy of a particular subset of U.S. culture.*

ON CREATIONISM IN THE ASTRONOMY CLASSROOM

Philip J. Sakimoto
Department of Astronomy, Whitman College, Walla Walla, Washington 99362 U.S.A.

Recently in the United States, a small but vocal group of citizens has been promoting a view known as "creationism," the assertion that the universe and everything in it was created in six days by divine fiat as suggested by a literalist interpretation of the Christian Bible. In a now common variant, "scientific creationism" or "creation-science," they assert further that this version of creation is proven true on "scientific" evidence alone, independent of any religious underpinnings. Legislation requiring the teaching of creation-science alongside of evolution in the public schools has been proposed in over twenty states; in Louisiana and Arkansas such legislation was passed, although it was later struck down by federal courts as a violation of the separation of church and state.[1,2,3,4]

Astronomers cannot afford to ignore the efforts of creation-scientists: like it or not, the movement has tainted the perceptions of science and religion that students bring to the classroom. Students from conservative Christian backgrounds are sometimes shy of legitimate science, fearing that science is somehow "anti-Christian." Students with strictly rationalistic/scientific world views sometimes scoffingly equate creation-science with all of Christianity, leading to potentially hostile relations in the classroom. In both cases, preconceived notions can hamper the spirit of free-flowing inquiry we seek in academe.

At Whitman College (a small, secular, undergraduate, liberal arts college in the Pacific Northwest), I have developed a *sotto voce* approach to addressing creation-science within the framework of a standard introductory astronomy class. With it, I hope to create a more honest inquiry into (and possibly acceptance of) legitimate scientific and theological perspectives.

On the first day of class, I assign a brief homework essay in which the students assess their religious or philosophical world views to see if they anticipate any conflicts between those views and concepts in astronomy. In making the assignment, which is taken directly from the textbook[5], I emphasize that the views expressed will have no bearing on the grade given; the grade is based solely on thoughtfulness and exposition style. I do not comment on the responses at this juncture; rather, I proceed with the class as if nothing significant has yet gone on. The students, however, have been forced to explicitly recognize their *a priori* prejudices; they must now wrestle openly with them as the class progresses, rather than holding them as unseen or unstated stumbling blocks to learning.

The next step is to teach my standard general astronomy class with an emphasis on the underlying methodology of the science: how do we know what we know? For example, I emphasize the classical "proofs" that the Earth is round, the connections between observations and models in stellar evolution, and the observa-tional evidence for the big bang. Exams are essay-style, forcing students to actively engage in the scientific process. The purpose here is to establish firmly the method-ologies of normative science; it is useless to talk about the methodological flaws in creation-science until the students have experienced normal science first-hand.

At the end of class, I give a single lecture on creation-science. The purpose of this lecture is not so much to "resolve the conflict" as it is to show by tracing roots of the arguments that there really is no need to have a conflict at all. I begin by quoting verbatim the tenets of creation-science as defined in Section 4a of the Arkansas Creation Science Statute,[3,6] reviewing the institutional structure of the so-called "creation research institutes,"[6,7,8] and quoting the statement of faith required for membership in the Creation Research Society in Ann Arbor, Michigan.[6] This unambiguously demonstrates the literalist Biblical origins of the creation-science movement.

I then trace the generally accepted Biblical origins of Genesis, beginning with the Babylonian mythopoeic tradition from which the creation story in Genesis is taken.[9] I emphasize that the intent of this story was to elucidate the relations between God, the created world, and human beings; it was never intended or seen as a scientific view of *how* the world was created.[10,11,12,13] I then review the history of the creationism movement, showing that it is a relatively recent and fallacious deviation from mainstream theology with roots in the "common sense" philosophy of John Witherspoon and Samuel Stanhope Smith.[14] My intent here is to show that (1) creationism has no basis in mainstream theology and (2) it is quite possible (and normative in theological circles) to believe in a world created by God without ascribing to the version of creation touted by creation-scientists.

With those theological aspects addressed, I turn to the scientific aspects of creation-science. I do not attempt a point-by-point rebuttal of creation-scientists' "scientific" arguments: life is too short to track down all of their fallacies. Instead, I criticize their methodology. They take a "two models" approach, insinuating that evidence *against* evolution is evidence *for* creationism. They use flawed argumen-tative techniques, including straw-man arguments, misapplication of physical laws, gross extrapolation, self-contradiction, and, in one case of which I am aware, false data.[15] They do not make their work available for critique by the scientific commu-nity: a recent survey of 135,000 manuscripts submitted to 68 scientific, technical, and education journals over a three-year period found only 5 manuscripts dealing with the creationist advances of which, due to poor presentation styles and failure to follow accepted scientific methodologies, none had been accepted for publication at the time of the survey.[16] In short, there is no point in considering the scientific arguments for creationism since there is no science in the arguments.

Together, these two lines of analysis show that creation science has no legiti-macy in either scientific or theological circles. To date, I have received no objections to my treatment of the subject, nor any direct rebuttals to my analyses. This is not surprising since my approach is identical to that which I use for all other sub-jects: I give a frank analysis based on the consensus of recognized authorities in the field. Students are by and large grateful to have creation science treated openly and honestly. In some cases, previously dogmatic Christians are led to a broader understanding of the role of science in modern theology. In other cases, previously anti-Christian students gain a greater sensitivity to the religious dilemma of strict creationists. Those who previously accepted both Christianity and modern science are affirmed in their beliefs. For myself, I am grateful for the opportunity to quiet some of the false antagonism between science and the Christian faith. Perhaps this will clear the path for those who might wish to pursue the wealth of scholarship now available on the legitimate relations between modern theology and the natural sciences.[17,18,19,20]

References

1. Roland Mushat Frye, "Creation Science Against the Religious Background," in *Is God a Creationist?: The Religious Case Against Creation-Science*, ed. R.M. Frye (New York: Charles Scribner's Sons, 1983), p. 1.
2. Joseph Palca, "No Go for the Creationists," *Nature*, **327**, 645, 1987.
3. Robert W. Hanson, ed., *Science and Creation: Geological, Theological, and Educational Perspectives* (New York: Macmillan Publishing Company, 1986), Appendices A & B.
4. Paul Reidinger, "Creationism in the Classroom," *Amer. Bar Assoc. J.*, **72**, Dec. 1, 1986, p. 66.
5. William K. Hartmann, *Astronomy: The Cosmic Journey, 4th ed,* (Belmont, CA: Wadsworth, 1987), p. 4.
6. *Science* (News and Comments section), **215**, 143, 1982.
7. Dorothy Nelkin, "Science, Rationality, and the Creation/Evolution Dispute," in *Science and Creation: Geological, Theological, and Educational Perspectives*, ed. R.W. Hanson (New York: Macmillan Publishing Company, 1986), p. 33.
8. Stanley L. Weinberg, "Creationism in Iowa: Two Defense Strategies," in *Science and Creation: Geological, Theological, and Educational Perspectives*, ed. R.W. Hanson (New York: Macmillan Publishing Company, 1986), p. 55.
9. Bernhard W. Anderson, ed., *Creation in the Old Testament* (Philadelphia: Fortress Press, 1984), especially for the articles by Hermann Gunkel and Gerhard von Rad.
10. Bruce Vawter, "Creationism: Creative Misuse of the Bible," in *Is God a Creationist?: The Religious Case Against Creation-Science*, ed. R.M. Frye (New York: Charles Scribner's Sons, 1983), p. 71.
11. Conrad Hyers, "Biblical Literalism: Constricting the Cosmic Dance," in *Is God a Creationist?: The Religious Case Against Creation-Science*, ed. R.M. Frye (New York: Charles Scribner's Sons, 1983), p. 95.
12. Nahum M. Sarna, "Understanding Creation in Genesis," in *Is God a Creationist?: The Religious Case Against Creation-Science*, ed. R.M. Frye (New York: Charles Scribner's Sons, 1983), p. 155.
13. Bernhard W. Anderson, "The Earth is the Lord's: An essay on the Biblical

Doctrine of Creation," in *Is God a Creationist?: The Religious Case Against Creation-Science*, ed. R.M. Frye (New York: Charles Scribner's Sons, 1983), p. 176.

14. James B. Miller and Dean R. Fowler, "What's Wrong with the Creation/Evolution Controversy?," *CTNS Bulletin* (Center for Theology and the Natural Sciences, Berkeley, CA), v. 4, Autumn 1984.

15. Examples of such flawed arguments are readily apparent to any practicing scientist who reads those aspects of creation-science literature touching on areas with which he or she is familiar. See, for example, *The Scientific Case for Creation*, or *Scientific Creationism*, both by Henry M. Morris and available from Creation Life Publishers, P.O. Box 15666, San Diego, CA 92115. The now-disavowed case of false data involved an alleged fossilized human footprint found next to a dinosaur footprint in a Texas riverbed. The human footprint turned out to be recently chiseled into the ground. I do not have references to this incident, and I would appreciate any documentation which readers can provide.

16. Eugenie Scott and Henry Cole, *Quat. Rev. Biol.*, **60**, 21, 1985; quoted by Roger Lewin, "Evidence for Scientific Creationism?," *Science*, **228**, 837, May 17, 1985.

17. Pope John Paul II, "Science and Christianity," in *Is God a Creationist?: The Religious Case Against Creation-Science*, ed. R.M. Frye (New York: Charles Scribner's Sons, 1983), p. 141.

18. Ian G. Barbour, *Issues in Science and Religion*, (New York: Harper and Row, 1971).

19. Ernan McMullin, ed., *Creation and Evolution* (Notre Dame: University of Notre Dame Press, 1985).

20. See also any issue of *Zygon: The Journal of Science and Religion*, or the *CTNS Bulletin* (Center for Theology and the Natural Sciences, Berkeley CA).

Discussion

L.A. Marschall: *What fraction of students seem "grievously wounded" by your course? What is their reaction?*

P.J. Sakimoto: I am not aware of any students that are "grievously wounded." This is because I am careful not to denigrate anyone's faith; rather, I encourage them into a deeper understanding of their faith. Judging from the responses to open-ended exam questions, a surprisingly large number of students — perhaps 20 to 30% — report that as a result of the course they are re-thinking their previous understanding of creation while holding firmly to their belief in God. Others — perhaps the majority — report no change in their views (they were mostly "evolutionists" already), but some of them mention a new-found sympathy for (but not belief in) the religious motivations of strict creationists.

SPATIAL THINKING AND LEARNING ASTRONOMY: THE IMPLICIT VISUAL GRAMMAR OF ASTRONOMICAL PARADIGMS

Lon Clay Hill, Jr.
Dept. of Physical Science, Broward Community College — Central, Fort Lauderdale, Florida 33314, U.S.A.

As several previous speakers have noted, astronomers are a rare breed. Even though we represent the oldest science, we are as rare as poets. The fact that our discipline is so distinct suggests that it should also have its peculiar pedagogical problems. I have been attempting to understand a pervasive element of all astronomical thinking — the use of spatial imagination to link celestial phenomena to terrestrial analogs. My work has two components — a theoretical explication of astronomical thinking and some practical use of these ideas in teaching situations.

If one looks at some of the fundamental "facts" discovered by astronomers, we will find that we often do not, in fact, directly verify our discoveries as scientists supposedly do. Primary astronomical "truths" such as the sphericity of the Earth, the heliocentric orbit of the Earth, the identification of the stars as distant suns, and the recognition that galaxies are enormous collections of stars, were not discovered by direct verification. Thus, we note that the Greeks discovered the Earth's sphericity nearly two millennia before its circumnavigation and that only a few thousand very bright stars in nearby galaxies have been resolved in our largest telescopes. Some of the primary objects of astronomical discourse, then, are not perceptible objects, but objects created by the spatial imagination.

Likewise, our methodologies are replete with analogies to perceptual experience that cannot be directly tested over the next few millennia. To cite a few, we assume that stellar parallax is similar to stereoscopic parallax, that the periodic diminution of the light of an eclipsing binary is due to occlusion and that the retrograde motion of planets and the proper motion of stars are due to differential-motion parallax. We are convinced of the appropriateness of these methodologies precisely because we see them as straightforward extensions of procedures taken for granted in our ordinary lives.

As astronomers, of course, we are successful precisely because we do not spend time wondering about these leaps of geometric and spatial imagination. However, as teachers of astronomy, we must make the spatial content of fundamental astronomical relationships *transparent* to students who do not possess unusual spatial thinking abilities. For my dissertation work at the University of Iowa, I am using both traditional and novel materials. This work includes extensive use of palpable models in both lecture demonstrations and laboratory exercises, an eclipsing-binary computer exercises in which students must choose pairs of stars that generate a set of given light curves, and a number of stereolessons that use stereograms to introduce students to important 3-D relationships. The underlying assumption is that if the students have a *direct image* of our "object" they can employ their full intuitive powers instead of merely those formal paradigms under their conscious control. If we as teachers thus make our spatial assumptions perceptually evident, we will find that our students will more surely use the seemingly technical tools of our trade.

References

Lon Clay Hill, Jr., "Astronomical Imagination and Stereoscopic Vision: Conscious and Unconscious Processing of Visual Data," *Southwest Regional Conference for Astronomy and Astrophysics* (Editors, Preston F. Gott and Paul S. Riherd). **9**, 67–76.

Discussion

Harry Shipman: *Physicists also face the problem of student learning styles and misconceptions that greatly interfere with the learning process and that are hard to deal with through simple lecturing. One solution they use is to use a faculty member in small discussion or problem sessions, either before or at the same time as the faculty member teaches a large lecture. When you teach in a large lecture — talking at rather than with the students — it's easy to remain blissfully ignorant of student misconceptions. Your paper, and the results of Project STAR, suggests that astronomy teachers, even if they have to lecture to hundreds of students, should make efforts to meet regularly with them in smaller groups.*

Chapter 9

High-School Courses

There is a long and honorable tradition of teaching astronomy in American high schools, a tradition whose history is discussed in the first two papers of the chapter. We read about the "ups" and "downs" of astronomy in the curriculum and some possible historical reasons. We see that it is important for astronomers to be included in broad curriculum planning. An interesting discussion follows this first pair of papers. The next chapter discusses a current project to improve high-school science teaching using astronomy as a base. Finally, we read about astronomy teaching on the high-school level in West Germany, Japan, and Bulgaria.

HISTORY OF THE TEACHING OF ASTRONOMY IN AMERICAN HIGH SCHOOLS

Darrel B. Hoff
Project STAR, Harvard-Smithsonian Center for Astrophysics,
60 Garden Street, Cambridge, Massachusetts 02138, U.S.A.

Early American colonists had a deep interest in astronomy. Between 1725 and 1764, Nathaniel Ames published the *Astronomical Diary and Almanac* at Cambridge, Massachusetts. This was a brief, widely-circulated source of astronomical information. It averaged an amazing 60,000 sales per year, while the better known *Poor Richard's Almanac* distributed only about 10,000 per year (Noble, 1970).

Practical skills were the dominant theme in early colonial schools. The astronomy that was taught dealt with natural phenomena such as phases of the moon, eclipses, and, for practical purposes, navigation and time-keeping. Astronomy was also frequently taught as a part of what we today would call physical geography courses. This practical nature of our culture dominated American schools until about the middle of the 1800's. Astronomy, as a separate subject, did appear in the curriculum of academies — forerunners of the modern American high school. A popular textbook of this period, *An Easy Introduction to Astronomy For Young Gentlemen and Ladies*, by James Ferguson, dated 1817, illustrates this point. It is almost completely devoted to the explanation of natural phenomena (Ferguson,

1817). It is a charming book, taught in classic dialogue form between the teacher and his students.

Specialized teacher-training institutions began in the United States in 1839. These were designed primarily to prepare elementary school teachers. A course of study included: orthography (spelling), reading, grammar, composition and rhetoric, logic, writing, drawing, arithmetic (mental and written), algebra, geometry, bookkeeping, navigation, surveying, geography (ancient and modern with chronologies, statistics, and general history), physiology, mental philosophy, music, constitution and history of the United States, natural philosophy and astronomy, natural history, the principles of piety and morality common to all sects of Christians, and the science and art of teaching. Commonly this was a one-year or, at most, a two-year program (Hinsdale, 1900).

The educational philosophy of "mental discipline" determined curriculum and teaching procedures during the second half of the nineteenth century, just as the influence of the common practical culture had exercised a dominant role during the first half. During the first half of the century, newer subjects such as astronomy entered into active competition with the classics. Conservative scholars decried these new subjects as too simple and poorly organized to meet the requirements of "mental discipline." This competition reached an acute stage near the middle of the century when a fierce controversy broke out in Great Britain between a group of brilliant scientists, on one side, and representatives of the great English public schools, on the other. The classicists maintained that the sciences were "shallow information subjects, lacking in organization, unsuited for discipline and altogether unworthy of the efforts of a high-minded school" (Noble, 1970). Faraday argued on behalf of the study of physical sciences for developing judgment skills and Lyell offered testimony to their contribution to the development of perception and reasoning powers. The scientists won, but the result was the emergence of science courses which began to take on "rigor" to the exclusion of simple phenomena. This "toughening-up" was designed to meet the requirements for "mental discipline." To illustrate the nature of the effect of this change, Burritt's popular *Geography of the Heavens*, published at the middle of the 19th century, still reflects a practical, observational approach (Burritt, 1854). Young's *Lessons in Astronomy*, published in 1895, takes on the rigorous appearance of more modern texts (Young, 1897).

I shall leave to Dr. Bishop (see the following paper) to discuss the effect of the "Committee of Ten," but I will share with you one set of science enrollment figures for American public and private high schools during the 1890's. These figures show that astronomy was taken by only about 4% of the students, geology by about 5%, chemistry by 9%, and physics by 20%. ("mental discipline," you know.) One should also keep in mind that many of the courses were not the standard year-long high school courses, as we know them today.

I personally believe that the figures generally represented the status of astronomy as a separate subject prior to the effects of the "Committee of Ten" and do *not* represent the immediate effect of the actions of the committee. I failed to locate any national figures prior to that year in the Harvard Library so I would be happy to be proven wrong on this point.

Table 1. Selected Science Course Enrollments in Public and Private
High Schools and Academies during 1897-1898[a].

Subject	Enrollment	Percent of Total
Astronomy	24,433	4.40%
Chemistry	47,448	8.55%
Geology	25,851	4.66%
Physics	113,650	20.48%

[a](Butler, 1904)

I do not dispute the ultimate effect of the committee's work. By 1920, astronomy had largely disappeared as a separate subject in American schools. A survey of high school curricula during the 1860's revealed that 14 schools in a sample of 20 taught astronomy. A similar survey done in 1915 showed only 1 astronomy course offered in the 40 schools sampled (Packer, 1924).

Dr. Bishop and I also differ strongly on the reasons for the re-emergence of astronomy as a separate subject in American high schools during the 1960's and '70's. I believe she would argue that the new availability of the relatively affordable planetarium dictated astronomy's re-emergence. I argue that cultural relaxation and the spirit of an "open society" during that period produced this re-emergence. While she may be right, I have the advantage of having been a member of the school and university setting at the time, and observed these changes first-hand. I do admit that I can supply no empirical evidence to support my view.

In the United States, Sputnik produced a cry for curriculum reform in schools during the early 1960's. One example of the nation's response to this crisis was the development of the Earth Science Curriculum Project, or ESCP, as it was called. This project, sponsored by the American Geological Institute and the National Science Foundation, developed and distributed a set of materials under the title of *Investigating the Earth*. The course is an amalgam of geology, astronomy, meteorology, and oceanography. Earth science, as a high school course, became very popular in the United States. A number of commercial texts were written, emulating its style and content. Currently earth science enrolls about 1.5 millions students annually in the eighth and ninth grades (Welch, 1984). An analysis of the content of the four most popular earth science textbooks currently in use indicates approximately 18% of a typical text is devoted to astronomy.

Prof. Owen Gingerich from the Harvard-Smithsonian Center for Astrophysics played a vital role in the development of another nationally-funded curriculum project, Harvard Project Physics. These materials were designed for a senior-high-school physics course and also included a section on astronomy. It is unfortunate that this text did not retain the popularity it deserved.

The sweeping curriculum reforms of the 1960's also reflected the "open society" of that time. Teachers were free to experiment with new courses; separate courses in geology, meteorology, oceanography, and astronomy (as well as other science courses) began appearing in the curriculum of American schools. Data obtained by a Project STAR survey reveals that many separate courses in astronomy had their beginning about twenty years ago. Project STAR's data suggests that about 15% of American schools offer a separate astronomy course. We know of approximately 1600 courses. I believe that this figure is a low estimate.

These courses, the survey revealed, are most often taught by a physics teacher. Earth-science teachers are the next most frequent teachers of these separate courses. I recently completed a survey of American earth-science teachers at the Harvard-Smithsonian Center for Astrophysics. Ten percent of the nation's 22,860 earth-science teachers were sent a short questionnaire inquiring about their teaching experience, course load, *etc*[1]. Twelve per cent of the respondents reported teaching a separate astronomy course in addition to the regular earth science course. If my results are valid, this would suggest that about 2700 earth science teachers teach a separate course in astronomy, in addition to those being taught by other science teachers.

Table 2. Separate Courses Taught by Earth Science Teachers.

Course	Percent
Astronomy	12%
Geology	10%
Meteorology	4%
Oceanography	3%
No Separate Course	71%

I have a great concern for the formal academic preparation in astronomy for our earth science teachers. I completed another survey three years ago in the state of Iowa. The results of that survey indicate that a third of Iowa's earth-science teachers have no formal preparation at all in astronomy. The average number of semester hours of astronomy in their college background is only about 3 semester hours — or one course (Hoff, 1988). Dr. Weiss' extensive work at the Research Triangle Institute suggests that nationally about 60% of the earth science teachers in the United States have no astronomy or space science college course work in their background (Weiss, 1987).

Clearly, there is considerable work to be done in the United States to improve high school astronomy instruction.

References

Burritt, Elijah H., *The Geography of the Heavens and Class Book of Astronomy*

---

[1]Survey conducted at the Harvard-Smithsonian Center for Astrophysics, Cambridge, Massachusetts, April 1988.

(1854), F.J. Huntington, New York, NY.

Butler, M.M. (ed.), *Monographs on Education in the United States* (1904), Louisiana Purchase Exposition Company, Albany, NY., p. 63.

Ferguson, James, *An Easy Introduction to Astronomy for Young Gentlemen and Ladies (1817)*, Benjamin Warner, Lancaster, England.

Hinsdale, B.A., *Horace Man and the Common School Revival in the United States* (1900), Scribner's, New York, NY.

Hoff, Darrel B., et al., "Earth science teaching in Iowa during the decade 1976-1986," *Journal of Geological Education,* **36**, 1988.

Noble, Stuart G., *A History of American Education* (1970), Greenwood Press, Westport, Connecticutt, p. 51.

Packer, Paul C., *Housing of High School Programs* (1924), Teachers College, Columbia University, New York, NY, pp. 8-9.

Weiss, Iris. *Report of the 1985-86 National Survey of Science and Mathematics Education*, Research Triangle Institute, November 1987.

Welch, Wagne L., et al. "How many are enrolled in science?" *The Science Teacher,* December, 1984.

Young, Charles A., *Elements of Astronomy* (1897), Ginn and Company, Boston, MA.

Discussion

C. Harper: *The renewed interest in astronomy in the 1960's would seem to be correlated with the development of the U.S. space program.*

THE COMMITTEE OF TEN

Jeanne E. Bishop
Director, Westlake Schools Planetarium, 24525 Hilliard Road, Westlake, Ohio 44145, U.S.A.

The college-educated populations of colonial 18th- and 19th-century United States were reasonably well-versed in principles of elementary astronomy — perhaps not fully in spatial concepts of seasons and lunar phases, though they certainly had some correct ideas about their causes. Astronomy had found a niche in academies and in the public high schools that succeeded them. In 1838 an observatory was installed in a Philadelphia high school. The states turned to the academies and early high schools for the majority of their elementary teachers, so the teachers who completed high school in the 19th century had been taught the reasons for common astronomical phenomena and conveyed these reasons to children. Parental teaching reinforced what was taught in primary schools. Until the late 19th century, colleges continued to offer astronomy or natural philosophy as part of the general curriculum. It therefore seems that a cycle of astronomy teaching and learning rudimentary astronomy was in effect until the late 1900's.

What caused the decline and virtual omission of basic astronomy from U.S. elementary and secondary schools for over fifty years, from about 1900 to 1950 and on? I conclude that it was the work of "The Committee of Ten."

In 1892, Charles W. Eliot of Harvard and other college administrators became concerned with inconsistent preparation for college; they requested that conferences be held by small groups of designated college and high school instructors in each of nine subject areas. The purpose of each meeting was to make recommendations for what should be taught in high schools that would be accepted for college admission. The resulting reports, published in 1893 by the U.S. Bureau of Education, became known as "The Report of the Committee of Ten."[1].

One conference or committee was designated, "Physics, Astronomy, and Chemistry," but the group of invited chemistry and physics teachers (no astronomy or natural philosophy teachers or professors) immediately changed the name to "Physics, Chemistry, and Astronomy." No reason for the change was given. The rest of this group's decisions also reflected the position that astronomy was of minor importance compared with physics and chemistry.

This group's most influential decision for astronomy was item #9 in its report: "That Astronomy be not required for admission to college." This followed "That Both Physics and Chemistry be required for admission to college." As with the name change, no rationale was provided for this set of declarations, such as those included with many of the important recommendations of the other subject conferences.

The "Physics, Chemistry, and Astronomy" conference was silent concerning what should be included in any astronomy course (if offered, the group said that it should be a 12-week elective) or unit offered in either high school or elementary school. In contrast, teams of researchers recommended 50 experiments for physics and 100 experiments for chemistry, which were printed with the conference report. The single astronomy teaching method advocated by the conference was #12: "That the study of Astronomy should be by observation as well as by classroom instruction." In harmony with this statement and as evidence of the widespread high regard for the Report of the Committee of Ten, several laboratory astronomy texts appeared soon after the Report was published.

Often, it takes quite a while for educational reforms to occur after recommendations have been made. But this was not the case with the report of the Committee of Ten. Comparisons of the percentages of students enrolled in high school astronomy between 1890 and 1950 show a very fast decline in the late 1800's and a slow continuing decline between 1915 and 1930, when the decline levelled off. The lack of recommendations for the study of astronomy in elementary school probably led to the disappearance of that subject from the elementary curriculum, except for what was taught about Earth motions under the heading of "Geography." The Geography subject conference did specify geography activities for elementary school, while the Natural History conference mapped out detailed elementary activities in botany and zoology.

I have concluded that the Committee of Ten orchestrated the disappearance of astronomy from U.S. schools and therefore from the common fund of knowledge. A review of astronomical knowledge of new college freshmen in 1940 showed that a large number of students did not know that the sun is a star, that the stars are farther apart than the Earth and sun, that the sun and stars move, that the planets move in noncircular orbits, that light takes time to travel through space, and that stars appear to move from east to west because of the Earth's rotation.

It seems that the Report of the Committee of Ten, whose precepts were followed in high schools, elementary schools, and colleges as if they were laws, was the cause of this chain of events: (1) Astronomy was dropped as a requirement for admission to college. (2) Students took chemistry and physics (and also biology), but not astronomy, since the former were required for college admission. (3) Astronomy was dropped from the list of courses of most high schools. (4) Astronomy was dropped from the list of required and offered courses in teacher-preparation institutions and many liberal arts colleges. (5) For over 60 years the average educated U.S. citizen largely was illiterate in astronomy .... In this period astronomy as a science was proliferating in leaps and bounds, and it was extremely exciting to those aware of the milestones: the source of solar and stellar energy, the extent of the Milky Way Galaxy, the existence of other galaxies, the expanding-universe interpretation of the red shift of galaxies, the birth of radio astronomy, and the dedication of the 5-m (200-inch) Hale Telescope on Palomar Mountain. But very little information filtered down to either high schools or elementary schools. Texts did not cover these areas, and teacher-preparation courses did not include them.

This brief discussion raises some questions ... Why were there no astronomers on the "Physics, Chemistry, and Astronomy" committee in 1892? Why did the recommendations of the Report of the Committee of Ten have so much influence? How might the effect have been avoided or reversed? How did people learn *any* astronomy in this "Dark Age of Astronomy" (first half of the 20th century)? And why was there a Renaissance in astronomy education beginning about 30 years ago? I have presented some of my ideas on these elsewhere.[2,3] I would welcome your insights on these questions, and on this topic generally.

[1] U.S. Bureau of Education. *Report of the Committee on Secondary School Studies.* Washington, D.C.: Government Printing Office, 1893.

[2] Bishop, Jeanne E. "United States Astronomy Education: Past, Present, and Future." *Science Education* **61** (3): 295–305 (1977).

[3] Bishop, Jeanne E. "Astronomy Education in the United States: Out from Under a Black Cloud." *Griffith Observer.* pp. 2–10. March, 1980.

Discussion

D. McNally: *One should, perhaps, not be too hard on the "Committee of Ten."*

They may have been responding (unconsciously?) to a wider trend. For example, in the UK, after the great era of 19th century popularization of astronomy by people such as John Herschel, Agnes Clerke, and Procter, astronomy in the early 20th century rapidly disappeared from the formal science educational curriculum. This pattern also seems to have appeared in other parts of the English-speaking world albeit at different rates. A possible underlying reason may be the excitement in physics research at the turn of the century.

J.E. Bishop: I agree that the perceptions of the Committee on Physics, Chemistry, and Astronomy may have been the same as groups elsewhere. However, I do think that other close countries and regions were influenced by the Report of the Committee of Ten. Possibly well-educated nonastronomers thought of astronomy as it was some 20 years before — positional astronomy with world-valued activities for time-keeping, navigation, and surveying. Astrophysics, by 1892, was earning a place as a proper field of astronomy (with much done in the United States), but few knew about it. You are right: physics may have drawn away the excitement.

J.M. Pasachoff: *The American Association for the Advancement of Science (1333 H Street, NW, in Washington, DC) has launched a major program, Project 2061 (named after the date of the next visit from Halley's Comet), to improve science and mathematics education across all school grades. Entitled "Education for a Changing Future," panels have prepared interdisciplinary statements of the kinds of knowledge that may be necessary in the future. Section D (Astronomy) of the AAAS (Jay M. Pasachoff, Retiring Chair; Virginia Trimble, Chair; and Arthur Davidsen, Chair-Elect) are among those representing astronomy. We will try to see that astronomy fares better than it did with the Committee of Ten.*

J.E. Bishop: It is important and admirable that the AAAS Project 2061 committee will work to avoid rejection of astronomy education in schools. I suggest that the group initiate and maintain channels of communication with school administrator organizations, teachers' organizations, and the National Science Foundation, which funds teacher workshops and programs. Hopefully scientist groups in other countries also will work with educational leaders and financial supporters to improve science and mathematics education generally and astronomy education particularly.

H.L. Neumann: *Astronomy education may have declined late in the 1800s because outdoor illumination was increased, so fewer people needed astronomical knowledge for everyday life.*

J.E. Bishop: It seems possible that some people felt so...maybe even Committee of Ten members.

W. Osborn: *This talk and the commentators have emphasized the importance of making one's views known when it comes to astronomy education. Our own state-specified curriculum in science for elementary schools is an example of what occurs when a committee specifies astronomy content in a curriculum without input from an astronomer or astronomy educator.*

J. Percy: *Astronomers should not* wait *to be invited to serve on curriculum advisory boards and committees. They should invite* themselves *to become involved as soon as they know that such boards and committees exist.*

J.E. Bishop: My reason for bringing this historical topic to the attention of the group is hope that awareness of it might motivate many to do exactly what Drs. Osborn and Percy recommend. Astronomers need to be better "cheerleaders" for their subject. I suggest these ways of becoming involved with pre-college science education: 1) offer to participate in programs or meetings for school administrators; 2) write enthusiastic articles about the nature and importance of science (with astronomy) for journals of school administrators, school boards, and educational regulation agencies; and 3) generally befriend and offer advisory support to officials in positions to make key educational decisions.

J.M. Pasachoff: *In American colleges and universities, at least, physicists often think that they can properly teach astronomy, and astronomy sometimes is forced to take a junior role in joint departments of physics and astronomy. Physicists, however, usually lack the breadth of astronomical experience to be able to tie together celestial topics from an astronomical point of view. Further, when astronomers lose their independence, their access to university administration and to funding is diminished. I feel that the smaller the department, the more important it is for the astronomers to be independent. After all, a relatively large group of astronomers can defend themselves. We astronomers should be emphasizing that astronomy is a separate science from physics, and that astronomy departments should be independent.*

SCIENCE TEACHING THROUGH ITS ASTRONOMICAL ROOTS

Philip M. Sadler and William M. Luzader, Project STAR,
Harvard-Smithsonian Center for Astrophysics, 60 Garden Street, Cambridge,
Massachusetts 02138 U.S.A.

1. Introduction

Project STAR (Science Teaching through its Astronomical Roots) emerged from growing concern about the decreasing enrollment in science classes, and lack of understanding of science and math concepts[1] (Figure 1). Using astronomy as a focus, Project STAR is trying not simply to increase the enrollment in high school science courses, but also to improve the students' understanding of science and its role in making sense of the world.

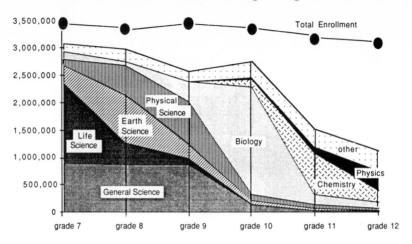

Fig. 1. Science enrollments in U.S. schools 1981-82 (from "How Many are Enrolled in Science?" The Science Teacher, NSTA, December 1984.

The educational approach of Project STAR is based on three principles:

- Mastery of a few ideas is more important for students than cursory exposure to many concepts.
- Students learn best through hands-on activities.
- Students enter the classroom with certain preconceptions, or "naive theories," about how physical systems work, without a sense of whether their understandings are accurate (often they are not).

To develop a curriculum around these principles, a team of scientists, high school teachers, and support professionals was assembled at the Harvard-Smithsonian Center for Astrophysics (CfA) in Cambridge, Massachusetts, and has been active since August 1985.

2. Goals of the Project

In the initial stage of the project, we have sought to develop three units of a high school course that would use astronomy as a basis for teaching principles of science and mathematics. Each unit involves approximately three weeks of classroom time. The materials developed and being tested are:

1. student activities and teacher guide,
2. computer software, and
3. video tapes and films.

3. Background

Before beginning the curriculum development process, we spent time looking into areas that could have a major impact on the effectiveness of our work. Much as a business extensively investigates a new product area, we explored opportunities to increase the probability of success.

A. Past Successes and Failures

During the last 30 years, there have been many attempts to produce popular and effective courses, through cooperation between university and public school educators. Some were successful and remain in use, such as BSCS Biology which is used in about 40 per cent of U.S high schools.[2] Some are out of print and no longer in use. Others, such as PSSC Physics and Harvard Project Physics, were initially successful but have declined in popularity. Some of the activities from these projects are still used as parts of courses.

These curriculum efforts used many different approaches to content, teaching style, laboratory activities, teacher training and dissemination. To benefit from their experiences, we conducted telephone and personal interviews with many of the principal investigators of these curriculum projects. These informal conversations revealed that:

1. Most teacher training concentrated more on the theoretical aspects of the curricula and proved to be ineffective.

2. There was little concern for the complexities of dissemination and marketing of the material.

3. Royalties from developed materials were returned to the NSF instead of to the project, making ongoing project activities, teacher training, and revision difficult.

4. Teachers had little time to assemble "hands on" materials and even less money for the repair and replacement of equipment.

5. The courses were directed at top students, thus neglecting the largest body of students.

6. New curricula were costly to develop.

B. Naive Theories

Recent educational research has revealed that students learn very little in their science courses.[3] Cognitive researchers have found that students learn science concepts much more slowly than their teachers believed. High-school science classes race through thousands of new words, symbols, definitions and concepts. Very few of these words, symbols, definitions, and concepts are actually understood by the students; almost none can be applied to new situations.

This research has shown that students have explanations for physical phenomena that are based on their own unique experience and logic. For example, when students and/or adults are asked about the cause of the Earth's seasons, most respond by saying, "The closer we are to the sun, the hotter it is." Very few even attempt an answer relating to the tilt of the Earth's axis, the varying length of daylight or the altitude of the sun in the sky, in spite of instruction to the contrary.

"Pre-conceived Knowledge" or "Naive Theories" can become critical barriers to the understanding of physics or any other science. Students' theories are notoriously resistant to change. Most students hold on to these explanations and never fully accept the more powerful scientific concepts. As a result, their intuitions about how things behave in the natural world never change and they cannot apply new concepts to their everyday experience.

Project STAR provides an opportunity to uncover the "naive theories" that students (or even adults) have about basic astronomical concepts. We attempted to develop materials and teaching methods that could effectively challenge these theories and replace them with more powerful, accepted ideas.

4. The Team

The Center for Astrophysics is a unique environment. At this one location, there are experts in almost every area of astronomy and astrophysics. Many are involved in state-of-the-art research and have distinguished teaching careers at Harvard. Several are accomplished authors as well. We have enlisted several of these individuals in an ongoing advisory capacity. This Science Panel has met weekly since the beginning of the project to discuss details of the program and to review materials that have been developed.

Local and national astronomy teachers have also been a key resource of Project STAR. Their experience and insights have helped move the project along more rapidly. In monthly meetings, local teachers provide feedback on what concepts students can learn and what materials work in their own classrooms. The teachers from outside Massachusetts keep records of their daily experiences with the students and the materials. These records are then sent to the project staff for evaluation and for incorporation in revised materials. During the 1986-87 academic year, nine Massachusetts teachers helped develop and then used our materials. During the 1987-88 academic year, they were joined by 17 additional teachers from around the U.S. in revising and trial teaching revised materials. Approximately 700 students used these new materials.

Project STAR has been fortunate in attracting teachers to take sabbaticals as project associates. In this way we have managed to have those with current classroom experience developing materials, leading workshops and working with CfA scientists.

5. Projects

A. Astronomy Education Survey

Before beginning our curriculum development, we wanted to find out how astronomy was being taught in high schools around the country. How much information existed regarding secondary level astronomy education? Where was it being taught? Who was teaching it?

Reference to astronomy in the nation's schools can be found in the *1977 National Survey of Science, Mathematics and Social Studies Education.* In schools with grades 10–12, astronomy was offered as a course in 6 per cent of the schools to 46,375 students. The number of schools with astronomy courses, 1,318, could also be interpreted as the minimum number of teachers who teach astronomy, since at least one teacher is needed to teach one course in a school.

An initial telephone survey of the school systems within the Greater Boston area (enclosed by Route 128, a circular highway about 20 km from downtown), showed that there are 6 high schools out of 54 school systems with separate astronomy courses. Extrapolating this ratio to the nation, the 16,000 school systems may have as many as 1800 astronomy teachers. While our sample was too small and non-random for any such conclusion to be reliable, we do infer that there are about 1,500 teachers and about 50,000 students involved in astronomy.

i) Initial Census — May 1986

To determine more accurately the extent to which astronomy is being taught in the nation's schools, we secured the names and addresses of 11,100 science department heads from the National Science Teachers Association and mailed each one a short questionnaire. The questionnaire included a very brief description of the project's goals and six questions related to the teaching of astronomy. These cards were mailed postpaid on May 15, 1986. Within four weeks, 23 per cent of the cards were returned and we learned that 15 per cent of the responding schools offer separate astronomy courses. The results have helped to develop a picture of the role that astronomy plays in secondary schools.

Fifty-six per cent of the responding high schools offer astronomy to their students as part of Earth Science, Earth and Space Science, or Physics. Over 15 per cent offer a separate astronomy course for one or two semesters. The large majority of Science Department Heads (76 per cent) would like more astronomy to be taught in their schools.

The national significance of these figures should not be underestimated. The number of qualified physics teachers in U.S. high schools has been dwindling for the last ten years. The best estimate from the American Association of Physics Teachers is that only about 8,000 remain.[4] By supporting and increasing the number of astronomy teachers in the nation's schools, it might be possible to attract more students to physical science in eleventh or twelfth grade. At present, only about 35 per cent of high school students take a physical science course (chemistry or physics) before graduation.[5] One approach to get more students into science classes is to increase their options.

ii) Teacher Survey — June 1986

One of the key questions on the census card was: "If an astronomy course is offered, who is the teacher?" Among the 2400 respondents, our census identified

408 teachers who taught courses of one semester or longer. The Project then mailed to these teachers a more detailed eight-page questionnaire.

Of the questionnaires sent to astronomy teachers, 62 per cent responded before school closed. We believe we gained thereby a strong sense of the attitude and needs of astronomy teachers. We estimate that 800 to 1200 more astronomy teachers remain to be identified.

The follow-up survey of the 408 identified teachers uncovered several remarkable facts. Most of these teachers are aware of the lack of curriculum materials and of any high school level textbooks. They have adapted by using college-level texts and developing their own courses and curricula. Today's astronomy teachers are a dedicated group of science educators. Most of them teach other science courses, and most of them have little or no interaction with other astronomy teachers elsewhere. Moreover, most of the astronomy courses are taught by the teacher who developed the course, and when that person no longer teaches it, the course is often discontinued. The most striking response to our survey was one of unabashed enthusiasm. "Finally," one of the teachers said, "we will have a forum."

These teachers had varied backgrounds. One might think that all astronomy teachers would also teach physics. Although a plurality did, all the other sciences were represented as well. Even some mathematics teachers are teaching astronomy (Figure 2).

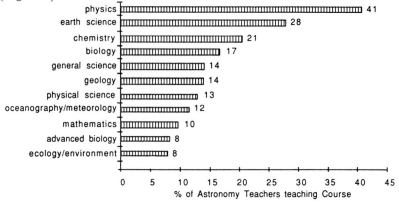

Fig. 2. Other Courses Taught by Astronomy Teachers

Most of these teachers (77 per cent) consider astronomy their hobby and many own telescopes and subscribe to popular hobbyist magazines. Thirty-six per cent of the astronomy teachers are members of astronomy societies or organizations. Thus, astronomy is not only a course they teach but also a lifelong avocation. By comparison, few high school science teachers consider themselves amateur physicists or amateur chemists. Many teachers have enough confidence in their knowledge/mastery of the subject matter to develop a unique astronomy course. Eighty per cent have written their own curriculum materials and laboratory activities. Only 14 per cent use a commercial text. As far as we know, no other subject generates this sort of grassroots interest. These facts demonstrate the interest and dedication of astronomy teachers.

Slightly more than half (57 per cent) of the teachers teach one astronomy class; the remainder teach from two to seven classes. Most courses are one semester long, but several schools offer full year courses. The average class size is 25 pupils.

The teachers identified four areas in which they were most dissatisfied. They want:

- Higher quality student and classroom activities and computer programs.
- A student workbook in astronomy (the need most often cited).
- Summer workshops in astronomy.
- Astronomy education newsletter or association.

In individual interviews with 25 astronomy teachers, we found that almost all thought they were the only teachers offering astronomy courses in their region of the country. Even teachers who taught in adjacent communities rarely knew of each other's existence. Their strong desire for a newsletter or association serving their needs reflects this feeling of isolation.

High-school astronomy teachers represent an extraordinarily dedicated group. Our goal has been to help support and expand the astronomy offerings. Moreover, teachers interested in teaching astronomy can be supported by these dedicated teachers. By sponsoring workshops, we are encouraging the teaching of more astronomy courses. Experienced and enthusiastic astronomy teachers are the natural choice for leaders of, and speakers at, these events. Many teachers have requested materials for integrating astronomy activities and topics into courses other than astronomy. These teachers could also help to lead the effort to incorporate more astronomy into earth science, chemistry, physics, and mathematics courses in the nation's schools.

B. Resource Review

Between 1956 and 1975, more than $387,000,000 was spent by NSF in efforts to produce and disseminate curriculum materials.[6] It has been very costly to write entirely new material, so we have worked to adapt existing materials whenever possible.

To aid this effort, Project STAR has assembled a resource library of astronomy education materials appropriate to our target audience. It includes teaching materials, articles, media, computer software, texts, and activity lab manuals. Research materials have been gathered, including dissertations, articles, studies, and reports on astronomy education. Teaching support materials have been acquired when practical, including posters, photographs, slide sets, simple models and equipment. Many of these materials were acquired at no cost by requesting them from publishers or individuals. These materials have been an excellent source of ideas.

C. Summer Institutes

i) July 1986

For STAR's first Summer Institute, a team of consulting teachers, along with the project's scientific and support staff, worked with consultants, authors, educators and workshop leaders for four weeks at the Center for Astrophysics. The goal was to establish a unified approach to the creation of the Project STAR curriculum. This Institute was intended to be the first of a continuing series of summer institutes aimed at further developing teacher competency in astronomy and space science.

The Summer Institute consisted of three integral parts:

1. Lectures and workshops on the latest developments in astronomy to provide the teachers with the opportunity to interact with researchers;

2. Workshops on astronomy education, centered on the materials and resources already developed by the Project STAR group at the Center for Astrophysics; and

3. Working sessions where new astronomy teaching materials were developed by the participants.

The primary goals were to help the participants to become more effective science teachers and to develop them as leaders to expand high school astronomy offerings in their locales. These participating teachers have served not only as resources to STAR but also as resources to their home communities by dispersing the ideas learned in the Summer Institute to fellow teachers in their local systems.

The teachers worked in three teams to draft the three curriculum modules: *The Solar System, Stars,* and *Galaxies.* They wrote activities and used overheads and slides to present their results at the last session of the Institute. The Institute was a great success in terms of building a team and establishing direction.

ii) August 1987

For the 1987 Summer Institute, 18 astronomy teachers from around the country came to the CfA for workshops about naive theories, STAR activities, and lectures on recent research on astronomy and astrophysics. They learned to teach the STAR curriculum materials and advised the staff on revisions. During the 1987-88 academic year, they used these materials in their own classrooms and administered tests to determine the materials' effectiveness. After using the materials they submitted detailed critiques and suggestions based on their experiences.

D. Computer Applications

i) Microcomputer-Based Spectrophotometer

A microcomputer-based spectrophotometer has been developed for Project STAR by the Technical Education Research Center (TERC). It has a broad appeal not only for astronomy classes but for physics and chemistry classes as well.

This real-time apparatus facilitates a wide range of experiments involving light. Using a linear photodiode array, a prism or grating, and some simple electronics, it generates a color display of the spectrum for many light sources: gas discharge lamps, filament lamps, flourescent tubes, the sky, the sun, filters, and fluorescing materials.

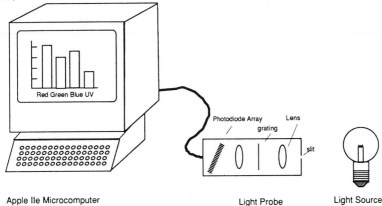

Fig. 3. Microcomputer-Based Spectrophotometer

ii) Spreadsheet Astronomy

We experimented with the use of computer spreadsheets and graphics to teach astronomical concepts. With only a few direct measurements on each of the 500 brightest stars in the sky, various graphs and charts could be produced.

A plot of right ascension vs. declination of these stars will produce an equatorial star chart. This map was used in an activity to determine the structure of the Milky Way Galaxy. This same information is used to produce charts that plot stars from a polar point of view for both the northern and southern hemispheres. These charts are cut, folded and taped together to create a spherical star map. They are used in the construction of the student celestial spheres, referred to the Low-Cost/High-Tech section.

iii) Desktop Publishing

Project STAR must produce hundreds of documents, involving text, tables and graphics. Previous curriculum projects have employed graphics artists, typists, typesetters and illustrators. Many have used outside services, which are costly and can slow production. Project STAR uses a modern and efficient desktop publishing system, the Macintosh Office with an Apple Laserwriter. This technology saves time and money, streamlines the production process and allows a much larger number of contributors to be involved in the writing process. Our newsletter, *STARnews*, is produced in-house.

iv) Animation Software

Microcomputers have tremendous power to simulate systems. Existing software can provide students with many experiences which are impossible or difficult to transmit in a laboratory, planetarium, or classroom. "Planetarium" programs can set up the sky for any season, time or location. Students can make measurements of the positions of stars, planets, the sun or the moon. Microcomputers can simulate gravitational systems from a single rocket in space to many-body problems. Drill and practice software can teach students basic measuring techniques used by astronomers.

For example, gravity simulations of galaxy "collisions" were created by generating individual frames with a Pascal compiler. The frames were then combined using software written to make computer movies. Students will be able to watch these simulations to get an understanding of gravitational interaction on a galactic scale.

v) Models of the Solar System

Below are frames from two animations showing the motion of Venus. The frames show the Earth-moon-sun system from above and the appearance of Venus with respect to the sun as seen from Earth. Students can compare a heliocentric model (Figure 4a) with a geocentric model (Figure 4b). Students can experience firsthand how observation can be used to test a theory. They can view these two simulations, then make direct observations to determine whether a theory is consistent with the data.

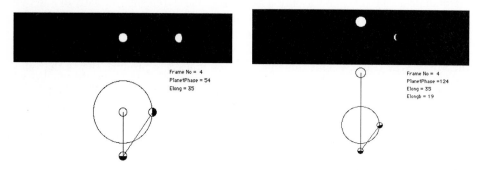

Fig. 4a (left): Heliocentric model of the Earth, Venus, and the sun; Fig. 4b (right): Geocentric model of the Earth, Venus, and the sun.

These simulations run very slowly, even on high performance microcomputers such as the Macintosh. Although they will run much faster on supermicros, and presumably on the next generation of normal microcomputers, interactive capability is not currently possible. Frame generation time is from 10 to 60 seconds. We have converted several simulations to videotape for classroom trials. Other possible simulations include moon phases, the altitude of the sun throughout the year, and the path and speed of comets.

E. Naive-Theories Film

Over the last few years, science teaching literature has produced many articles about student misconceptions and naive theories. We are convinced that students' prior beliefs about the world play a major role in the learning of science. Many researchers have found that paying attention to the preconceived ideas of students is a very effective strategy for science teachers. Many teachers, however, are unaware — or at least tend to minimize — the role of these ideas. They often attribute much more understanding to students than they possess, seemingly because students use scientific terms and phrases without really understanding them.

Dr. Matthew Schneps has produced a video entitled "A Private Universe" that demonstrates the powerful role that these preconceived ideas play in learning science. In interviews with college seniors at graduation and with high-school freshman before being taught about the seasons, it was found that both groups have the same misconceptions about the cause of the seasons on the Earth. The high-school students are interviewed again after instruction. The film illustrates the role that model building and discussion play in student learning.

"A Private Universe," which is being distributed by Pyramid Films, has won a Gold Medal at the Houston International Film Festival and a Silver Apple at the National Educational Film and Video Festival in Seattle, Washington. It has been used extensively for teacher workshops during the 1987-88 school year, where it stimulated interesting discussions about the learning and teaching of science.

6. Curriculum Development

A. General Outline

Project STAR has developed a unique set of materials for high school astronomy teachers. Built around the idea that students best learn science by confronting their models of scientific concepts and applying them in both astronomical and terrestrial situations, we have concentrated on the following conceptual areas *Orientation in Time and Space, Laws of Nature and Nature of Light*. Each conceptual area is discussed from three perspectives: *The Solar System, The Stars*, and *The Galaxies*. Each perspective of a conceptual area constituted a single STAR Module (creating nine STAR modules). STAR Modules are materials which can be taught in two to four weeks, and include teacher notes, laboratory supplies and students workbooks.

i) The First Three Modules

The first set of activities to be prepared related to *Orientation in Time and Space*. Each activity begins with a prediction in which the students make a commitment to their misconceptions about nature. This prediction is addressed during the activity so the students will discard their misconceptions and accept more powerful concepts. The specific modules developed for this unit were:

The Solar System: The concepts of measuring distance and size are developed by angular measurement of nearby objects and astronomical objects using similar triangles. Students used this information to build scale models of the solar system.

The Stars: This module begins with an exploration of how the intensity of light decreases with distance. Using simple equipment, students measure light in the classroom, then progress to measuring the luminosity of the sun and stars. By performing simple calculations, they then estimate the distance to nearby stars. Using geometric methods developed in *The Solar System* module, students compare the results of distance measurements made with different methods.

The Galaxies: By classifying various deep-sky objects and plotting these objects on a map of the sky, the students construct a model of our own Galaxy. The sizes of, and distance to, various galaxies are estimated using the methods developed in the first two modules. From reproductions of Palomar and Lick Sky Surveys and photographs from the Harvard Observatory plate stacks, students make calculations and build a model of our local group of galaxies.

ii) Future Modules

Activities for the other two units, *Laws of Nature* and the *Nature of Light*, were outlined by the teachers at the 1987 Summer Institute:

Laws of Nature: orbital motion and gravity, gas laws, nuclear energy;

Nature of Light: spectral analysis, temperature and color of stars, sizes of stars, stellar evolution

The activities for these units will be produced under the proposed extension of the project.

B. Student Journals

In keeping with the "learning by doing" philosophy of Project STAR, student journals were used as a tool for promoting students' understanding of science concepts. Project STAR consulting teachers used two such techniques: Moonwatch and Sunwatch exercises and Moon Journals. In the Moonwatch and Sunwatch, students draw the surroundings of their observation spot and record the position of the sun or the moon over a period of two weeks to three months. They acquire an intuitive understanding of the paths of the sun and/or the moon before clarifying it intellectually through class discussion. The Moon Journal is a record of daily observations of the moon — not just its position, but other details as well — shape, color, thoughts about past observations, questions, *etc.*

The more detailed their observations, the better. This exercise stimulates them to think more carefully about the causes and effects of the phenomena they observe. They notice things that they would not have noticed before. They understand more about how things work and they have more questions.

C. Low Cost Equipment for Activities

One of the responses to the teacher questionnaire was the lack of student activities involving observation and prediction. Thus the project designed several inexpensive and versatile pieces of equipment that have been tested in participating schools and at teacher workshops. These activities have become the most popular parts of the curriculum.

i) Celestial Spheres

Astronomy is a very "spatial" science. However, most students have great difficulty mastering the spatial relationships posed in most textbooks. STAR has developed a very inexpensive (< 50 cents) celestial sphere that students construct themselves.

The celestial sphere is assembled from two vacuum-formed clear plastic hemispheres on which the students trace the position of the stars from a spherical star map. A reformed metal coat hanger supports a small earth globe inside the hemispheres to form the completed sphere.

As the sphere is built, the students practice constellation identification. After it is built, a pin, representing the sun, is placed at various places along the ecliptic to model the changing position of the setting sun for a period of a few months. It may then be used to predict future positions of the sun and the visibility of certain constellations at various times during the year. A single hemisphere is used to track the daily motion of the sun at different days during two seasons.

ii) Telescope

An essential part of any introductory astronomy course is an explanation of how a telescope works. For under 50 cents, each student can make a telescope comparable to that which Galileo used to make his first astronomical observations.

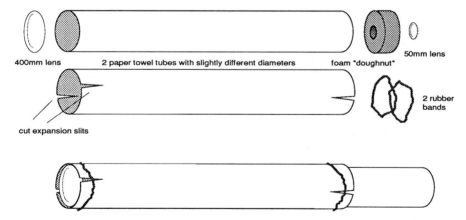

Fig. 5. A simple, inexpensive telescope for an introductory astronomy course.

The molded plastic lenses and foam insert are inexpensive, while the rest of the materials may be collected in the student's home. Also, by adding a simple reticle, measuring angular sizes becomes a straightforward task. Using this instrument, the student can make quantitative observations of terrestrial objects and the moon's surface.

iii) Luminosity of the Sun

How bright is the sun? How far away are the stars? The inverse-square law is a fundamental tool in answering these questions. Some interesting and useful activities can be carried out with an easy-to-build device called a *null photometer*, which was used in the PSSC Physics lab exercises (Figure 6). Using this device, students observe that nine lights are needed to illuminate a surface three times the distance of one light, thus verifying the inverse square law. Given the Earth-sun distance, measuring the photometer-bulb distance, and then using the inverse square law, the students can also determine the power output of the sun by comparing it to a known source such as a 200-watt light bulb (Figure 7).

Fig. 6. A null photometer. When the photometer is illuminated equally from both front and back, the aluminum foil "disappears."

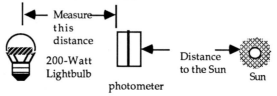

Fig. 7. A simple experiment to measure the power of the sun, using a null photometer.

The inverse-square law is also used to determine the distance to a bright star. For this activity the students build an artificial star from optical fiber. This device, when held at the proper distance, matches the apparent brightness of some star in the sky. They use an ordinary flashlight (with two D-cell batteries), a piece of plastic thread (a length of optical fiber), some clear tape, and a small sheet of aluminum foil) (Figure 8).

By assuming that a star emits the same amount of light as the sun, the student can estimate the distance to some of the brightest stars. Even though the numbers were not exact, this approach gave results to the correct order of magnitude.

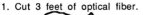

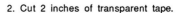

3. Tape fiber across face of flashlight.

4. Cover end of flashlight with aluminum foil so that no light escapes.

Fig. 8. An experiment to estimate the distances to some bright stars using an "artificial star."

7. Evaluation

A. Formative: trial schools

Feedback from teachers and from students has been essential to the improvement of our curriculum materials. Local astronomy teachers met one Saturday each month during the school year to discuss the units and activities that they tried in the classroom. At these meetings, they shared new ideas for presenting the materials as well as the difficulties encountered by their students.

The national teachers kept records of how the presentations went with their students. The journals contained comments on how the students reacted to the activities, suggestions on improvements to the activities, and the time it took to complete an activity. The following are quotes from the teachers journals:

> "Project STAR does what it was intended to do, *i.e.*, to provide average to above-average 11th and 12th grade students with meaningful astronomy lab activities."

> "...students, as a whole, enjoyed the STAR program..."

> "What is being done in these chapters is better than the usual textbook..."

> "...my reaction ...is a positive one. Student response ...was also very encouraging."

The most popular activities were:

1. building scale models of the Solar System;
2. building the Celestial Spheres and using them to explain the seasons; and
3. using the Inverse-Square Law to determine the distances to the stars.

The least popular were:

1. passing around an Earth globe to demonstrate the seasons;
2. using the North Star to determine latitude on Earth; and
3. classifying astronomical objects

Some of the comments from teachers concerning desired improvements were:

"...text was missing ...the math will need to be reduced or explained more simply."

"...hard to do activities all of the time."

And of course, one student stated:

"...I wasn't interested in knowing about how astronomers have found out things."

B. Summative: pre-tests and post-tests

Both before and after a class had worked through Project STAR materials, the students were tested on their general astronomy knowledge. The pre-test and post-tests are the same. This method allows us to represent our effectiveness statistically.

The data for the pre-tests and post-tests are still being collected. Those received so far suggest a slight increase in the scores from before to after. However, the statistical analysis of those data will not be straightforward. For example, there are questions about the validity of the control groups. A more complete analysis of these data will have to wait for the extension of the project.

8. Dissemination

A. Conferences

Members of the STAR staff, Science Panel, and individual consulting teachers have presented 45 papers and workshops. We estimate that a minimum of 3,000 teachers have participated in these various workshops and paper presentations.

B. Newsletter

STARnews, the project newsletter, keeps us in touch with our constituency. Each of the four issues was distributed to approximately 2500 people, including astronomy teachers on the STAR mailing list and to 300 employees of the Center for Astrophysics. The newsletters included:

- articles by staff and scientists of the CfA
- news of Project STAR current projects
- an "Activity of the Month" for teachers and students

- profiles of the staff and others involved with Project STAR
- classified ads which solicited help from others

C. Publishers

Wide adoption and use of the materials we develop will depend to a large extent on our ability to publish and market a successful and popular new textbook. We are exploring the opportunities for a collaborative effort with an appropriate publisher of national standing. Project STAR has contacted the major publishers of science textbooks in the United States. A new text, backed by the marketing expertise of a knowledgeable publisher, would help to achieve widespread use of these new materials.

In all, two dozen publishers were contacted. Several have given positive reactions. Negotiations are continuing with Kendall-Hunt Publishing Co. of Dubuque, Iowa, and the Smithsonian Press in Washington, D.C.

References

[1] Welch, Wayne W., Linda J. Harris, Ronald E. Anderson, "How Many are Enrolled in Science?" *The Science Teacher*, National Science Teachers Association, December 1984.

[2] "What are the Needs in Pre-college Science?" National Science Foundation, 1979, p. 90.

[3] McDermott, Lillian C., "Research on conceptual understanding in mechanics," *Physics Today*, July 1984.

[4] Private conversation with Jack Wilson, executive director of the American Association of Physics Teachers, July 1986.

[5] Welch, W. W., *et al.*, *op. cit.*

Discussion

B.W. Jones: *Have any follow-up studies been done on the students' understanding, by comparing those who have been through STAR with a control group that has not? Such a follow-up should be done well after completion of the course. It would be alarming to find that the STAR program had little long term impact on understanding.*

P. Sadler: Follow-up studies have only been made just after course completion: the STAR group performed significantly better. It would be difficult to do a longer-term follow-up, because the classes are dispersed.

G. Verschuur: *A general measure of angle that is useful for everyone is that a fist held out at arm's length subtends an angle of 10°. This fact is useful since the sun moves 15° per hour, that is, one and a half fists' width, a useful way to figure the amount of daylight before sunset. It may be worthwhile using both a fist's width and a 2° finger's width (for smaller angles) to allow students to estimate angles.*

Comment by J. Percy: I have found that, even with university non-science students, angular measure is one of the most difficult concepts for students to understand.

R.J. Dukes, Jr.:

1. *Until all college astronomy students have been exposed to your materials in high school, these materials are useful at the college level.*

2. *We need to work on astronomy (and all science instruction) at the elementary level. Elementary students are fascinated by science. Between 4th and 8th grades, we (teachers) kill this interest. This is directly responsible for the decrease in astronomy enrollment after 10th grade. We need projects similar to STAR for elementary schools.*

G. Vicino: *I agree with all the ideas on which Project STAR is based, and I invite all the Project STAR crew to know the work we are doing in Uruguay. We have a full-year course of astronomy in our high schools (for pupils 15 or 16 years old) and our program is one of active learning, with a course full of experiments and observations to teach the pupils to think, not to get the knowledge given by a teacher and received in passive form.*

N. Sperling: *Project STARs sense of proportion seems stuck at the astronomy of 100 years ago — all positional. Most astronomy these days reveals the physical nature of objects, which my students find much more interesting than their locations. Surely modern science principles can be revealed in a superior manner by examples drawn from modern science! Confine the positional astronomy to exercises already developed, and develop physical models next. Try differentiation (a grand unifying principle of planetology). Try stellar evolution from protostellar nebulae to protostellar nebulae. Try nucleosynthesis. Or, for instrumentation, try the tradeoffs in telescope design (see my article "Of Pupils and Brightness," Griffith Observer, January 1985). Try variable stars — apply to classrooms the participatory planetarium program from Lawrence Hall of Science, University of California, Berkeley. Try atmospheric evolution (of the Earth, probably).*

D. Hoff: In response to Mr. Sperling's remarks, your concern for inclusion of more modern topics is desirable. Project STAR is concerned with modern topics. We do not propose a modern course. We propose to teach *Science Through its Astronomical Roots*. That is our name. As a result, our material will not be a fact-laden course. We believe that less may be more.

W. Luzader: Research indicates that most students are not learning positional astronomy by using the "exercises already developed." Phases and seasons should be a part of an introductory astronomy course because people do have strange (if not incorrect) models of how these work and they are natural phenomena that can be easily observed. Any activity that encourages students to look at the real sky and follow real events in nature should be used in an introductory course.

How can students "understand" differentiation if they do not know the sequence of planets from closest to farthest from the sun or how that is important to any model of solar system formation? In addressing the concept of nucleosynthesis, is it more important that the student "understand" the difference between fusion and fission and how that relates to the power plant down the road or "learn" the proton-proton or carbon-nitrogen cycles?

Some of the topics mentioned by Mr. Sperling will be addressed in later activities. It should be kept in mind that the existing activities comprise about a fourth to a third of the total materials that will eventually be developed by Project STAR for a full year, high school level, introductory astronomy course. Topics will be chosen and activities written in a way such that their everyday applications can be presented and that certain basic scientific concepts be understood.

R.R. Robbins: *The training of teachers to use the materials is especially critical. I recall the death of excellent materials written by Stanley Wyatt and others at the University of Illinois, because the teachers didn't know how to use them.*

By the way, I do not agree with the idea that the Wyatt materials were not appropriately designed. They did many of the same things.

W. Bisard: *An interesting extension of interviewing students about misconceptions is to interview teachers in general. My experience with "teaching teachers" says we must help them and future generations of teachers.*

B. Riddle: *We heard a general discussion about modeling and other activities to prepare students at high school level. Comments made concerning need for lessons at elementary school level were also voiced.*

Project STARWALK (see poster display) is a program that involves elementary school teachers and their students in lessons centered around three seasonal visits to a planetarium facility. Lessons are designed as classroom activities to prepare students for planetarium visits, and as followup exercises to reinforce and consolidate planetarium lessons. The visit to the planetarium is designed as a laboratory exercise based on concepts dealing with seasons and time. Seasonal stars and constellations are also studied during each visit.

G.A. Carlson: *Many textbooks I have used stress that the Earth-moon system is a "double" planet system, and that the moon's orbit is always concave to the sun, never convex. Should I, in an introductory service course, be presenting this concept? From your brief presentation, I got the impression that the concept of "the moon goes around the Earth" is the way to go. How much of the Earth-moon concept do you feel we should present, realizing that 20 per cent of my class are future elementary/secondary teachers?*

S.R. Prabhakaran Nayar: *Most of the students have the misconception that the waxing and waning of the moon and eclipses are both due to shadowing by by the Earth. It would be worthwhile to discuss both simultaneously and to discuss the differences between them.*

W. Luzader: Future teachers should "understand" the basics so they can "teach" the basics. I feel that it is more important for a student to understand the phases of the moon and how they relate to eclipses than understanding the orbit of the moon with respect to the sun. The convexity of the moon's orbit would be a good way to test or apply the students' understanding about the Earth-moon-sun system.

N. Pasachoff: In addition to this admirable attempt to restructure astronomy courses, it is also very useful to work within the system. J. Pasachoff and I teamed up some years ago with one of the largest of the Amercian publishers of books for elementary schools and high schools: Scott, Foresman and Co. (Glenview, Illinois). We have provided them with new, expanded, exciting, up-to-date material on astronomy for EARTH SCIENCE *(commonly 8th grade, about age 13) and* PHYSICAL SCIENCE *(commonly 9th grade, about age 14). The first editions of these books, published in 1983, have so far sold over a million copies. Since copies are commonly bought by school systems, and used for a period of at least 5 years, the books are reaching over 5 million students. Thus the opportunity to improve interest in astronomy through normal publishing channels is significant. The second edition is in press (1989 publication), and we have recently contributed astronomy and other material to grades 3 to 6 of Scott, Foresman's elementary school series* DISCOVER SCIENCE *(1989).*

A. Fraknoi: *I very much agree with the need for workshops to help teachers do more and better astronomy in the schools. The Astronomical Society of the Pacific has been running 2-day intensive national (and this past year international) workshops for teachers in grades 3 to 12 each summer for the last seven years. We have been especially gratified to see teachers (in elementary and junior-high school) taking these workshops who have absolutely no background in astronomy at all. About 150 to 200 teachers come to take our workshop each year.*

Anyone who would like more information on these workshops can feel free to write to the A.S.P.

ASTRONOMY AND ASTROPHYSICS IN THE CURRICULA OF THE GERMAN GYMNASIUM

Hans L. Neumann
Friedrich-Dessauer-Gymnasium, Frankfurt/M, and Frankfurt Public Observatory,
Federal Republic of Germany

1. Introduction

The school system in the Federal Republic of Germany is generally of the classical European type as defined by D. Wentzel in his paper for this colloquium. Children enter school at age 6. After 4 years of primary school they enter secondary schools. The gymnasium type lasts for 9 or 7 years, depending on the state. The final exam (Abitur) entitles the students to go to university.

The interest of pupils in learning about astronomy is generally very high. But the usual science or physics teacher's education in most of our federal states does not include any thorough knowledge of astronomy; physics curricula are overcrowded with traditional matter and more and more modern technical applications, without reducing the time allotted to old topics. Practically speaking, there are very few lessons left that might be used for astronomy teaching.

The current situation reflects the contents not only of the recent standard curricula but also of curricula for many decades in the past. Only during the last 10 years could a slight trend for the better be recognized. But the situation has worsened again, as it has been demanded that Computer Science become a new subject of rather high priority for schools.

Generally, in junior high school only rudimentary aspects of astronomy are included in the standard syllabi, namely: phases of the moon, eclipses, and some basic optics of telescopes in physics courses; seasons and time in geography courses. The discussion of concepts of energy supply may lead to including some more astronomy or astrophysics beyond only solar energy. There are only a few exceptional states in which an 8-to-12-lesson course of Space Science is included in the physics curriculum. Optional courses in an additional subject are possible in some states.

For senior high school, the situation is varied, because of different state regulations and organizational models. For students taking physics during grades 12 and 13, it depends on state, school, and teacher whether or not the student learns astronomy. In an elective physics course, astronomy may be included as a topic for one or two semesters (out of four). But none of the federal states has astronomy or astrophysics as a standard topic in the curriculum for physics majors.

2. Astronomy Curricula

Curricula are developed separately in every federal state. A first project was carried out in Baden-Württemberg in close cooperation with the Astronomische Gesellschaft during the early 1970s. The resulting curriculum led to the introduction of astronomy as an additional separate subject for senior high schools there. It can be chosen for grade 13, and it is rather popular with students. The curriculum (Table 1) has become a standard basis for curriculum development in some other federal states.

Table 1. Curriculum in Baden-Württemberg

I.	Apparent and True Motions in the Sky
	Daily and Yearly Motions
	Gravitation and Keplerian Laws
	Bodies of the Planetary System
	The Sun as a Star
	Solar Properties
	Internal Structure
	Solar Activity
II.	Properties and Physics of Stars
	Hertzsprung-Russell Diagram
	Color-Magnitude Diagram
	Stellar Evolution
	Properties of the Milky Way System
	Properties and Physics of Galaxies
	Cosmology

About the same time, a group from Bochum University used a different approach in developing a curriculum for North-Rhein-Westfalia. Their basic concept is to teach physics from an astronomical or astrophysical point of view (Table 2). Two semesters deal primarily with electric and magnetic fields and electrodynamics or atomic physics, respectively. Astrophysical aspects (like radioastronomy or stellar spectroscopy) are used here as applications of general physics only.

Table 2. Curriculum in North-Rhein-Westfalia

11/I	Kinematics of the Planetary System
11/II	Gravitation and Space Flight Physics
12/I	Analysis of Optical Stellar Radiation
	Hertzsprung-Russell Diagram
12/II	Electric and Magnetic Fields
	Radioastronomy
13/I	Atomic Models
	Radiation in Stellar Atmospheres and Space
13/II	Nuclear Processes in Stars

As an example for the curriculum of a one-semester course, I give the newly proposed syllabus of Hessen (Table 3). It will presumably become effective in 1989.

Table 3. Curriculum in Hessen (proposed)

The Planetary System
Motions on the Celestial Sphere
Gravitation, Keplerian Laws and Applications
Properties of Planetary System Bodies
The Stars
Distance and Brightness
Important Physical Properties
H-R Diagram and Stellar Evolution
Galaxies
Physics of the Milky Way System
Galaxies and Cosmology

3. Books

Members of the working group in Baden-Württemberg have published a matching two-volume textbook of rather high standard. It is one of four textbooks on the German market, intended for schools but of very different levels. The lack of adequate teaching material may be in some cases one more reason not to do astronomy at school, for the individual teacher will have to develop most of the material for his lessons on his own.

4. Teachers' Education

At the university, astronomy is almost completely absent in physics teachers' education. During employment, teachers can participate in continuation courses to keep up with the evolution of science and with didactics. Courses with astronomy or astrophysics as topics are also offered by the regional institutes responsible, and they are usually strongly overbooked. An "index" C, which shows the average number of astronomical courses per year for a federal state, is mostly $0.1 \leq C \leq 1$, but would be $C \approx 2$ if demand were followed. This lack underlines the low priority given to astronomy by state officials in education.

Fortunately the associations of professional and of amateur astronomers both usually dedicate part of their convention programs to teachers and school astronomy. Over a long time span, these conventions contribute remarkably to providing better information about astronomy to science teachers.

Discussion

M. Dworetesky: *Would you agree that syllabi and levels of the German gymnasium courses are more or less comparable to North American university courses for physics students, and possibly at a level comparable to that of the UK introductory course described in my paper?*

H. Neumann: As I'm not familiar with the US system, I rather would like to compare the German 13th (perhaps even 12th) grade to introductory college levels. Comparison must be checked from the details, as for the UK system. But the similarity with the UK system might be higher.

TEACHING ASTRONOMY AT KEIO SENIOR HIGH SCHOOL, JAPAN

Yukimasa Tsubota
Keio Senior High School, 4-1-2 Hiyoshi, Kouhoku-ku, Yokohama-shi 223, Japan

1. Introduction

The major problem in teaching astronomy in our senior high schools has to do with the nature of the Japanese educational system. The typical science curriculum consists of physics, chemistry, biology, earth science, and general science I & II. The Japanese Ministry of Education allows General Science I to fulfill the minimum high-school graduation requirement in science. General Science I covers the basics of earth science. Astronomy has been taught as a part of General Science I and Earth Science.

Table 1. The Japanese Ministry of Education's
Current Educational Guideline

Subject Area	Subjects	Credits[a]	Statistics[b]
Science	Gen. Science I	4	Required Course
	Gen. Science II	2	Uncommon
	Physics	4	22%
	Chemistry	4	39%
	Biology	4	32%
	Earth Science	4	7%

[a]One credit = 35 classroom hours of lessons. One classroom hour = 50 minutes.
[b]From text sales.

Many Japanese high schools do not offer earth science because it is not covered in the college entrance exams (Table 1). Moreover, teachers usually spend many hours with the students memorizing the basics of earth science rather than allowing some time in the laboratory.

Keio Senior High School has a unique and creative academic curriculum (Table 2) that is not influenced by the Japanese college entrance exam because the graduates are admitted to Keio University without the entrance exam. The school was founded in 1886, has 130 faculty members, and 2567 students (all boys) in grades 10 to 12. It is located on the campus of Keio University, 20 minutes by train from Yokohama, and 30 minutes by fast train from Tokyo.

2. Lesson Plans

Keio's tenth-grade earth-science schema are as follows. The teaching materials cover the Earth, sun, solar system, and stellar evolution. The objectives of these plans are: 1) to have students understand nature via experiments and exercises, and 2) to have them learn scientific methodology and to acquire deductive reasoning through these materials. The meaning of the different kinds of brackets, and of the use of italics, is given at the very end of this section.

Table 2. Keio's Science Curriculum

Subjects	Credits	Grades	Statistics (1988)
Biology	4	10	Required
Earth Science	3	10	Required
Chemistry	3	11	Required
Physics	3	11	Required
Advanced Chemistry	3	12	196 (23%)
Advanced Physics	3	12	196 (23%)
Projects on Physics	2	12	288 (33%)
Projects on Earth Science	2	12	75 (9%)

Lesson Plan 1: The Earth (22 lessons)

Earth dimension (3)

- *Earth's shape*: evidence
- [Earth ellipsoid]: Eratosthenes' method, oblateness, curvature

The motion of the Earth (16)

- [Positional astronomy]: horizontal coordinates and equatorial coordinates, meridian, zenith

- {Solar time and sidereal time}: definition of time, local time, universal time, time difference, equation of time
- *Apparent motions of celestial objects*: diurnal motion, altitude of meridian transit, circumpolar star
- [Foucault's pendulum]: rotation of the Earth, coriolis effect
- [Apparent motion of sun and planets]: direct motion, retrograde motion, stationary, from geocentric theory to heliocentric theory
- *Revolution of the Earth*: annual aberration, annual parallax, distance of fixed stars
- {Synodic period and sidereal period}: conjunctions: inferior conjunction, superior conjunction, greatest elongation, opposition
- [Orbit of Mars]: ecliptic coordinates, Kepler's laws
- Moon's motion: law of universal gravitation, circular motion, {altitude of artificial satellites}

Our solar system (2)

- *Comparative planetology*: radius, mass, density, materials, gravity, rotation period, sidereal period, atmosphere
- {Length of the day}: effect of sidereal motion.
- <Planetary exploration>

Recognizing the peculiarity of the Earth (1)

Lesson Plan 2: The Sun (23 lessons)

The surface of the sun (9)

- [Sunspot and facula]; <the sun>; revolution period of sun, sunspot distribution, 11-year period of sunspot and climate change, (introdution to statistical analysis)

Energy source of the sun (5)

- [Direct solar radiation]: solar constant & atmospheric transmissivity
- Speculation on solar energy source: comparison of energy sources, chemical energy, gravitational energy, atomic energy
- {Determination of the solar surface temperature}; Stefan-Boltzmann law, albedo, effective temperature, and greenhouse effect

Composition of the sun (4)

- [Spectroscopic observation of the sun]: continuous spectrum, emission spectrum, absorption spectrum, Fraunhofer lines, <solar activity>

- [Spectroscopic astronomy]: spectral types, stellar surface temperature, Wien's law, determination of solar surface temperature

The sun as a star (4)

- {Absolute magnitude and visual magnitude}; luminosity, magnitude, distance
- [Hertzsprung-Russell diagram]: classification of stars, main-sequence stars, giant stars, supergiant stars, white dwarfs, spectroscopic parallax
- Stellar evolution: solar evolution

Introduction to modern astronomy (1): <x-ray astronomy>

( )   school hours
[ ]   laboratory works
{ }   calculation exercise
< >   audio-visual material
discussion lab

3. Discussion

I believe that astronomy is a suitable subject to teach scientific methodology and to introduce deductive reasoning. However, some students have been disappointed in our methods of teaching astronomy. Perhaps it is the result of the discrepancy between the material we teach and the students' interest, which run more to constellations and astrology. Students do not seem especially interested in astronomy as a laboratory science. Learning astronomy can be very difficult without the fundamentals of physics and advanced mathematics.

So is astronomy needed at a senior high school? My answer is "yes." Conceptual astronomy should be developed for senior-high school students. Some laboratory work might be delayed until the college level.

A PROJECT OF ASTRONOMY TEACHING IN THE SECONDAR SCHOOLS

N.S. Nikolov and T. Stefanova
Department of Astronomy, University of Sofia, A. Ivanov Str. 5, 1126 Sofia, Bulgaria

At present, astronomical teaching in the Bulgarian secondary schools is incorporated in the subject of physics. This situation dates from about 10 years ago. Before that, astronomy was a separate school subject with 15 to 30 hours in the last (10th) year of school. Now a new reform aroused by the rapid social and economic progress is in operation. During the course of the reform, one not very definitive decision was taken by the Ministry of Education to again separate astronomy, with about 30 hours in the 11th year of a school system which now has 12 years. In connection with this reform, the Ministry asked the Department of Astronomy at Sofia University to work out a project for developing a didactic system for teaching astronomy in the secondary schools. The elements of the system are: alternative variants for the curriculum and the respective textbooks; advisable teaching methods with the appropriate didactic means (diagrams, models, slides, films, *etc.*); computer programs to demonstrate astronomical objects and phenomena and for pupil computer dialogues, including means for controlling the knowledge level.

In order to find out what was essential:

1) We examined the teaching of astronomy in some other countries.

2) We examined past teaching of astronomy in our country and especially the school textbooks used during the last century and a half (Bulgarian schools started to operate more than a century before liberation from the Turkish yoke).

The studies from these items allowed us to discover a tendency to increase the astrophysical content of school curricula over time. This conclusion led us to pay special attention to the formation of basic astrophysical concepts in school courses in astronomy. For the formation of these concepts, we decided to apply the concept of theoretical generalization from the Soviet pedagogical literature.

3) We carried out a pedagogical experiment for settling the level of astronomical knowledge in the conditions of the present-day school system as well as finding the typical mistakes that pupils make in assimilating astronomical lessons. This establishing experiment was the first stage of a whole pedagogical experiments, respectively.

Because of the similarities in the methods of these three experiments or three stages of our pedagogical experiment, let us now concentrate on some details:

1) The experiment covered about 1200 school students from 12 schools of all kinds all over the country.

2) As a main subject of examination, we chose the stars and, more concretely,

their essential characteristics — temperature (spectra), luminosity (absolute magnitude), mass, *etc.*

3) For each characteristic, we gave three stages of problems graded in difficulty. The first stage of problems could be solved with only a simple *reproduction* of the perceived essential knowledge of the suitable lesson. The second stage — normal problems — one could solve by *using* the perceived knowledge. The third stage of problems could be solved only by means of creative reflection. That is why we call these *creative* problems.

4) In order to provide numerical evaluations, we decomposed each problem and its solution into their basic elements. We defined a coefficient κ (the relation between the number of true solved elements in a concrete solution and the total number of elements) as a grade for each problem given to the pupils.

The internal experiment held in 1985 (stage 1) showed the following average coefficients for the reproductive and creative problems in the characteristics luminosity, temperature, and mass, respectively:

	Luminosity	Temperature	Mass
Reproductive	0.73	0.63	0.36
Creative	0.30	0.23	0.16

These results show that it is necessary to use as much as possible methods that require the continuous active participation of pupils, often placing them in situations that need creative reflection. In addition, the experiments showed that pupils do not perceive sufficiently deeply the physical essence of stellar characteristics and of astronomical notions in general.

With respect to these conclusions, we worked out two variants of the system for the educational experiment. The variants concern mainly the structure of the educational material. We divided each variant into subvariants, differing predominantly in the method of teaching and respective visual means, devices, *etc.*

In Fig. 1 we show partially the results:

a) The coefficient κ for the temperature (spectrum) of stars from the educational experiment held in 1986/1987;

b) The mean κ for the three years.

For the rest of the characteristics, the results are similar. The quantity η gives the relation of κ for the experimental classes to κ for the reference classes. We believe that the increase of η from the reproductive to the creative level is due at least partially to the system developed by us.

In the forthcoming third stage — the definitive experiment that follows from the three-year educational experiment — positive elements would be used, such as suitably thought-provoking problems given by the teachers to students, selected by experimentation in previous years. After that, we would recommend the system to the Bulgarian Ministry of Education.

In conclusion, it is necessary to mention that the system for teaching astronomy at the secondary school level in the People's Republic of Bulgaria developed by our Department would certainly permit corrections, improvements, and ameliorations.

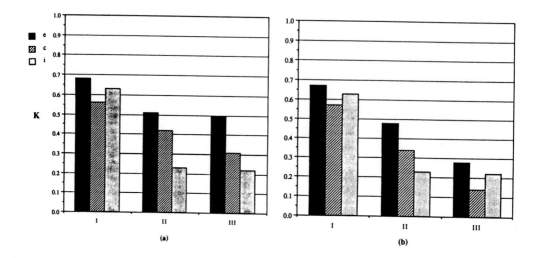

Fig. 1. Some results of the experiments about the knowledge about the concept star temperature *(spectral type, color) (a) from the school year 1986/87; (b) the average from three years of experiment. The number* κ *shows the ratio between the correctly solved elements and the total numbers of elements for solving the problems given to pupils. The symbols denote: i - inference experiment; e - experimental classes; c - control classes for reproductive (I), normal (II), and creative (III) levels of knowledge, respectively. The number* η *shows the ratio between* κ *for the experimental and control classes.*

The values of η *are: (a)* η *(I) = 1.21,* η *(II) = 1.21,* η *(III) = 1.65 (b)* η *(I) = 1.20,* η *(II) = 1.53,* η *(III) = 2.00.*

Chapter 10

Teacher Training

Training school teachers to teach astronomy is one of the most impor-
tant steps in improving astronomy education. This chapter discusses the
activities of societies and universities in France, the U.S., Mexico, and
Italy to train elementary and high schools teachers. It concludes with a
discussion of an American society that contributes to astronomy educa-
tion both schools and universities.

CLEA: AIMS AND ACTIVITIES

L. Gouguenheim[1,2], L. Bottinelli[1,2], J. Dupré[2], M. Gerbaldi[2,3]

[1]Observatoire de Paris, Section de Meudon, 92195 Cedex, France
[2]Université Paris Sud, Centre d'Orsay, Bât. 470, 91405 Cedex, France
[3]Institut d'Astrophysique de Paris, 98 bis, Bd. Arago, 75014 Paris, France

The Liaison Committee between School-Teachers and Astronomers (in French,
"Comité de Liaison Enseignants Astronomes" or CLEA) was officially created about
ten years ago under its present form, but its story began in 1970. In that time, there
was no astronomy at all in French school programs, neither in elementary nor in
secondary schools.

A discussion was beginning about introducing physics earlier in the curriculum,
with a specific purpose, the main ideas being (i) to avoid too much formalism (and
formalism is a strong general characteristic of French science teaching) and (ii)
to concentrate on experimentation and on the various representations of natural
phenomena.

Of course, astronomy enters these schemes quite well, but several arguments
were developed against its introduction. These arguments resulted mainly from (i)
strong competition between the various fields of physics and (ii) the ignorance of
the school teachers: the large majority of them had never learned about astronomy
in their studies.

Anyway, young children and teenagers were obviously considerably interested
and we were a few astronomers who decided to push the projects of (i) introducing
the teaching of astronomy at school together with (ii) training school teachers well.

A first association, called "Groupe Spécialisé Enseignement," was then set up in 1970 by the French Astronomy National Committee. It is now a section of the French Astronomers Association (Société Française des Specialistes d'Astronomie, SFSA).

The importance of meeting school teachers and discussing with them was soon realized, and the first national meeting took place in 1976 in Grenoble, during the IAU General Assembly, under the responsibility of IAU Commission 46, whose president was Dr. D. McNally. The meeting was very successful. It was attended by about 150 school teachers who expressed strongly their need for contact, information, and organization. The extensive report was published and widely distributed; it included, in particular, the following conclusions adopted by the participants:

"A committee including both astronomers and school teachers should be created, its main objectives being the introduction of astronomy in school programs together with the training of school teachers." The participants underlined the importance of organizing astronomical summer schools, publishing a newsletter, and circulating teaching material.

The first summer school took place in 1977, the first issue of our newsletter *Les Cahiers Clairaut* was published in 1978 and CLEA was born. It has today about 1000 members; Profs. J.-C. Pecker and E. Schatzman are our Honorary Presidents. The 40 members of the Council are mainly representatives of the different French educational districts.

1. CLEA: Main Educational Objectives

CLEA's main educational objectives are the following:

(1) to give access to theoretical knowledge through practical activities;

(2) to increase the mood of observing and experimentation;

(3) to urge the various kinds of teachers, working at different levels, on different subjects, to exchange their experiments and to hold a dialogue. This is particularly important to overcome the barriers between disciplines and teaching orders;

(4) to produce and circulate good quality educational material (textbooks, slides, video, scale models...) of low cost, easy to use, not too time consuming, and well tested from an educational point of view. The feedback comes from the network of CLEA members.

2. CLEA: Methods and Results

The methods must be appropriate to adult education. We favor a friendly and non-hierarchical atmosphere, and a good mixing of theoretical and practical activities.

We are also open to collaboration with foreign groups with similar activities. This is the case, for example, for the Italian group "Associazione Casa-Laboratorio del Censi": Nicoletta Lanciano has participated in several CLEA General Assemblies and we have published in *Les Cahiers Clairaut* the French translation of one of their experiments.

Our main results, up to now, concern first the effective introduction of some astronomy at various levels in school programs, at elementary school (ten-year-old children), and secondary schools, for all 13-year-old and for 16- and 17-year-old "non-scientific" students. For the first time, astronomy will be introduced in 1989 for "scientific" 17-year-old students.

The training of school teachers has been developed through astronomy summer schools: up to three different schools have been organized each year, for the past 12 years (see the paper by Gerbaldi *et al.*). A large number of 3- to about 10-day sessions during the year are organized by CLEA members in various educational districts (Appendix 1).

Fig. 1. Milky Way gossip, by G. Paturel.

A large number of astronomical clubs and educational projects are developed by teachers, and astronomical educational material has been produced at low cost and is circulating among CLEA members (slides, video, written material, stellariums, scale models...).

3. Les Cahiers Clairaut

Les Cahiers Clairaut are named after the French mathematician Alexis Clairaut, who measured with Maupertuis the length of the terrestrial meridian in Lapland and thus confirmed Newton's view. We chose his name because he has written in the foreword of his book "Eléments de géométrie": "J'ai pensé que cette Science, comme toutes les autres, devait s'être formée par degrés; que c'était véritablement quelque besoin qui avait fait faire les premiers pas et que ces premiers pas ne pouvaient pas être hors de la portée des Commençans, puisque c'étaient les Commençans qui les avaient faits." We have the strong feeling that the best way of teaching astronomy in particular and science in general, is to go step by step, following the steps of discoverers, which is a good way to simplify the problems.

Four issues of *Les Cahiers Clairaut* are published each year, each of them being published at the beginning of each season. The contents include:

- book reviews (G. Walusinski, A.M. Louis...),
- CLEA activities (G. Walusinski),
- letters from the readers,
- scientific news and notes: "Milky Way Gossip" (Fig. 1) (L. Bottinelli),
- history of astronomy (a large effort is made in this field, for which school teachers have marked their strong interest, and many astronomers have contributed; examples include not only Dr. C. Iwaniszewska but also several school teachers, G. Walusinski, J. Ripert, J. Vialle),
- astronomy and computer science (M. Toulmonde, J.C. Allard, C. Dumoulin...),
- interpretation of an observation (J.P. Rosenstiehl, G. Paturel, F. Suagher...),
- astronomy at elementary school (L. Sarrazin, V. Tryoën, A. Delavergne, D. Vallarché...),
- astronomy at secondary school (J. Ripert, B. Sandré, J. Chappelet...),
- making an observation (D. Bardin, J. Heidmann...),
- physics (H. Gié, J. Dupré, M. Gerbaldi, B. Leroy...),
- astrophysics: (many astronomers are contributing: P. Léna, A. Brahic, E. Schatzman, J. Schneider, D. Alloin, E. Gérard, S. Collin, R. Hakim, H. Andrillat, A. Acker, P. Boissé, J.P. Parisot, J.P. Zahn, C. Vanderriest, A.C. Levasseur-Regourd, L. Celnikier, L. Nottale, and many others...),
- constructing an instrument (D. Toussaint, D. Bardin, J. Ripert, M. Jonas, A. Dargencourt...),
- constructing a scale model (B. Sandré, C. Dumoulin, J. Ripert, C. Piguet, J.L. Fouquet, V. Aguerre, A. Delavergne...).

4. Examples of Demonstrations Made by CLEA Members

Celestial sphere

This big sphere (Fig. 2) helps to illustrate celestial coordinates and the various celestial motions. It was first constructed by C. Piguet, who published the plans in *Les Cahiers Clairaut*. Two of them, constructed during summer schools, are circulating: they can be used in a classroom, because of their large size. However, they are rather expensive (about 200 U.S. dollars); much smaller ones (9 cm in diameter) have been constructed by B. Sandré, at a very low cost of about 5 dollars.

The star box

Constructed by J. Ripert, who published the plans in *Les Cahiers Clairaut*, it is a "light box" which helps to recognize the constellations.

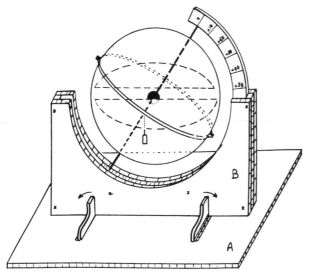

Fig. 2. A big celestial sphere: (a) the plans; (b) the realization

An armillary sphere

Constructed in cardboard according to the plans made by J.L. Fouquet, in about 6 hours (Fig. 3).

Movable viewgraph illustrating the motions of the moon

Designed by J. Ripert, it needs 4 sheets of viewgraph and 6 snaps; it illustrates the rotation of the Earth, the motion of the moon around the Earth, and the fact that the same face of the moon is seen from the Earth because the moon is also rotating, the phases of the moon¡, and the solar and lunar eclipses. It is distributed by CLEA and costs about 10 U.S. dollars.

Fig. 3. An armillary sphere under construction

5. Financial Support

Teaching astronomy from a practical point of view is one of CLEA's major objectives; however, this requires some material. Many efforts are being made for developing low cost (but good quality!) educational material and giving a chance for the teachers to get this material, or to be able to make it themselves. Some financial support is thus needed. *Les Cahiers Clairaut* is self supported, but all the practical work, including typewriting, is done free of charge by G. Walusinski, a retired school teacher. The educational material is sold at cost.

The summer schools and the annual sessions are supported mainly from the French Department of Education and from the Astronomy National Committee. They are warmly acknowledged.

Appendix 1
Examples of CLEA activities in the various educational districts

Besançon (J.P. Parisot, F. Puel, F. Suagher...): production of videotapes, slides, annual sessions

Dijon (R. Hernandez): various conferences

Grenoble (A. Omont, A. Richelme, C. Barathon, M. Bonneton...): circulating planetarium, annual sessions

Limoges (L. Sarrazin, C. Dumoulin...): annual sessions, videotapes...

Lille (M. Laisne, C. Mossler): annual sessions for elementary school teachers

Lyon (G. Paturel, C. Piguet...): annual sessions, educational material

Marseille (M.F. Duval, J. Donas, D. Bardin...): slides, circulating planetarium, annual sessions, and summer schools

Montpellier (H. Andrillat, H. Reboul, A. Cordoni, F. Gleizes, M.O. Mennessier...): educational material, annual sessions

Nice (J. Chappelet, J.L. Heudier, J. Ripert, V. Tryoën...) planetarium, annual sessions, educational material

Paris (L. Bottinelli, M. et M. Bobin, A. Brahic, A. Dargencourt, J. Dupré, M. Gerbaldi, L. Gouguenheim, E. Hadamcick, A.C. Levasseur-Regourd, A. et M. Rivière, B. Sandré, C. Vignon, G. Walusinski...): annual sessions and summer schools, circulating planetarium, educational material

Poitiers (J.L. Fouquet, J. Gagnier, J. Vialle, J. Vallantin...): annual sessions, educational material

Reims (D. Toussaint, G. Bazin...): planetarium, annual sessions

Rennes (F. Dahringer): circulating planetarium, annual sessions

Strasbourg (A. Acker, E. Legrand, J.M. Poncelet...): production of slides, videotapes, educational material; planetarium, annual sessions, and summer schools; a large activity is being developed around Strasbourg planetarium

Toulouse (J.P. Brunet, Talon, S. Vauclair...): circulating planetarium, annual sessions

Discussion

J. Fierro: *Is there any way one can get* Les Cahiers Clairaut? *and/or any of the materials (simple ones) by mail?*

L. Gouguenheim: The CLEA secretary is: Mr. G. Walusinski, 26, Bérengère, 92210 Saint Cloud (France). Any information concerning *Les Cahiers Clairaut* or CLEA publications can be obtained either from him or from me.

A. Fraknoi: *I found this talk very interesting. In the U.S., we are some years behind our colleagues in France. The U.S. professional societies began a newsletter for teachers only in 1984, but the first year we received 12,000 requests for the newsletter from teachers. Today, we have over 20,000 readers, with more coming in. Naturally, since we provide these newsletters free, we are concerned about the ongoing cost of the program. Thus I am interested in how CLEA and* Les Cahiers Clairaut *are funded?*

L. Gouguenheim: The funds that we receive from either the French National Astronomical committee or from the Ministry of Education are used mainly for Summer

Universities. The school teachers pay a subscription of 60 French francs ($\simeq$ 10 U.S. $) a year (4 issues).

D. McNally: *I recommend* Les Cahiers Clairaut *highly. Although they are written in French, they are written clearly, so that readers with a liminted knowledge of the French language can still understand.*

AN EXAMPLE OF CLEA ACTIVITIES IN THE TRAINING OF SCHOOL TEACHERS

L. Bottinelli[1,2], J. Dupré[1], M. Gerbaldi[1,3], L. Gouguenheim[1,2]

[1]Astronomie, Bât. 470 Université Paris Sud, 91405 Orsay Cedex, France
[2]Observatoire de Paris, Section de Meudon, 92195 Meudon, France
[3]Institut d'Astrophysique, 98 bis Bd. Arago, 75014 Paris, France

Various simple activities are being developed by the French Comité de Liaison Enseignants Astronomes (CLEA) in the training of school teachers. We give in the following one example of a very simple instrument which we call "Alphonse's box," after Alphonse Delavergne who invented it. It is also called a *heliograph*. This very simple and inexpensive instrument enables (1) the plotting of the daily apparent path of the sun, (2) the determination of the duration of sunlight, and (3) the determination of the declination of the sun. The following description is due to Maryse Jonas:

Material

1. photosensitive paper (used by architects) that can be handled in semi-darkness; its sensitive face is yellow and it is developed by the vapor of an ordinary ammonia solution in about ten minutes;

2. a simple can, with a cover that will shut tight; approximate dimensions: diameter 10 cm, height 15 cm, with a hole about 0.5 mm in diameter on one side;

3. two pieces of wood, 15 x 15 cm and 15 x 20 cm; one butt hinge; one threaded rod with two nuts and two washers.

Orientation of the Box

The first piece of wood is horizontal, and the second one is in the equatorial plane. The axis of the box is in the direction of the polar axis, so the angle α between the two pieces is 90°- the latitude. The butt hinge allows the heliograph to be used at different latitudes. The cover of the box is fixed to the small equatorial board, and a plumb line, together with a protractor, enables the setting of the latitude (Fig. 1). The photosensitive paper must be placed inside the box (in semi-darkness) with the sensitive side exposed, using either magnets or Scotch tape to hold it in place. Its bottom edge must be flat against the bottom of the box. The 0.5 mm hole must not be covered by the paper. The box must be placed exactly on the meridian. This can be done at night, by observing the pole star through a thin tube put along the axis of the box. Figure 1 shows the assembly and orientation of the box, and Figure 2 shows the box assembled.

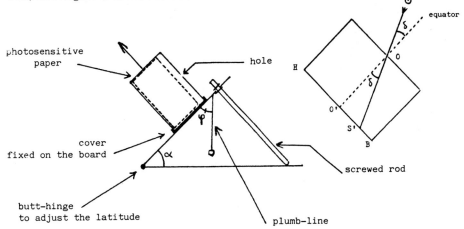

Fig. 1: *The assembly and orientation of the heliograph.*

Use of the Recording

Figure 3 shows the records obtained in different seasons. The general shape of the curve gives the sign of the declination of the sun. Noon (local solar time) is given by the straight-line tangent to the curve and parallel to the edge of the sheet. Note that this determination is not possible at the equinoxes. The time scale can be obtained by noting that the scale of the record is R x 2θ, where R is the radius of the box and θ is the angular velocity of the sun in the sky (15°/hour). From this, we can determine the duration of sunlight, the times when the sun was hidden by clouds, and the times of sunrise and sunset. The declination δ of the sun is also easily obtained (Fig. 1) by measuring, at noon, $O'S' = HS' - HO'$ and noting that $\tan \delta = O'S'/2R$. Finally, the equation of time (observed solar time - mean solar time) can be obtained from the difference between the noon observed and noon, mean solar time. The heliograph must be placed in exactly the same place for one year, the hole being unmasked for only one minute each 15 days at the same civil time (noon, for example).

Fig. 2: A photograph of the heliograph assembled.

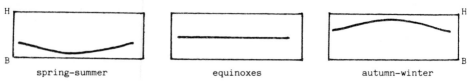

Fig. 3: Different shapes of records obtained at different seasons.

TRAINING OF SCHOOL TEACHERS AT FRENCH ASTRONOMY SUMMER UNIVERSITIES

M. Gerbaldi[1,2], L. Bottinelli[2,3], L. Gouguenheim[2,3], F. Delmas[1], J. Dupré[2,4]

[1]Institut d'Astrophysique, 98 bis Bd Arago, 75014 Paris, France
[2]Université de Paris XI, Centre d'Orsay, Lab. d'Astronomie Bât 470, 91405 Orsay, France
[3]Observatoire de Paris, Section de Meudon, DERADN, 92195 Meudon, France
[4]Université de Paris XI, Centre d'Orsay, Lab. Infrarouge Bât 350, 91405 Orsay, France

1. Introduction

In September 1976, at the end of the IAU General Assembly held at Grenoble (France), a one-day meeting concerning the teaching of astronomy was organized by Commission 46. It was decided during this symposium, which brought together 150 French school teachers and 50 astronomers, among other things, to organize a summer school of astronomy the following summer. Since then, such a school has been organized every summer. These astronomy Summer Universities are one of the activities developed by the non-profit organization CLEA (Comité de Liaison Enseignants Astronomes) whose activities are discussed elsewhere in these proceedings. In astronomy, children are always very curious. Because of this interest, in the 1970's some French astronomers applied pressure on the Education Ministry to introduce the subject in schools, and they were successful. However, astronomy was not introduced as a new separate subject, but rather as a part of another subject, mainly mathematics or physics.

Since few secondary school teachers receive instruction in astronomy during their university education, they need to be trained in astronomy. It is well known that one achieves better understanding of a subject by concentrated involvement. The solution adopted for training was therefore to be together, school teachers and astronomers, for 10 days in an isolated place. Table 1 lists the Summer Universities held to date.

In what follows, the methods used during these Summer Universities and the results will be discussed.

2. Objectives of the Summer Universities

The objectives of these Summer Universities are those of the CLEA. The aims can be expressed on one hand concerning *content* and on the other hand concerning *pedagogical methods*; both sides are equally important.

Table 1. Summer Universities Held in France.
Number of participants (not including the pedagogical team members)

Year	Organizers				Total
	Bottinelli, L. Delmas, F. Dupre, J. Gerbaldi, M. Gouguenheim, L.	Acker, A. Parisot, J.P.	Celnikier, L.	Duval, M.F. Donas, J.	
1977	69				69
1978	63		55		118
1979	67		50		117
1980	55				55
1981	61				61
1982	80		55		135
1983	70				70
1984	91				91
1985	82	50			132
1986	98	70			168
1987	79			15	94
1988	61	50		25	136

The content varies from one year to another and covers a wide range. It can be related to a specific program, for example:

– in elementary classes (pupils from the age of 7 to 9): *notions of time and space*
– in secondary school programs:

     pupils between the age of 13 to 14: *physics program: optics*

     pupils between the age of 16 to 17: *physics program: stellar evolution*

     pupils between the age of 17 to 18: *mathematics program: history of ideas.*

It can also be related to some specific astrophysical concepts, for example:

– *the various coordinate systems*
– *the use of the Universe as a unique laboratory considered as an extension of any terrestrial laboratory, to be able to have a better understanding of the approximations of the physics laws deduced from experiences on Earth.*

Concerning *methods*, the goals are for any participant:

– to be able to choose from all the activities proposed, the one best suited to him
– to be able to develop his critical awareness, elaborating on what can be read in books

– to be willing to ask questions

– to be willing to explain his pedagogical experiences in a classroom

– to learn to use pedagogical tools such as: overhead projector, models, planetariums, microcomputers, other audiovisual equipment

– to be able to communicate with other teachers who attend the same Summer University but who are teachers of different subjects. This point is important: the school teachers who attend a Summer University have various backgrounds. Table 2 divides the participants of the previous Summer Universities by their backgrounds. The participation in a Summer University is not restricted by the background of the school teacher: the only rule for registration is: "first come, first served." The mixing of school teachers with different backgrounds enriches everybody.

Table 2. Typical Backgrounds of the Participants.
(1988 Summer University)

Teaching Level		Background		
Primary School Teachers	8 %	Mathematics	10	%
Secondary School Teachers	50 %	Physics	71	%
(1st level)				
Secondary School Teachers	42 %	Physics-Chemistry	8	%
		Earth Sciences	7	%
		Non-Science	4	%

We also want to push very hard the following ideas:

– we are very much in favor of using the history of science to acquire knowledge: astronomy is very suitable for that purpose

– we wish to show that recent developments in astrophysics can be introduced in classrooms.

3. Pedagogical Strategy

During a Summer University there are 3 categories of activities:

– lectures (9:00 – 12:00 noon)

– small workshops and practical classes (2:30 – 4:30 and 5:00 – 7:00 p.m.)

– tutorials (9:00 – 10:00 p.m.)

– night observations (10:00 p.m. –)

a) *Lectures*

The lectures are always in the morning. They last 3 hours and all the participants attend them. During the last 5 years, the lecturers have been L. Bottinelli,

J. Dupré, F. Durret, M. Gerbaldi, L. Gouguenheim, M. Gros, G. Paturel, J. Ripert, B. Sandré, V. Tryoën, and G. Walusinski, as well as A. Acker and M.F. Duval. The aim of these lectures is to give all the participants some basic facts. For example:

– the analysis of light in astrophysics

– a reflection of epistemological type (that is, the study of the nature, sources, and limits of knowledge), for example: the representation of the solar system during the 17th century

– modern astrophysics (cosmology).

Of course not all these subjects are developed during the same year.

b) *Small workshops and practical classes*

The activities are organized into small workshops or into practical classes of 10 persons according to the subject: theoretical or more practical. Everyone can choose from all the activities proposed. Several workshops and classes run in parallel. The small workshops and practical classes are always during the afternoon and are directed by a tutor.

The overall aim of these activities is to illustrate an astrophysical concept that may have been presented during the lectures. Therefore, during the afternoon, activities considered include:

i) *the analysis of a collection of data or of a series of photos (generally obtained by professional astronomers)*

The exercises cover a broad range of subjects, for example:

– analysis of light:
  - motion (Doppler effect)
  - magnitude determination
  - temperature determination

– coordinates and time

– history of astronomy (Kepler's laws)

During a Summer University about 30 different such small workshops are available.

ii) *the making of models*

During a practical class, the participants make themselves a model or a small instrument that can be used during the following nights to make an observation.

For the construction of these small instruments out of cardboard or wood, we intend to allow every participant to make his own instrument and to take it home. We therefore prepared ready-made pieces to avoid wasting too much time. Further, everybody can make his own instrument even if not skilled at using materials and tools.

Among the models and instruments so far made were:

- an astrolabe
- a model of the solar system (to scale)
- an heliograph
- a simplified equatorial mount
- a refractor (diameter 5 cm)

25 such projects are carried out during a Summer University (see Bottinelli *et al.*, in these same proceedings).

iii) *carrying out an observational sequence; for anyone studying astronomy it is necessary to be practical and to observe the heavens with the naked eye, with binoculars, or with a small telescope*

These activities are not designed for future use in the classroom. Participants must adapt them to the level of their pupils. We do not want to give packages of "ready-made" lessons.

During the Summer University, there are some informal sessions in which the participants discuss how to adapt the activities they have learned to their pupils. Moreover, every evening, tutorial sessions are designed to clear up any aspects that were not understood during the lectures. These tutorial sessions are organized at the demand of the participants themselves, to counterbalance the great variety among the backgrounds of the participants. The instructor for any of these activities can be either an astronomer or a school teacher who has himself acquired sufficient knowledge in some astronomical field.

During the last 5 years, the main instructors were: R. Arhel, D. Bardin, C. Canard, F. Dahringer, A. Dargencourt, Ch. Dumoulin, J.L. Fouquet, M. Jonas, R. Gouguenheim, C. Piguet, J. Ripert, A. Rivière, M. Rivière, B. Sandré, D. Toussaint, V. Tryoën, and C. Vignon.

The methods used in a small workshop are direct and concise because we want to have a high efficiency level in a short time. But each instructor is free to choose any method he wishes to use. Because of the fact that there are many instructors in a Summer University, the methods used can vary greatly. In any case, each participant is absolutely free to choose any activity he likes.

4. Organization of these Summer Universities

The organization and the putting into practice of a Summer University of 80 participants is done by a pedagogical team that consists of professional astronomers and school teachers. Such a team consists typically of 6 astronomers and 10 school teachers. The members of this team share the organization of the Summer University.

a) *Pedagogical organization*

Many activities are available for the participants. Before the beginning of the Summer University, each participant receives a short description of all the activities offered.

On the first day of the school, one of the organizers explains the program in detail so every participant can make his own daily plan.

Of the activities put forward, some had been developed during the preceding school year by school teacher members of the pedagogical team. In this way, the content of the Summer University is enriched by feedback.

This pedagogical organization requires a certain environment, such as:

− a library

− a small workshop

− a small photographic lab as much for the development of photographs as for their printing on paper

− a small color lab to develop slides.

b) *Practical organization*

The practical organization is as much concerned with housing the participants as with putting together all the equipment needed.

First, we must find a location where the Summer University can take place. Since it must be favorable for night observation, we must be far from a city. Further, housing for 100 people and both board and catering must be available. Moreover, a conference room for 100 participants and several (6 to 7) small rooms must be available for practical classes, since the latter run in parallel sessions during the afternoons. We need also a place to organize the workshop as well as suitable rooms for the photographic labs. A place that fulfills all those requirements is generally a summer residence.

To take care of the registration of the participants as well as of their housing, a team member is fully in charge, some weeks before and during the Summer University, of the lodging and its financial counterpart. Before the Summer University, some amount of work is necessary to put together all the equipment needed. We have to buy wood, cardboard, some tools, nails, screws, and so on...used in the practical classes. We must also reproduce all the documents needed during the small workshop and pre-cut pieces of wood or of cardboard for the making of the models.

Concerning the funds necessary for such a school, the CLEA succeeded in obtaining financial support from the CNRS (Centre National de la Recherche Scientifique), the CNFA (Comité National Français d'Astronomie) and for the past 4 years from the Educational Ministry, but in the last year we had to be selected from many proposals covering a wide range of subjects.

The participants need only pay for their board and meals. All the funds collected are used for the equipment needed. None of the lecturers receives any fee. In order that the summer residence can function as a Summer University, some team members must be there two days in advance to put everything in order.

The total volume of all the equipment needed is 12 cubic meters for a Summer University of 80 participants. To transport the equipment from our University (Paris XI), where it is used also with the students, we used to rent a small truck and drive it ourselves from Paris to the location (roughly 900 km). This summer we contracted a moving company to transport all the equipment.

5. The Feedback

How to evaluate what the participants had acquired during the school?

We evaluate it throughout the courses by informal reports or during the tutorial sessions in the evening. But the major evaluation is done the last day during a poster session, during which small groups of participants (one for each practical class) present how a model is made, how it functions, and what its use is.

The last day of the Summer University, we distribute to all the participants a questionnaire concerning the content, the methods, the pace, and the way the Summer University was carried out. This questionnaire is of great help for organizing the next Summer University.

Moreover after the end of the Summer University, some months later we used to publish the proceedings, including the text of all the lectures as well as that of the small workshops. We did so 11 times, but as it became more and more expensive and time consuming, and because during the 11 years all the major subjects were described in one way or another in one of the previous proceedings, we decided this year not to publish that way. Instead, during the Summer University, we distributed to every participant the full text of all the lectures as well as the reports written down for the poster session, which were photocopied during the Summer University.

We try also to keep in touch with participants during the year that follows, through the non-profit organization CLEA.

6. Conclusion

Figure 1 shows that these Summer Universities contribute to the training of the school teachers from every region in France. The number of teachers who have been trained since 1977 varies greatly from one region to another, but is in fact, proportional to the number of pupils in each region.

We conclude that — despite the fact that, in France, astronomers are concentrated in a limited number of observatories and institutes that are not linked with the universities — the method of school teacher-training by Summer Universities is powerful in the sense that it reaches school teachers far from astronomical centers.

On a different level, another success of these Summer Universities is that some of the teachers, after attending several times, can then become tutors. Not only do they propose new experiments and test them with their colleagues, but they also organize local training in astronomy during the school year, in their districts.

Acknowledgment

We would like to express our deepest thanks to F. Warin for all the photographic work she does for the Summer Universities.

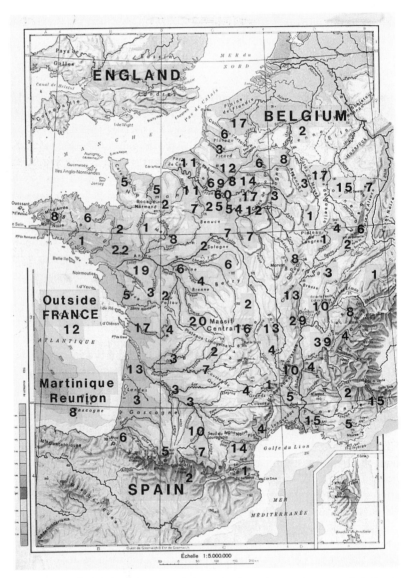

Figure 1: Geographical distribution of the school teachers who attended the Summer Universities, organized by L. Bottinelli, et al., since 1977. "Outside France" means teachers who are teaching in French Lycées located in foreign countries.

Figure 2: Cover of the Proceedings of a Summer University
(drawing by D. Bardin)

Discussion

H. Shipman: *First, let me say how impressed I am with the number of teachers involved and the variety and depth of activities. My question is how do the elementary teachers react to your program; are they overwhelmed by the program, or are they as pleased as the high school teachers?*

M. Gerbaldi: We are very careful to ask participants about their backgrounds, and structure activities accordingly. The participants were very satisfied.

C. Iwaniszewska: *I would like to comment about the publication of the big volume after the Summer University; it is a lot of extra work for the organizers, but is so very useful for participants.*

M. Gerbaldi: This year, in 1988, we reproduced the text of all the lectures before the beginning of the Summer University in order to give them immediately to the participants. For the results of the small workshops and the practical classes, small groups of participants wrote down, in a few pages, the results they obtained and how the scale models or the small instruments were made. We rented a photocopy machine to reproduce all these reports and they were distributed to all the participants. Thus this year, every participant went home with the proceedings.

THE UNIVERSE IN THE CLASSROOM: A NEWSLETTER ON ASTRONOMY FOR TEACHERS

Andrew Fraknoi
Astronomical Society of the Pacific, 390 Ashton Ave, San Francisco, California 94112, U.S.A.

Since 1984, the Astronomical Society of the Pacific (ASP) has sponsored a quarterly newsletter on teaching astronomy in grades 3–12. Cosponsored by the American Astronomical Society, the Canadian Astronomical Society, and the International Planetarium Society, the project now reaches over 20,000 teachers and schools in the US and Canada.

We would like colleagues overseas to examine the newsletter, and, if they find it useful, to consider translating and distributing it. We will be glad to make it available without charge to anyone who is interested.

1. The History of the Newsletter

Since 1978, the summer meetings of the ASP — held in a different city each year — have featured one to three days set aside for workshops on astronomy for teachers. In recent years, between 150 and 200 teachers have come each summer — at their own expense — to take these credit workshops and learn more about astronomical topics, activities, and resources.

Participants in these workshops asked the ASP year after year if we could set up a way for the "graduates" to keep in touch with us and continue to get updates, especially about new discoveries and new teaching resources. In late 1984, with a grant from the US National Academy of Science's Slipher Fund, we were finally able to respond to their requests by starting a newsletter on teaching astronomy.

We set up a distinguished Board of Advisers for the newsletter, consisting of noted astronomers and educators in the US and Canada, and sent a news release to our workshop "graduates" and some teachers' and astronomy journals. We thought that in its first year, announcements of the newsletter would probably generate several hundred requests, mainly from alumni of our workshops and other astronomy workshops around the country. We were shocked to receive about 10,000 requests for *Universe in the Classroom* during the 12 months after the project was announced.

These requests came from schools and teachers throughout the US and Canada and from grade levels ranging from first grade through community college. A significant number of requests were received from professors of science education in universities with a special interest in teaching teachers.

We had already asked the American Astronomical Society to cosponsor the project and their Council gave it an enthusiastic endorsement and its financial assistance. We then began to look around for other sources of support and assistance. Starting at home, the ASP Board decided to devote most of the interest from the new Bart J. Bok Memorial Endowment to help the newsletter. The Canadian Astronomical Society stepped in to help support the mailing of issues to Canada. Associated Universities, Inc., and the National Radio Astronomy Observatory gave us some welcome support. And the Slipher Fund has generously continued its financial assistance to the project for several years. A number of other foundations and companies have also helped with printing, equipment, and other forms of assistance.

2. The Newsletter Today

The newsletter is offered free of charge to teachers, librarians, and youth-oriented-group leaders. Only written requests for the newsletter are honored and subscribers are asked to write in on school or youth-group stationery, to make sure that we are not sending copies to students or the public.

The issues so far have been written by Andrew Fraknoi and Sherwood Harrington (ASP) with contributions by Dennis Schatz (of the Pacific Science Center) and John Percy (of the University of Toronto). Each issue features astronomy information, one or more classroom or observing activities, and a section on teaching resources.

We write each issue of the newsletter with a varied audience in mind. We want to please not only the experienced science teacher but also the beginner with very little science background who may be facing his or her first-grade class. Many of these teachers have told us they don't cover topics in modern astronomy because they are afraid their students will know more than they do. The newsletter tries to address such concerns by giving them information and tools that they can put to immediate use in the classroom, as well as resources for further information for both students and teachers.

The Universe in the Classroom

A Newsletter on Teaching Astronomy

Sponsored by the:

Astronomical Society of the Pacific

American Astronomical Society

Canadian Astronomical Society

International Planetarium Society

Number 12   © copyright 1989, Astronomical Society of the Pacific   Winter 1988 - 89
                              390 Ashton Ave., San Francisco, CA 94112

Activity Corner

1. How Soon Can You See a Crescent Moon?

An interesting activity is to see what the youngest crescent Moon is that you can see. Some very experienced observers are able to see the Moon less than 24 hours after it's new. How well can you do?

[For an article on an international effort to see a very young crescent, see: "Moonwatch — July 14, 1988" by L. Doggett, et al., in *Sky & Telescope*, July 1988, p. 34.]

2. When Is the Moon Visible?

It is often surprising to youngsters that the Moon sometimes can be seen in the *daytime* sky. Just when the Moon rises and sets depends primarily on its phase, and only at full Moon does it behave as some might expect — that is, to rise at around sunset and be up all night long.

Keeping track of the visibility of the Moon as it cycles through its phases can be a fascinating (and instructive) thing to do. Moon-tracking activities could follow a wealth of different paths — from precise timing of moonrise and moonset to a more general noting of where the Moon is (and what phase it's in) each time you see it.

To help you devise a program of Moon-viewing, the accompanying table charts the times when the Moon rises, sets, and so on during its different phases. We should note that the times in the table are *very general* (correct to within an hour or two). The precise times of moonrise, moonset, and so on depends on a number of factors besides the Moon's phase — your location on Earth (latitude and longitude) has a major effect, for example. (Exact local times of moonrise and moonset are often printed in large daily newspapers, usually in the weather section.)

Notice that knowing when the Moon rises and sets in its various phases allows you to tell time (roughly) whenever it is visible! For example, if you're awakened in the night on a camping trip and you notice a third quarter moon high in the eastern sky, then you know that sunrise is coming soon. On the other hand, a full moon high in the sky would reassure you that you have plenty of time for more sleep — the full moon is highest around midnight.

3. Lunar Eclipses

Observing a lunar eclipse (which — because it can be seen over a much wider area — is much more easily seen than a solar eclipse) is a safe and enjoyable family activity. Upcoming eclipses are listed in astronomy magazines such as *Sky & Telescope* and *Astronomy*, as well as in the *Abrams Planetarium Sky Calendar* that comes with membership in the Astronomical Society of the Pacific. No special precautions need to be taken in viewing such an eclipse, and it's fun to organize family, friends, and neighbors when you know such an eclipse is coming. ■

Moon Phases and Time of Day					
Phase	**Rises**	**In Eastern Sky**	**Highest In Sky**	**In Western Sky**	**Sets**
New	[~sunrise]	[morning]	[noon]	[afternoon]	[~sunset]
Waxing Crescent	[just after sunrise]	[morning]	[just after noon]	[afternoon]	just after sunset
First Quarter	~noon	afternoon	~sunset	night (pm)	~midnight
Waxing Gibbous	afternoon	~sunset	night (pm)	~midnight	night (am)
Full	~sunset	night (pm)	~midnight	night (am)	~sunrise
Waning Gibbous	night (pm)	~midnight	night (am)	~sunrise	morning
Third Quarter	~midnight	night (am)	~sunrise	morning	~noon
Waning Crescent	just before sunrise	[morning]	[just before noon]	[afternoon]	[just before sunset]

Times in brackets [ ] indicate that the Moon can't be seen because it's too close to the Sun on the sky.

A typical item from a typical issue of *The Universe in the Classroom*.

This year, the International Planetarium Society has joined the Societies involved with the newsletter as a sponsor. Because planetariums represent the largest and most effective interface between professional astronomy and teachers, we are especially pleased to have them as partners in this endeavor and are working with them to encourage even wider redistribution of each issue through local planetariums.

3. Looking to the Future

At present the direct circulation of the newsletter is over 20,000 and still climbing. In many schools and school districts, the initial recipients make dozens or hundreds of additional copies for further distribution. In addition, articles from our newsletter have been excerpted and reprinted in many dozens of local and national magazines and newsletters read by teachers.

We would like to encourage our colleagues from around the world to translate and distribute the newsletter in their own countries, if it is appropriate. To obtain some recent issues and more information, please write the author at:

Teachers' Newsletter Distribution
Astronomical Society of the Pacific,
390 Ashton Ave.,
San Francisco, California 94112, U.S.A.

Whether or not you wish to (or need to) translate, please feel free to duplicate copies to teachers and others who can use them. (Each issue contains a paragraph giving educational institutions blanket permission to make additional copies). The only thing we ask is that you either reproduce the issue in full or — if you only use specific articles — that you give full credit and copyright information on each copy.

A SCIENCE AND MATHEMATICS TEACHING CENTER

Walter Bisard
Central Michigan University, Mt. Pleasant, Michigan 48859, U.S.A.

1. A Crisis in Science Education?

It is widely known that a crisis in science and mathematics teaching exists in the United States. This crisis has reached all levels of education, from elementary to secondary to colleges and universities. The problem, which is easy to define but difficult to resolve, is trifold: there are not enough high-quality science and mathematics teachers; present teachers are teaching out-of field or are out-of-date and in need of subject updating; and the average education graduate is only minimally qualified to teach science and math. These factors have caused national and state "alarms" to be published that point out the increasing mediocrity and an overall lack of science education in our nation's schools.

2. Science and Mathematics Teaching Center

Central Michigan University, long among the nation's leaders in the training of teachers at all levels, is taking steps to remedy this alarming situation. A Science and Mathematics Teaching Center, designed to confront the problem head-on, has been founded. The Center's major purpose is to improve the quality of science and math teaching. This goal is being accomplished through a series of workshops, seminars, in-service programs, conferences, and teacher outreach programs.

3. What the Center Can Do For Teachers

Through a series of seminars, workshops, and mini-courses, the Center is well-prepared to instruct the teacher and prospective teacher in the latest developments in science and mathematics education. Even though these functions cover a wide range of topics, all have common elements, including:

- the emphasis upon hands-on processes or activities and the development of higher level thinking skills;
- the application of science and math to current events and technology;
- the introduction of new instructional materials and techniques; and
- the covering of interdisciplinary topics which do not fit well in existing courses.

For instance, some of the seminars/workshops presented by the Center have included series on Halley's Comet; Mathematics and the Gifted Student; Utilizing the Apple Computer in the Physics Lab; Keeping Reptiles in the Classroom; and National Science Teachers Association (NSTA) Materials For Teachers. Future programs will be devoted to the Michigan Educational Assessment Program (MEAP) science tests, problem solving in math, influencing females to study science/math, weather instruments, and examination of moon rocks. Major science conferences, such as the 35th Edison Science Institute, featuring many of the nation's best science speakers, also are planned.

Continuing relationships with the NASA educational materials network via the NASA Regional Teachers Resource Room at the Center and the National Science Teachers Association allows firsthand examination of relevant materials and classroom science activities/experiments. Just as science has made tremendous strides over the years, so have the materials and equipment that accompany the teaching of science and math. The Center will serve as a centralized source of materials that are vital to the continuing education of teaching professionals. This function is a team effort with the adjacent Instructional Materials Center.

There are many new materials in science and math education, many of which remain unknown to local school districts. Often these materials require special maintenance, care, and instruction in their use. The Center, with its facilities and expertise of its faculty members, will address those special needs. An example of this new and promising technology is an interactive laser-video-disc-system that is being developed in particular subjects.

The Center will provide a science/math lab for student microteaching. This will allow for the videotaping of a teacher's performance for evaluation at a later date.

The Center has worked very closely with hundreds of Michigan school districts and other universities in an effort to improve the quality of science and mathematics teaching. Along with arranging for mini-courses, workshops, and seminars for teachers, the Center also will assist in the evaluation of a school district's curricula and make recommendations for improvement. This effort includes extensive and cooperative grant development with school districts for the improvement of science/math education. Recently, a major grant helped train master teachers of science and mathematics at the middle school level who then returned to their school districts to conduct professional development activities for their fellow teachers.

THE WEEK-END ASTRONOMY CLASS FOR TEACHERS

James H. Hensley
Physics Department, University of Wisconsin–Platteville, One University Plaza, Platteville, Wisconsin 53818-3099, U.S.A.

1. Introduction

Astronomy is an integral part of many high-school science programs. Project STAR and the Science Assessment and Research Project at the University of Minnesota have recently recognized this. In addition, astronomy is a part of most elementary and middle-school science programs. In the Platteville, Wisconsin, school system, the solar system is a unit of study for all third grade students and a study of the stars is a part of the eighth grade science program. This is also true for other school systems in this area, in the Chicago area, and I would suspect, across the nation.

However, most elementary school teachers have had little science course work and none in astronomy. Middle-school and high-school teachers have better backgrounds for teaching science but little or no astronomy course work. Some of those who teach astronomy are active in local astronomy groups and read *Astronomy* or *Sky and Telescope* magazines, but this is the exception rather than the rule.

2. The One-Credit Solution

Teachers return to the university to take courses for a variety of reasons. Some take additional courses to advance on the salary schedule, some to meet accreditation requirements, and others to make up deficiencies. Since the fall of 1981, one-credit, Saturday courses in astronomy have been offered through the Office of Conferences and Workshops at Governors State University, located in University Park, a suburb of Chicago on the far south side.

The courses meet on two consecutive Saturdays, from 9:00 a.m. to 5:00 p.m., with an hour break at noon for lunch. The students are in class for fourteen hours and receive one credit hour upon successful completion of the course. The Saturday format allows teachers to complete the course as conveniently as possible. The courses have attracted teachers at grades K through 12, students completing degrees at Governors State University, and others interested in learning about astronomy.

No general source of outside funding is available for the teachers to attend. Many school districts will provide reimbursement for tuition and fees for teachers to take additional course work in approved areas of study. Some of the teachers were given support by their local school district.

Governors State University has supported the weekend astronomy courses by providing the teaching staff.

The first course in the series is *Astronomy in the Classroom*, a course designed to give the participants some updated knowledge in selected areas, such as results from the Voyager Spacecraft, the Supernova 1987A, and the Giotto Flyby of Comet Halley; hands-on experience with selected astronomy activities, such as the pocket sun clock, phase of the moon with small spheres, and elliptical orbits; and sources for astronomy resource materials: the Astronomical Society of the Pacific, Hansen Planetarium, and NASA.

It became apparent after *Astronomy in the Classroom* was completed in the fall of 1981 that a need existed for additional astronomy courses. Subsequently the following courses were developed: *Terrestrial Planets; Jovian Planets; Comet Halley; Stars and Nebulas; Galaxies and Quasars; Survey of the Night Sky; Introduction to Archaeoastronomy; Telescopes and Observatories; Intelligent Life in the Universe.*

An extensive set of class notes was developed for each course. The notes are reproduced and given to each student. These notes give the student a structure to follow as the day progresses. Breaks are taken every 50 to 60 minutes, depending on the material and activities of the course. Whenever possible, activities illustrating the subject matter are included. As an example, an incandescent filament controlled by a rheostat is viewed through a diffraction grating to illustrate a continuous spectrum and Wien's Displacement Law. Students draw several ellipses with string, thumb tacks, pencil, cardboard, and paper, and then determine the index of flatness for each ellipse.

3. Conclusions

The one-credit, two-Saturday, format for astronomy courses for teachers has worked well in the south suburban Chicago area. The course offerings are rotated so that each course is repeated once every two years. The class notes are revised each time a course is offered. New activities are integrated into the material whenever possible.

The student response to the courses has been excellent. Most students who complete one course come back for additional courses, and many have completed the entire sequence. The enrollments have ranged from twenty to sixty. The courses will be offered again during the 1988–89 school year on selected Saturdays.

Daily schedules for each class may be obtained by writing to the author.

THE TRAINING OF PRE-COLLEGE TEACHERS THROUGH WORKSHOPS IN ASTRONOMY

Mary Kay Hemenway
Astronomy Department, University of Texas, Austin, Texas 78712, U.S.A.

Most pre-college teachers in the United States have not been trained in astronomy. In Texas, astronomical concepts enter the pre-college science curriculum at all levels.

During the last six years, I have presented two different types of workshops: intensive three-week summer institutes for rather small numbers of teachers, and shorter 6- to 18-hour workshops for larger groups.

The University of Texas Astronomy Department and McDonald Observatory have hosted five of the intensive institutes. The institutes were established to offer an opportunity to learn astronomy through hands-on activities (Texas education regulations mandate that 40 per cent of all pre-college science classes should be laboratory oriented), and to introduce the teachers to modern astronomical research and facilities.

The teachers bring a variety of scientific experiences to the institute. Unlike most teachers in Texas, 52 per cent of them have had an astronomy course. Since most pay all their own expenses, they demonstrate a high interest in astronomy. In five years, a total of 44 teachers have attended a three-week institute with six teachers attending more than once. Their background preparation showed that only 12 per cent had had an astronomy laboratory course, while 30 per cent had attended astronomy workshops. Astronomy as a separate subject was taught by 18 per cent.

The institute is an intensive experience with about 100 contact hours spread over the three weeks. Although the participants attend some lectures on current research by other faculty or research staff, the main focus of the institute is to provide them with hands-on activities and experiences that they can use in the classroom. The activities performed by the participants are taken from *Modern*

Astronomy (Robbins and Hemenway 1982).

Participants begin by learning about measuring errors by building and using a cross-staff and quadrant. They use these instruments to measure the Earth's rotation rate through night-time observations of stars. Depending upon their own interests and scientific training, they choose from a variety of other experiments, with each participant receiving individual attention from the instructor. The most frequently chosen activities concern lunar surface features, the apparent motion of the sun, optics, using small telescopes, photography, and spectroscopy. For the past three years, the students have had an observing run at McDonald Observatory near Fort Davis in west Texas during the third week. Part of the two weeks in Austin is spent preparing to use the 76-cm Cassegrain reflector, with a 4" x 5" (10 cm x 12.5 cm film size) film camera. In west Texas, besides observing and developing film, they enjoy tours of the observatory complex and an opportunity to stay with professional astronomers. Thus the institute offers the pre-college teacher both classroom oriented exercises as well as a chance to share the experiences of research astronomers.

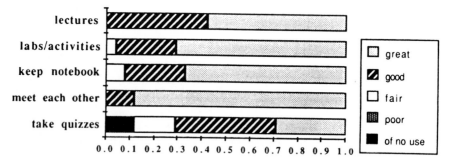

Fig. 1: *Participants perceptions of summer institutes, 1984–87, based on 71 % — complete survey of 34 participents.*

Results of a follow-up survey showed very positive perceptions of the institute. Participants especially valued the opportunity to meet other teachers with similar interests, and liked the discipline of maintaining a lab notebook. More importantly, 97 per cent of the participants used information from the institute in their classes. The activities themselves were used by 87 per cent in their classes.

With shorter workshops and different audiences, time constraints alter the structure and content of the workshop. Over six years, I gave 9 workshops to about 200 secondary-school teachers and 16 workshops to almost 450 elementary-school teachers.

Elementary-school teachers often have had poor preparation for science teaching. In Texas, prior to 1984, they needed no science courses to receive their certification, and now only need two courses. Science teachers major in science in preparation for teaching at the secondary level, but often teach outside their field. Workshops are often the only formal training in astronomy that a typical teacher receives.

In a lecture-demonstration format, a lot of information is conveyed but little real understanding occurs. Changing adults' pre-conceptions is difficult. Pre-tests show that most elementary-school teachers can't explain ideas like the seasons, lunar phases, apparent motion of the sun, or how a telescope works (see also a paper by Sadler, this volume). In order to improve understanding, the workshop emphasized activities rather than lectures. Even in one day, teachers can measure solar shadows and angles, demonstrate phases of the moon with balls, compare relative sizes of solar system objects, and classify lunar surface features or galaxies. Activities for secondary teacher workshops were taken from *Modern Astronomy*, while elementary teachers' activities came from *Astronomy Adventures*. During a recent elementary teacher workshop, the pre-test grade on such concepts was 50 per cent, which improved to 68 per cent after the workshop. A shorter workshop allows larger numbers of teachers to learn some astronomy and to observe teaching by the discovery process. The cost per participant is quite low compared to a three-week workshop. Unfortunately, the size of the larger group precludes individual attention. Although elementary-school teachers teach 22 students per year (while secondary-school teachers may teach up to 300), the elementary teachers probably have a stronger influence on their students' attitudes toward science.

There is a need and demand for short workshops for teachers of all grade levels, and for intensive institutes.

References

Robbins, R.R. and Hemenway, M.K. *Modern Astronomy: An Activities Approach*. University of Texas Press (Austin, Texas 78712), 1982.

Astronomy Adventures. Ranger Rick's Nature Scope, vol. 2, #2, National Wildlife Federation (1412 16th St., Washington, D.C. 20036, U.S.A.) 1986.

Discussion

D. McNally: *Could you please tell me the price a teacher has to pay for a 3-week summer institute — U.K. teachers are most unwilling to pay anything for any form of course.*

M.K. Hemenway: Tuition and fees, text, notebook = $130 (out of state students pay an additional $300; only 2 of the 44 have been from outside Texas). Room and board for 2 weeks in Austin range from $240 to $300 (some stay with friends or relatives in the Austin area, or live there). Room and board at McDonald Observatory (3 nights) ≈ $110. Transportation varies. Most participants carpool to the McDonald Observatory and pay a portion of the gasoline (about $25) to the driver. Six of my participants have attended more than once, so they paid double the above estimates. Summary: minimum $200, maximum $860 + transportation to and from Austin.

AN ASTRONOMY COURSE FOR MEXICAN HIGH-SCHOOL-LEVEL TEACHERS

M.A. Herrera

Instituto de Astronomia — UNAM, Apartado Postal 70-264, 04510 México

1. Introduction

In 1985, the National University of Mexico (UNAM) created a series of updating courses for high-school level teachers, both as an answer to an explicit demand of the teachers themselves, who asked for "fresh" information in their fields, and as an effort to improve the general level of undergraduate students, whose grades in the entrance examinations to the University had been secularly decreasing for years. To date, three "packages" of intensive 50-hour courses have been offered (during the three summer vacation periods) and Astronomy has been present in all of them. However, it should be mentioned that this presence was due only to the personal interest of the director of the project, since the subject "Astronomy" is *not* included in the official programs. In the following, we present a brief description of the courses and some interesting results.

2. Program and Bibliography

The original purpose of the course was to provide a more or less complete perspective of the most recent and spectacular results of astronomical research, such as accretion disks around black holes, the early universe (strings and axions), active nuclei, *etc.* However, it was soon found that the astronomical background of all the "students" was essentially zero, a fact that prevented them from following even the most elementary arguments if astronomical concepts were involved. It was thus decided to abandon the original program and to give, instead, a "canonical" introductory course on astronomy. At this point, a new problem arose when the students rejected the proposed bibliography (Abell, Zeilik, Shu, *etc.*) because "they are in English," and a suitable bibliography in Spanish had to be looked for. The final decision was to lecture without an "official" text, although references to "Cosmos" (Sagan), "The First Three Minutes" (Weinberg), "Introduction to Science" (Asimov), *etc.*, were liberally made.

3. Attendance and Interest

The original registration and the actual attendance in the courses were as follows: in 1985 — 8 and 6; in 1986 — 17 and 11; in 1987 — 22 and 16. It is evident that the numbers are rather poor, particularly if we consider that there are some 5000 high-school teachers in the University system. This problem, however, has not been restricted to astronomy; in fact, it has been general. The most "crowded" course, for instance, has always been computer science, and its "record" is 40 participants. Also, an average of 7 courses, out of a total of 25, have been canceled every year because their registration was precisely zero. Taking this into account, and remembering that astronomy is the only course whose subject is not taught in high schools (*i.e.*, there are no astronomy teachers), our average attendance of 15 turns out to be excellent.

The basic preparation of those attending was: in physics, 19 per cent; in chemistry, 53 per cent; in other subjects, 28 per cent. It is interesting to note that about one-fourth of the students do not teach subjects directly related to astronomy. We have had biology teachers, philosophy teachers, aesthetics teachers and even a French teacher. This shows, again, that astronomy (as we already knew) is as appealing to the scientific mind as it is to the artist or philosopher.

The lack of previous preparation in astronomy posed no problem with regard to interest. In fact, there were so many questions and discussions that the proposed program has never been completely covered.

4. Method and Evaluation

All the lectures consisted of an oral presentation accompanied by numerous slides. The main idea was to motivate the students to participate, and questions were not only allowed at any moment, but also were stimulated. The dialogue, once established, was used to extend, to complement, or to deepen the concept under discussion.

It was very interesting to discover that teachers hate to be evaluated. In spite of their resistance, a couple of tests were prepared by the instructors and left as "homework," with a formal promise from the students that they would do them alone (some didn't). A final result, which was either "yes" or "no", was derived from the examinations and from their participation and interest during the course. There were only a few failures in each course.

5. Results and Conclusions

First of all, it should be mentioned that we all enjoyed the course. There was always lots of discussion and interest didn't seem to decay, even when the lectures lasted for three straight hours. These long discussions also revealed some unexpected facts, the most interesting (in our opinion) being the following:

a) All the teachers, independently of their specific subject, were equally ignorant about the simplest motions of the celestial sphere. In fact, they had never noticed the diurnal motions of the stars (they knew, of course, that the Earth rotates, but they had never associated this motion with actual observable motions). Most of them also thought that the phases of the moon were eclipses and, as a final example, not a single one understood the mechanism that produces the seasons. (Ed. note: *See also Sadler paper, this volume*).

b) A biology teacher turned out to be the most ardent enemy of Darwin I have ever met. We had many of the typical (and useless) discussions that occur when faith confronts scientific fact and which always end with each protagonist believing the same thing he believed before the discussion started.

   c) Two chemistry teachers (chemists themselves) were absolutely convinced of the influence of the moon on the growth of plants and hair.

   d) About 50 per cent of the total attendance were firmly convinced of the validity of astrology (and, of course, they also accepted pyramidology, the Bermuda Triangle, *etc.*).

   e) UFOs were accepted, and defended, by some 25 per cent of the total attendance. After a discussion on relativity and interstellar trips, there were still lots of phrases like "it could be, anyway," and "scientists have been wrong before," *etc.*

   A final comment, which follows from the preceding results, is that this kind of course should be given as frequently as possible and that, in addition, it should be compulsory for all high-school level teachers to attend.

THE DIDACTIC ACTIVITIES OF THE ITALIAN ASTRONOMICAL SOCIETY

M. Elena Dilaghi Pestellini
Responsabile Didattica Societá Astronomica Italiana
Osservatorio Astrofisico di Arcetri, Largo E. Fermi, 5, Firenze Italia 50125

   The Italian Astronomical Society is one of the oldest Italian scientific associations: founded in Rome in 1871 as the Society for Italian Spectroscopists, it includes professional as well as amateur astronomers and teachers among its 800 members.

   In the last 10 years, the Society has been concerned with the problems of teaching, as a result of the enormous demand from schools where astronomy is often not well taught at any level because of a lack of suitable qualified teachers.

   To meet this need, the Society publishes a quarterly journal entitled *Il Giornale di Astronomia*, which includes news of astronomical events, informative articles and accounts of didactic experiences as well as other information that may be helpful to teachers of astronomy.

   Issues on single topics have also been published, such as *Observation of the Sky, Astronomical Exercises, The Planetary System, Astrophysics, Analysis of Electromagnetic Waves in Astrophysics, etc.*, as well as a series of teaching aids that are easy to consult and practical to apply.

   The Society periodically organizes refresher courses for teachers in schools of all levels. These courses are held in cities throughout Italy by professional astronomers and experts in educational problems. Summer residential courses were held, in collaboration with the Ministry of Public Education: in Asiago in 1986 for physics teachers with the topic "Analysis of Electromagnetic Waves in Astrophysics," and in Potenza in 1987 for teachers of Natural Science with the topic "Positional Astronomy and the Planetary System." The local courses or summer schools consist of theoretical lessons, experiments, and daily and nightly observations of the sky. During the development of courses, teaching equipment, duplicated lecture notes, slides, audio-visual material, *etc.*, are assigned; later, teachers who have attended the lessons experiment with new methods of teaching and curricula together with the pupils.

Fig. 1. Exhibition of material for the teaching of astronomy.

The Astrophysical Observatory of Arcetri has prepared an exhibition of basic teaching equipment in positional astronomy in collaboration with the Society, which undertook to set up the exhibition in various Italian cities. The exhibition was prepared by the Astrophysical Observatory of Arcetri during the didactic performance "Scuola come" in Florence in 1984; the I.A.S. presented the exhibition in many Italian cities, together with lectures, lessons for the teachers, and visits for pupils. An astronomer assisted. Besides a short history of astronomical research, the exhibition included basic experimental instruments.

The exhibition was created by P. Ranfagni (Institute of Astronomy — Florence University) and E. Brunetti (Astrophysical Observatory of Arcetri).

The Society has announced a biennial prize for teaching research in astronomy named after Prof. G. Tagliaferri; the prize is open to teachers and researchers. Topics include teaching experiences from various schools, educational aids, basic computer programs, slide series, teaching grants, *etc.* P. Ranfagni, E. Brunetti, and

P. Stefanini made a series of three hundred slides; these are of special educational value and they have been used during numerous refresher courses on the subject "sky observation."

The Society is also represented in certain commissions of the Ministry of Public Education for the compilation of new programs for physics and natural science.

ASTRONOMY ACTIVITIES OF THE AMERICAN ASSOCIATION OF PHYSICS TEACHERS

Robert J. Dukes, Jr.
Physics Department, The College of Charleston, Charleston, South Carolina 29424, U.S.A.

The American Association of Physics Teachers (AAPT) was founded in 1930 by members of the American Physical Society whose main interest was the teaching of physics. Since then its primary goal has remained the improvement of physics teaching at all levels. The Association publishes two journals (*American Journal of Physics* and *The Physics Teacher*) and a bulletin (*The Announcer*). The *American Journal of Physics* is intended for papers concerning physics teaching at the level of colleges and universities in the United States while *The Physics Teacher* is concerned primarily with introductory physics, including high-school courses. Two national meetings are held annually as well as numerous regional meetings.

Fortunately for astronomy education, AAPT's definition of physics has from the first included astrophysics. Early papers in the *American Journal of Physics* (then called the *American Physics Teacher*) included "Epsilon Aurigae, Colossus Among Stars, a Story of Co-operative Research in Photometry, Spectroscopy, and the Theory of Gases" by Struve and Lemon (1938) and "Mnemonic for Bethe's Solar Energy Reactions" by Randall (1948).

In 1983 the AAPT Executive Board established a Committee on Astronomy Education whose primary duty was to provide information for physicists who are teaching astronomy. This committee has sponsored a number of symposia at various meetings of the Association as well as working to publicize activities of other societies concerned with astronomy education among AAPT members. A survey of AAPT national meetings from 1971 through 1983 (Dukes and Strauch, 1984) showed that approximately five per cent of all papers presented at national AAPT meetings concerned astronomy (approximately 170 papers on astronomy out of a total of 3000 papers). This ratio has remained approximately constant to the present when 20 of the approximately 400 papers presented at the two 1988 national meetings concerned astronomy. Also in 1988 five papers on astronomy were published in *The Physics Teacher* and approximately 18 papers in the *American Journal of Physics*.

Finally the *American Journal of Physics* publishes a series of annotated bibliographies called "Resource Letters." Two have been directly concerned with astronomy education. The first of these was an extensive listing of astronomy education resources (Berendzen and DeVorkin, 1973). The second listed laboratory exercises for astronomy (Kruglak, 1976). Others that are of interest to astronomers include those on cosmology (Ryan and Shepley, 1976), black holes (Detweiler, 1981), x-ray astronomy (Canizares, 1984), and cosmology and particle physics (Lindley, Kolb, and Schramm, 1988).

In summary, the American Association of Physics Teachers through its various activities is continually working towards improving astronomy education in this country. Its journals are receptive to manuscripts on topics similar to many of those presented at this meeting. I encourage you to submit papers on astronomy education to them.

References

Berendzen, R., and DeVorkin, D. 1973, "Resource Letter EMAA-1: Educational Materials in Astronomy and Astrophysics," *Am. J. Phys.*, **41**, 783–808.

Canizares, C.R. 1984, "Resource Letter XRA-1: X-Ray Astronomy," *Am. J. Phys.*, **52**, 111–19.

Detweiler, S. 1981, "Resource Letter BH-1: Black Holes," *Am. J. Phys.*, **49**, 915–25.

Dukes, R.J., Jr., and Strauch, K. 1984, "Trends in Physics Teaching," *Physics Teacher*, **22**, 446–448.

Kruglak, H. 1976, "Laboratory Experiences for Elementary Astronomy," *Am. J. Phys.*, **44**, 828–33.

Lindley, D., Kolb, E.W., and Schramm, D.N. 1988, "Resource Letter CCP-1: Cosmology and Particle Physics," *Am. J. Phys.*, **56**, 492–501.

Randall, C.A. 1948, "Mnemonic for Bethe's Solar-Energy Reaction," *Am. J. Phys.*, **16**, 56.

Ryan, M.P., Jr., and Shepley, L.C. 1976, "Resource Letter RC-1: Cosmology," *Am. J. Phys.*, **44**, 223–30.

Struve, O. and Lemon, H.B. 1938, "Epsilon Aurigae, Colossus Among Stars, a Story of Co-operative Research in Photometry, Spectroscopy, and the Theory of Gases," *Am. J. Phys.*, **6**, 123.

Comment

Jay M. Pasachoff: We have proposed in the AAPT Astronomy Education Committee, that the organization's name be changed to "The Association of Astronomy and Physics Teachers." This choice would retain the current initials and logo. Though the request did not pass at the highest levels, it is being repeated and may well be granted eventually.

Hopkins Observatory 1837

Stereoscopic view of the Hopkins Observatory, photographer unknown. Williamsiana Collection, Williams College. Further information on page 432.

Chapter 11

Popularization

Astronomy is fortunate in being one of the most attractive sciences to the general public. We hear first the personal story of a prolific author and television popularizer of longest standing. A group of three papers discusses amateur astronomy and its role in education. A series of papers then considers astronomy for the general public in Mexico, India, Scotland, West Germany, New Zealand, and Australia.

Several of these papers discuss public observatories, which are particularly numerous in Europe. Finally, we read about efforts in Australia and the United States to bring astronomy to students older than the traditional student population.

THE POPULARIZATION OF ASTRONOMY

Patrick Moore
Farthings, West St., Selsey, Sussex, U.K.

I suppose it is inevitable that astronomy should be one of the easier sciences to "popularize." The sky is all around us; even our remote cave-dwelling ancestors must have looked up into the sky and wondered at what they saw there, even though they could have no idea of the nature or scale of the universe. Naturally, they believed the Earth to be supreme, and to have everything else arranged around it for our special convenience. Believe it or not, this point of view is not quite dead even now — and this brings me on to my first point.

Some time ago I attended a meeting of the International Flat Earth Society, held in London. Its members believe that the world is shaped like a pancake, with the North Pole in the middle and a wall of ice all around. The meeting was quite remarkable, and participants were totally sincere. Later, I rather ill-naturedly put them in touch with a German society whose members maintain that we live on the inside of a hollow sphere, and I understand that they are still fighting it out; but of course this is quite harmless — and as I have often said, the world would be poorer without its "Independent Thinkers." But other aspects of eccentric thought are less laudable, and of course I am thinking of astrology, which has experienced a curious revival in recent times. Regrettably, it has even been given tacit approval by a few professional astronomers who certainly ought not to be lured into making unwise statements. I will not take this any further; suffice to say that in my view, it is essential at the outset to make a clear differentiation between astronomy and astrology. I maintain that astrology proves only one scientific fact — "There's one born every minute" — but it can do harm.

I come now to pure astronomy. And I think I must ask your indulgence if I spend a minute or so in a personal view, because unlike most of you, I am not a professional astronomer, and have never been. My doctorate of science is an honorary one. What should have been my university years were spent flying; this was the period from 1939 to 1945, and again there is no need to say more about that aspect now.

My first interest, and still my main interest, concerns the Moon, and this was also the subject of the first book I ever wrote under my own authorship (it had been preceded only by a translation from the French of Dr. Gérard de Vaucouleurs' admirable little book about Mars). In 1957 I began my BBC television series *The Sky at Night*, which still continues; I have not missed a month since then, which I understand is something of a record. But the longevity of the series is not due to me; it is due entirely to the subject; each program is watched by around 4,000,000 viewers, so it may have been some help in "spreading the word."

There are some obvious traps into which it is too easy to fall, and into which I have no doubt fallen on many occasions. First, astronomy is a spectacular science, and there is an obvious tendency to overdramatize it. In a popular book, or for that matter a television program, it is simple to show view after view of "what Mars may be like," "an alien civilization on Beta Cygni C," or "astronauts touching down on a planet of a sun inside the Hercules cluster." In moderation this is all very well, but if taken too far it can lead to a completely false impression of what it is all about.

Secondly, you need to maintain the novice's interest, and to delve too deeply into technicalities straight away means that those who are not genuinely fascinated will drift away. It means steering a middle course.

When I started my television series, the Space Age had still not begun; it was six months later that it was opened by Sputnik 1 — not with a whimper, but with a very pronounced bang. The change in outlook was evident at once, and it was sometimes assumed that the only object of studying astronomy was to send men to the Moon. This was also the time of the popular awakening to the value of radio astronomy, with the inauguration of the 250-foot (76-meter) "dish" at Jodrell Bank now known, very aptly, as the Lovell Telescope. It has been said to me that radio astronomy makes all optical work obsolete — and one has to explain that all branches of the science work together rather than in separate compartments.

But perhaps the most hackneyed question of all, asked at many popular lectures, is: "What is the use of studying the stars, and sending men into space, when there is still so much to be done down here on Earth?" Some political extremists are only too eager to further this impression. One has to explain that it is no longer possible to separate any branch of science from any other, any more than one can divorce arithmetic from algebra, but when people are sufficiently indoctrinated it takes a good deal of persuasion to drive one's point home. Not everyone can readily appreciate the close links among astronomy, physics, chemistry, medicine and biology, for example, and I am bound to say that these links are not always emphasized as forcefully as they ought to be.

Another question, which has been put to me more times than I can count, concerns the methods of "starting out." The usual letter begins: "I think I am interested in astronomy, but I do not see how I can make a start without spending a large sum of money on an expensive telescope, and this is something which I cannot afford." Here I feel that books that show nothing but amazing pictures taken with the world's great instruments are somewhat misleading. I doubt if any Voyager view of Saturn, for example, can be as telling as one's first actual view of the planet through a 3-inch (7.5-cm) telescope.

On average, I receive around forty letters per day, many of them from young enthusiasts, but also many from adults. (I answer them all, but it does take time; my ancient typewriter works overtime, particularly when I have been away for a week or two.) And my answers to the standard questions are always more or less the same. I think it may be worth my summarizing them here, though others may have different approaches:

1. *Do some reading from a suitable elementary book.* This is what I did myself, at the age of six (which takes me back to 1929). The book was called *The Story of the Solar System,* by G.F. Chambers, and it had been published in 1898, so that even then it was somewhat out of date — but the essentials were there.

2. *Obtain a simple star-map, go outdoors on a dark night, and start learning your way around the constellations.* As we all know, this is not nearly so difficult as might be thought — ask the absolute beginner how many stars he can see on a clear night, and he is apt to reply, "Millions" — and I remember making a pious resolution to identify one new constellation every night. It soon worked. The method is to select one or two obvious groups, beginning probably with Orion (if visible) and Ursa Major, and use them as guides to the rest. In fact, starting with these two only, one can in time work out all the rest. I know, because I have done it myself. The old cliché about an ounce of practice being better than a ton of theory is true in astronomy as it is true in everything else.

3. Next, if you are still interested, *consider some optical aid.* This is where the trouble often starts. Buying a very small telescope is a recipe for disaster. These tiny instruments have poor optics, small fields of view, and mounting about as firm as blancmanges[1]. Unfortunately, advertisers can make them seem very attractive. For years I have been waging a war against them, and I think with some success; the obvious alternative is to buy binoculars — which are much more useful to the beginner than he or she will appreciate at first.

I have a standard "telescope letter" that I send out, and I have distributed thousands of copies, mainly to young inquirers. Again, not everyone will agree with me in saying that the minimum useful aperture for a refractor is 3 inches (7.5 cm),

[1][Ed. Note: A blancmange is a dessert, made from gelatinous or starchy substances, and shaped in a mold, particularly popular in England.]

and for a Newtonian reflector 6 inches (15 cm), but I feel that it is at least of the right order.

4. *Join a society.* In Britain, as in the United States, there are many local societies. We also have the national British Astronomical Association, which is mainly amateur though with many professionals, and which has an outstanding observational record. There is no actual age limit — I joined when I was eleven, though I think this was a record at the time. (I became President exactly fifty years later!)

By then I think the beginner will have settled firmly into one of two categories. He or she may simply prefer to remain an "armchair astronomer," following what goes on and taking an intelligent interest; or he and she may want to undertake practical observing — and here I feel it is important to stress that astronomy is one of the very few sciences in which the amateur can still play a useful rôle. I need not elaborate this here; all of us know how valuable are the contributions from, for example, the comet-hunters and the nova and supernova hunters, and from those who follow variable stars and time-dependent planetary phenomena.

For your armchair astronomer, it is surely important to maintain a flow of information which is technical enough to be worth while, but not too technical to "lose" the enthusiast who has only a limited amount of time to spare for his hobby (as is almost always the case). This is where books, broadcasts, and television programs come in. When I started out, there were very few popular books either for adults or for juveniles; the book I happened to read was an adult one, but it was simply written. Nowadays, the choice is very wide, and it has to be said that the standard is very variable. There is not much that can be done about this; one has to use one's own instinct; to sort out books that look attractive, but soon lapse into astrology and flying saucers, is not always easy.

The vital thing is that early enthusiasm should not be killed. And this is where I am bound to be slightly controversial, even though my remarks apply to Britain and may not be applicable to the United States. School subjects can be made dull. Sadly, they frequently are — and this is why I have never been in favor of making astronomy a school subject "on its own." Of course, it is part of science classes, and must be so; but I would be unhappy to see it taught as a special subject divorced from other science classes. As things are, the would-be enthusiasts gravitate to it naturally. If it is forced down their throats, they may recoil. The ideal is to have school astronomy clubs, which are numerous and which do a splendid job.

There are, too, many adult classes that are equally useful. They can, obviously, be more specialized, and this is where I feel able to say a little more about television, simply because I have been involved in it for so long. I try to give *Sky at Night* programs at various levels — sometimes very elementary, others deeper; and where we go more deeply into a subject, I am always anxious to involve an expert who can speak with authority, as opposed to my doing it second-hand. I may add that those who have appeared on *Sky at Night* programs with include Harlow Shapley, Bart Bok, Clyde Tombaugh, Neil Armstrong, Dale Cruikshank, Fred Hoyle, Fred Whipple, Jan Oort, Yuri Gagarin, Alla Masevich — quite a galaxy! I find that the programs are watched not only by beginners but also by those who are experts in their own fields and want to be kept abreast of what is happening elsewhere. As I have often said, your cosmologist need not be aware of the latest investigations into the behavior of clouds on Mars.

Finally, one must cater to the young enthusiast who is interested enough to want to follow astronomy as a profession. Again I have had many inquiries; I do my best to help, and I must admit that it gives me immense pleasure to find those who are now eminent in their fields, carrying out work of which I would never be capable, but who first contacted me when they were about to "start out." If I have any rôle to play in the realm of astronomy, it is in urging on others to do things that are far beyond my own capabilities.

There was a time — and I do remember it — when astronomy was regarded as a study separate from anything else, and when the popular image of an astronomer was that of an old man with a long white beard sitting in a lonely mountain-top observatory, night after night, watching the stars. Untrue thought this was, it was a deeply-rooted picture. Of course it is not so today, and I think that most people are aware of what astronomy means. What we have to continue doing, I am sure, is to make certain that we present the right views — and this is being done, thanks to the many professional astronomers who realize the value of popularization. After all, the beginner of today is the researcher of tomorrow.

Discussion

J. O'Byrne: I agree with the comments about telescopes, and especially binoculars. However, there is a place for some smaller refractors if only because people insist on buying them. There are a few good quality 60-mm refractors available that will show the rings of Saturn, the bands and satellites of Jupiter, and the Moon reasonably well. They also provide a view of some double stars and clusters. Often a child wants a telescope and nothing larger is possible. Providing it is good quality, such a telescope can play a rôle.

P. Moore: Saturn is a special case. With the other planets, a small telescope will show little more than good binoculars, and this also applies to the moon. In my view, there are no real advantages in a tiny telescope as compared with binoculars.

Ed. Note: In small telescopes, the quality of the mounting is often more of a problem than the quality of the optics. No telescope of any size is of use if the mounting does not hold steady or permit setting on an object.

S. Isobe: There are many good astronomy books that give a whole view of astronomy. From pictures in these books, the general public sometimes thinks that the larger or the more expensive the astronomical instruments such as telescopes are, the better view of astronomical objects they can get. This is not true and we should teach what astronomical view one can get with different sizes of astronomical instruments.

THE CONTRIBUTION OF AMATEUR ASTRONOMERS TO ASTRONOMY EDUCATION

Cecylia Iwaniszewska
Institute of Astronomy, Nicolaus Copernicus University, Torun, Poland

I would like to dedicate this paper to the memory of my husband, Henryk Iwaniszewski, an astronomer working in radio astronomy and electronics, who until his untimely death seven years ago had been very active as president of our local branch of the Polish Amateur Astronomers Association. He was especially keen about introducing astronomy to the general public.

I want to speak here mainly, but not exclusively, about the IAU Colloquium No. 98, "Contribution of Amateur Astronomers to Astronomy," which was held in 1988 in Paris. First of all, some definitions. Thomas Williams of the American Association of Variable Star Observers (AAVSO), from Houston, Texas, introduced at the conference several criteria for identification — first of astronomers, and then of professionals and amateurs. According to Williams:

- astronomers must display a serious intent to contribute to the advancement of astronomy,
- they must produce results over an extended period of time,
- their work must be conducted using acceptable methods or techniques,
- the results obtained should be communicated to other astronomers.

The distinction between professionals and amateurs is as follows:

- a professional astronomer is a person who practices the science of astronomy for a livelihood with great skill,
- an amateur astronomer does astronomy for pleasure rather than for money, and derives his or her income from other means than astronomy.

A distinction should be made here also between amateurs and recreational sky observers, who, while appreciating the beauties of the night sky, are not motivated to use their time and skill to contribute to science.

Taking into account the above criteria, Thomas Williams produced a world list of amateurs, or amateurs who later became professionals, comprising nearly 700 names, from the mid-sixteenth century to the present. When looking over the astronomical interests of the persons on this list, I found out that in seventy cases, just ten per cent, the amateurs had been interested in education or popularization, although not exclusively. This was the past; what are the interests of present day amateurs?

At the Paris Colloquium, we saw that the amateurs' contribution to astronomy might be divided into three parts, directed towards three aims:

— to astronomical observations, where they can contribute valuable data for further research work, and indeed they are a great help for professionals,

— to the past, the history of astronomy, where they show more patience in studying old documents and taking care of historical instruments and buildings,

— to the present and future, to the education of contemporary society and generations to come, introducing sometimes quite spectacular forms of popularization.

When listening for five days to the accounts of the activities of amateur astronomers, I began to wonder why they display much more enthusiasm than professionals?

Is it actually because, for this group, astronomical work is a hobby, and not work one has to do for one's salary?

Is it because amateurs usually work in local associations, ordinarily independent of teaching institutions, schools, universities, which may in turn have no institutional interests in popularization at large?

Or is it because of the equal status of every member in an amateur association, independent of their professional position and education but dependent only on their abilities and amount of work they want to spend on the association activities and projects?

I think that in smaller towns, communities, or even provinces, a local association of amateurs may act for the astronomical education of the general public in a way similar to the activity of a planetarium staff, whose tasks are not restricted exclusively to arranging sky-shows under the planetarium dome.

Here are some examples of the actions of amateurs towards raising the level of astronomical education in their respective countries, examples taken from different parts of the world.

My first case is a small, or medium-sized country, with two universities, no observatory, and only one professional astronomer. How can one improve the local astronomical knowledge? By taking advantage of special events, like the coming of Comet Halley, which happened to have in the southern hemisphere especially favorable observing conditions, and . . . founding an amateurs' club. The members begin by taking photographs of the Comet; then become interested in other celestial bodies; then they want to know more, so they organize astronomical lectures and special courses; then they want to communicate with each other so they publish their own periodical, a trimestrial magazine, and run astronomical columns in daily newspapers and patronize TV programs. These people become the nucleus of astronomical activity aimed at reaching a higher level of astronomical education in the capital, in the whole country. The amateur group has been named "Club de Astrofísica del Paraguay."

It seems a pity that one cannot change the 76-year period to have Comet Halley visiting us more frequently, perhaps every 30 years, once in every generation?

In another, northern country, 7000 amateurs are members of some 30 amateur associations, the most important being the Ursa Astronomical Association. Ursa plays an important role in public education, publishing books, star maps, magazines, organizing Astronomy Days, *etc.* However, in my opinion, the most important step towards helping astronomy education in Finland has been the donation of wall star maps produced by the association to all, more than 5000, schools in the country. In that way the teachers are obliged to learn some basic astronomy to be able to comment on the gift; that would be the least they can do. Or will the maps be carefully stored in school cupboards?

One must stress here the importance of the existence of small astronomy clubs in schools. They must of course gain some more experience, a longer tradition like, for instance, the Pleiades Club in Nice, France, founded more than twenty years ago. The Club activities, like observations, exhibitions, and travel to big observatories, were supplemented a few years ago by a project of active education: the construction of a school planetarium and observatory, a building of a sort of metal-grid covered by a concrete layer and situated in the school-yard. It took about four years of children's work during school-free days, sometimes also during vacations, to have the building ready. Now they are still working in the interior. In that way the club members acquired not only astronomical knowledge but also got to know about real hard work, work done collectively, work done for a group, for others who will learn in College Valeri in the future. And that is also education.

After hard work, people like to have some recreation, to take walks through parks or woods. Why not teach people some astronomy during their leisure time instead of in the classroom? The Swiss Amateur Association designed a set of Solar-System Promenades. Taking the diameter of the Sun equal to 1.4 m, and that of Mercury equal to 5 mm, models of every planet in suitable size and distance were placed over a considerable area. Information about every planet is given next to the model. To reach Pluto from the Sun one has to stroll over 6 km, and that is a fairly good walk.

Big cities have other problems: how to introduce some astronomy to a capital with much traffic, with people rushing all the time? How to reach large groups of people waiting for the next car, the next bus? In 1985, the year of Comet Halley, a group of young French amateurs arranged for astronomy presentations on the platforms of 16 stations of the Paris underground. For two weeks, they patiently explained exhibitions, commented on movies, answered questions. And, since in the end many millions of visitors attended the shows, I suppose this was one of the biggest popularization enterprises.

When talking of large groups of people one ought to mention amateur associations, who invite the public to their yearly conventions. It happens in Japan, where the twentieth anniversary of amateurs gathered 1000 visitors; in Venezuela at the tenth anniversary of amateurs; and at the Thai Astronomical Society meeting celebrating 300 years of Thai astronomy, since the first observations of the solar eclipse were made in 1688 by King Narai.

In the National Reports on Astronomy Education sent by the National Representatives to IAU Commission 46, I have read about amateur societies and their activities in nearly every country. Their work and help in popularization and education are everywhere deeply appreciated.

Now I would like to turn to a small country, my own country, to our Polish Amateur Astronomers Association, more than sixty years old, which publishes a monthly magazine, "Urania"; which has organized national seminars on teaching problems, taking together teachers and professionals; and which organizes various forms of activity for pupils and students. Our local Torun branch offers interested secondary-school students activities in a club conducted by two professional astronomers, outside their normal duties. During the school year, they meet every week at the Observatory; during vacations they have astronomical summer camps in the mountains or at the sea. During winter, our amateur branch organizes a series of lectures for the general public, where we try an interdisciplinary approach, to show the connection of astronomy with other sciences. The titles of lecture series of the past are as follows: *Time in Nature, Energy, Visible and Invisible Light, Magnetic Fields in Nature, The Turbulent Universe,* and *Cosmic Chemistry,* the lecturers being professional astronomers, or physicists, geologists, geophysicists, or even biologists, generally from our Torun University. The lectures take place in a modern, well equipped lecture room of the old, 14–16th century Town Hall of Torun, right in the middle of the old city, a few meters from the house where Nicolaus Copernicus was born 515 years ago.

And I would like in the end to mention another solar-system model, a model for the Torun audience. The idea of preparing models based on the local highest buildings, which I use in popular lectures, came from the paper of Rainer Gaitsch from the German Federal Republic presented at the Societá Astronomica Italiana conference in 1985. Taking then the 53-m tower of the old St. John's Church of Torun as the diameter of the Sun, we get a model in which the Earth is 51 cm and the Moon is 12 cm, and all planetary orbits can be drawn on the map of Poland: the first in close vicinity to Torun, Neptune's orbit passing through our capital, Warsaw, and Pluto still farther out.

And then, let me end with the words describing the reason why amateurs are interested in the sky, the words of our Torun patron, Nicolaus Copernicus, from book I of *De Revolutionibus*: "...and what is there more beautiful than the *HEAVENS*, which indeed embrace everything that is lofty and beautiful?"

Additional Comments

C. Iwaniszewska:

1. Yesterday I received some comments about the name of Commission 46, and I see that in my talk I used the word "Education" — not "Teaching" as Prof. Sam Okoye has proposed. I think it could not be possible to change the name of the Commission, but indeed we think about *education*.

2. Since there has been some discussion about the introducing of astronomy as a separate subject, I would like to announce that in Poland a new subject has been introduced in secondary schools, "physics with astronomy," where astronomy has to be taught as another part of the subject, and not as something

left over.

3. Looking over last evening's posters, I felt much moved by one of them, about introducing astronomy to the visually disabled. I think we usually do not remember about those who cannot see the beauties of the sky.

Discussion

A.E. Troche-Boggino: *I am grateful to you for your mention of a club from my country, the* Club de Astrofísica del Paraguay *and its educational magazine* Astrocosas. *I am sure all its members are going to be pleased with what you have said of us.*

J.-C. Pecker:

1. *Even* professional astronomers *do work for the* excitement *of astronomy: if they had been looking for money, they would have selected some other occupation — banking, business ...*

2. *The "Club des Pléiades" you mention as having built an observatory and a planetarium at school by themselves, spent only slightly more than $1000 for the building. Any school can afford that small expense!*

3. *There is a serious need for some international link (union, federation, ...) among amateurs (individuals, clubs, associations, ...).*

4. *I do not believe that IAU Commissions 38 and 46 should merge: Commission 38 addresses "grown-up" qualified astronomers. Young astronomers need to travel, even after being educated or after getting their Ph.D.'s.*

5. *At school, should astronomy be taught as a part of some "physics and astrophysics" course, by a physicist? I doubt it: physicists have a tendency to describe astronomical problems as* applications *of physical questions (plasmas, spectra, solid state, thermodynamics); but then, one never sees the "unity" of the astronomical realm.*

C. Iwaniszewska: Teachers must be suitably prepared but I still think it is better to have physicists instead of biologists, *etc.*, doing astronomy teaching.

P. Moore: *I was glad that the speaker referred to the popularization of astronomy for handicapped people. I am associated, unofficially but deeply, with handicapped children, and when I was recently at a school for blind children, I was asked to prepare a text that could be produced in Braille. It is not easy, but I hope that it can be done usefully. Also, I am sure that the speaker, like myself, always warns beginners about the dangers of observing the sun. In my view, direct observations of the sun with any telescope, with or without a dark filter, should always be avoided.*

EXTRAGALACTIC OBJECTS: AN EXAMPLE OF POPULARIZING ASTRONOMY FOR THE AMATEUR ASTRONOMER

J.V. Feitzinger
Observatory of the City of Bochum and Astronomical Institute,
Ruhr-University, Postfach 10 2148, D-4630 Bochum 1, Federal Republic of Germany

1. Introduction

The handbook *Astronomy* (*Handbuch für Sternfreunde*), edited by G.D. Roth, will soon appear in its fourth edition. Newly written by 25 authors, the two volumes contain an additional chapter: Extragalactic Objects. The intention of my 50-page contribution to these volumes is to show the amateur astronomer that extragalactic work is not disappointing, ending in the northern sky with a fuzzy picture of the Andromeda Nebula.

The former editions of Roth's handbook were mainly characterized by the statement that the observation of extragalactic objects is beyond the capabilities of the amateur. In the last ten years the situation has drastically changed, especially because of new photographic techniques and sensitive emulsions and the availability of CCD's even for amateur astronomers. Small telescopes are no hindrance to doing interesting extragalactic work.

2. Simplification

To lower the threshold fear for the beginner, illustrative galaxy pictures are used. The pictures were taken by many different telescopes and cameras. Active German amateur astronomers contributed exclusively to this picture gallery.

A further way to lower the threshold fear follows the path of scientific research. In simple experimental steps, the amateur should learn how successful observations are done.

For example:

Observing	$\longrightarrow$	inventing concepts making experiments controlling variables	photography, processing
Classifying galaxies	$\longrightarrow$	questioning interpreting data using numbers	

Since the level of a broad, mixed educated readership has to be met, simplification can be used in a ninefold manner: 1) restrict to qualitative statements 2) idealize 3) simplify the statement 4) simplify the reasons for the statement 5) simplify by omission 6) generalize 7) particularize 8) use models with some special components 9) go back to the historic evolution; this offers often a simplified, naive picture.

Working with "simplification", the permissible level has to be controlled. The three controls are: 1) the simplification must be appropriate to the mixed levels of education of the readership 2) the simplification must continue and not block further, more difficult steps 3) the simplification must be technically relevant. Technical relevance and false information must always be distinguished. Technical relevance is most important.

3. Organization and Examples

The approach — "extragalactic objects" — is subdivided into the following sections: 1) catalogues and picture material 2) classification of galaxies 3) interacting galaxies and galaxies with peculiarities 4) structure of galaxies 5) general characteristics 6) brightness and color 7) stars and gas 8) mass and luminosity 9) structure formation in galaxies 10) spiral structure 11) cosmic cycles and energy distribution 12) distances 13) active galaxies and quasars 14) the universe 15) amateur techniques and amateur tasks. The emphasis of the different sections is: Sections 1–3: "find and look"; 4–8: "gather facts and think"; 9–14: "understand"; and 15: "work."

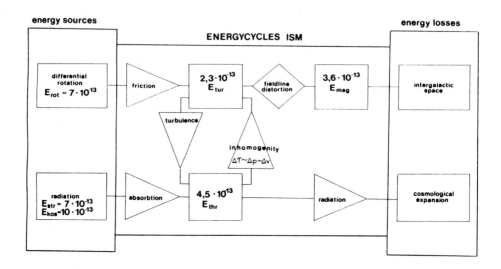

Fig. 1: *A flow diagram of the energy equilibrium in the interstellar medium is presented. The energy sources and the energy losses are linked with different reservoirs and dissipation processes. This visualizes the galactic ecosystem. The mean energies are given in erg/cm^3; E_{rot} = rotation, E_{str} = radiation, E_{kos} = cosmic radiation, E_{tur} = turbulence, E_{mag} = magnetic, and E_{thr} = thermal.*

Discussion

J. Percy: *In textbooks and other general astronomy books, there are several advantages to using photographs taken by amateurs:*

  i) They are often of very high quality.

  ii) They are often of "everyday" objects that are discussed in school textbooks.

  iii) Amateurs are usually pleased to contribute their work for publication.

  iv) Readers can "identify" with photographs taken by people like themselves.

TAKE THE "A" TRAIN[1] TO THE STARS

John Pazmino and Sidney Scheuer
Amateur Astronomers Association, 1010 Park Avenue, New York, NY 10028, U.S.A.

Astronomers, in addition to their scholarly and academic functions, have the mission to bring enlightenment to the people. In the City of New York, astronomers fulfill this mission through the Amateur Astronomers Association. Over the decades, the Association, or AAA, evolved a multi-faceted scheme of public enlightenment in astronomy. Under this scheme, astronomy in New York City has become a free-standing cultural amenity on a par with streetfairs, artshows, plays, and parades.

Once a month during the school year, the Association presents a formal public lecture on astronomy. These are convened in the American Museum of Natural History, the ancestral birthplace of the AAA. Occasionally, lectures are featured at a large university in the City for time and place variety. At these lectures, a professional astronomer explains some contemporary topic on a first-year college level, illustrated by slides and viewgraphs. The lectures — and all public activities of the AAA — are free of any charge. Area high schools and colleges employ the AAA lectures as an extra-curricular activity for their students.

In the summer, the AAA stages public stargazing in Carl Schurz Park, along the East River in Manhattan. Though located in the dense Upper East Side, Carl Schurz Park offers clear views of about two-thirds of the sky with adequate shielding from nearby lights. These sessions, convened monthly in clear weather, feature the celestial sights of the season: the moon, planets, clusters and nebulae, and double stars. Telescopes and charts are provided by the AAA.

The Amateur Astronomers Association operates the astronomy program at Gateway National Recreation Area under contract with the U.S. National Park Ser-

---

[1] *Ed. Note:* The "A" train is one of the New York subway routes that serves Harlem; it was made especially famous by Duke Ellington's jazz piece "Take the 'A' Train." The authors can provide detailed information about the "A" train for any readers who have a special interest in this topic.

vice. The parklands and preserves of Gateway stretch along the southern frontier of New York and offer clean dark skies for the urban dweller. The programs include popular-level slideshows, equipment demonstrations, skywatching tutorials, and clear-weather star-viewing. Staff for the Gateway activities is drawn from the AAA's Brooklyn and Staten Island Chapters.

National Astronomy Day is a theme day celebrated in April or May each year. In New York, National Astronomy Day is hosted by a museum, park, or school while the AAA provides the astronomy program. In conjunction with the host facility, this program includes slidetalks, equipment and project exhibits, panels and seminars, fleamarkets, and viewing of sunspots by day and the stars by night. Sometimes movies, videos, and planetarium shows round out the day's activities.

The AAA maintains a panel of experienced speakers available for other clubs, museums, schools, and social and civic groups. Speakers may be AAA members or patron astronomers. The panel can supply a single person to give a simple slidetalk or it can cater to an all-day astronomy fair with a corps of astronomers. Speakers-panel services are customized to suit the client. Fees for speakers-panel service are quite attractive to even the smallest client.

The Association serves the newsmedia by explaining and interpreting astronomy events like comets, novae, and eclipses. Newsmedia obtain quick authoritative answers to their questions and they can engage the AAA for press interviews or radio/TV appearances. The press, both national and local, routinely carry notices of AAA public activities. The Associations's interpretation mission extends also to consultations for outside authors and exhibitors.

Astronomers who serve with the Amateur Astronomers Association directly fulfill one major goal of their profession — to bring enjoyment and edification in astronomy to the people. If you have this goal, too, do consider membership in the Amateur Astronomers Association. For membership details — or for exploring other ways to practice public service astronomy through the AAA — contact us.

STRATEGIES FOR PRESENTING ASTRONOMY TO THE PUBLIC

James S. Sweitzer
The Adler Planetarium, 1300 S. Lake Shore Dr., Chicago, Illinois 60605, U.S.A.

Imagine an American trying to explain a triple play to someone who understands English poorly and who has never seen a baseball diamond, much less a game. Yet this is the sort of challenge astronomy educators face every day both in the classroom and in more informal settings.

In a college classroom, the professor has some time to teach students the language of science and astronomy. As the students gain proficiency, they can then comprehend increasingly complex concepts. Through repeated classroom exposure and study, eventually most students learn the basics of astronomy.

In a public setting such as a planetarium, or in a work of journalism, the constraints on the educator are very different. The contact time is often so brief that it is measured in minutes rather than hours. Familiarity with basic background concepts and the language of science is lacking or absent in most audiences. And finally, the public consists of voluntary learners, rather than grade-motivated students.

The purpose of this paper is to suggest three educational strategies that are very effective for public learners. The paper will finally present examples of applications of these ideas at The Adler Planetarium.

1. Basic Strategies

Strategy 1: Create pictorial models. Pictures are vital to helping people understand concepts rapidly, but not all pictures are equally useful. Some of the hardest images to understand are false-color computer images. Although they make perfect sense to their creators, they can be meaningless to those who have no prior mental concept of what they are looking at.

Cognitive scientists have long known that the mind looks for edges and shapes when trying to comprehend something new. Thus, simple line drawings are probably the most useful means for making a concept understood rapidly.

Before showing actual photographs of the real astronomical object, present a wire-frame drawing — the more generic and symbolic the better.

Although the idea upon which this strategy is based seems obvious, planetariums often violate it in the way they present complex astronomical images. When the photo is presented first, the audience never listens to the explanation because they are concentrating on their primary need to understand visually. Line drawings presented previously can avoid this.

Strategy 2: Use common language. Most scientists, when listening to accountants use financial jargon, will understand how astronomers sound to the general public. Even using an occasional word such as "excitation," "evolutionary," or "conservation" can be confusing and misleading in a brief educational experience.

Clear and simple language is best. Unfortunately, this limitation may require simplifying the subject and ignoring the complexity of the scientific issues, so it is essential to plan carefully the goal and scope of the message.

Effective language tactics include three important ones frequently used by journalists:

1. Begin with a brief "hook" to capture the audience's interest. If the message starts with a difficult concept or begins with jargon, the audience will be lost.

2. Explain concepts using analogies. This is one of the only ways to teach people about large scale phenomena quickly and effectively.

3. Use colorful, descriptive language that actively paints a picture in the minds

of the audience. Avoid the experts' natural tendency to over-explain concepts.

Strategy 3: Take humanistic perspectives. Even though the universe is vast and depersonalizing, the public does not want to feel small and insignificant. As with the previous two strategies, taking into account human perception and understanding requires the use of humanistic considerations to capture the audience's attention effectively and communicate the message. Here are some well-known techniques:

1. Make the subject concrete. Most people cannot rapidly reason abstractly. They can, however, clearly picture concepts that are made concrete.

2. A human story or human conflict can be a very powerful method to capture an audience's interest. Care must be taken, however, not to mislead or lose sight of the science when applying this technique.

3. Address important human concerns such as: life, death, and one's place in the universe to capture peoples' interest directly. Luckily, these are central topics in astronomy and motivate all who contemplate the universe.

2. Application of Strategies at The Adler Planetarium

The strategies presented above are used actively at The Adler Planetarium. Here are three programs that have been improved by these ideas.

Video telescope observing. Since 1987, The Adler Planetarium has conducted public observing programs using a computerized telescope with narrow-band filters and CCD imaging. When a picture is recorded, it is transferred live to our sky theater and projected on the dome for the audience. Any one of hundreds of possible objects can be located and projected within seconds. This opportunity seriously challenges the lecturer to explain rapidly what people are seeing.

The Adler has solved this challenge by showing audiences a simple line drawing of what they will be viewing. Often these pictures are more symbolic than realistic — a star in a bubble to represent a planetary nebula. Nevertheless, they help by giving people a mental concept in a pictorial form they can readily understand.

Exhibition language. In late 1988, The Adler mounted a small exhibition of a type of meteorite that is believed to have come from Mars. This exhibit faced two problems: first, verbal communication through exhibits is notoriously difficult because people walk away from long or complex descriptions; second, the scientific evidence and argument that the meteorites are from Mars are multifaceted and laden with jargon. Thus, for this exhibit, special attention was focused on the words. The result was an attractive title — "Mars Rocks" — and a jargon-free set of short labels presenting the important evidence in the audience's language.

Scale-of-the-solar-system show. One of the best examples at The Adler of "Strategy 3" is a short multi-image show now being produced to help people understand the scale of both the solar system and the planets. This will be done by using the old technique of creating a concrete model on a human scale. The scale for this model of the solar system was chosen to relate, in a memorable way, familiar objects with familiar places — fruits and vegetables with local landmarks. With 1 A.U. equal to 1 kilometer, and the sun at The Adler Planetarium, the Earth is the size of an apple atop the Sears tower and the other planets can be located at other famous landmarks. This model will be presented as a slide show for visitors to The Adler.

3. Conclusion

There are dangers inherent in the journalistic approach implied by these strategies. If misused it can lead to an emphasis or a simplification that misrepresents science.

Used properly, however, these techniques can help the public comprehend the exciting astronomical discoveries of our age. And if this approach is ignored, the danger is that the public will believe that astronomical knowledge is arcane and only for the elite — a message no one wants to send.

THE POPULARIZATION OF ASTRONOMY IN MEXICO

Julieta Fierro
Instituto de Astronomia, UNAM, Apartado Postal 70-264, 04510 México

Mexican professional astronomers have a great task ahead if they want to popularize astronomy. We are very few and the country has great educational needs, with 4 years of elementary school being the average educational level.

In order to overcome this challenge, the Institute of Astronomy of the National University has popularized astronomy using the following resources: 1) publications, 2) public lectures, 3) advice to museums and planetariums, 4) radio and television interviews, and 5) courses.

1. Publications

a) *Books*: A few years ago, Mexican astronomers started writing books in Spanish, from astrophysics for college students to general themes intended for the layman, and children's books. The number of books now published in Spanish is 25, and several more are being written.

b) *Yearbook*: The Institute of Astronomy has been publishing a yearbook for the past 50 years. It includes ephemerides, positions of cities, lists of Messier objects, and important astronomical events.

c) *Periodicals*: The Institute of Astronomy publishes a monthly newsletter, *Orion*, that is distributed to newspapers, planetariums, and several magazines (which may reproduce freely the material therein). It is only a few pages long, and includes at least three astronomical developments, a book review, astronomical predictions for the coming months, an astronomical cartoon, and a poem.

2. Public Lectures

We have found that one of the most effective ways to popularize astronomy is through public lectures. The Institute of Astronomy staff gives some 200 lectures per year to people of all age levels and all over the country. This is very time consuming, considering there are only 40 professional astronomers in the country.

3. Counselling

The Institute of Astronomy has helped small museums with their astronomy exhibits, by updating them, remodeling, *etc.* At present, it is active in setting up temporary science exhibits at a subway station where 200,000 people commute daily. The Institute also checks the programs written by the planetariums whenever this is requested.

UNESCO has an office in Mexico that produces audiovisual material for schools. The Institute also checks their material and proposes alternative ones. (After being at the Williamstown Colloquium (these *Proceedings*), which was so stimulating, we proposed the construction of several new projects).

4. Radio and Television

The Institute of Astronomy gives periodic interviews to several radio and television stations. Our National University radio and television systems have a monthly interview program on astronomy. Once in a while, the Institute of Astronomy organizes special events like night observing through a telescope or a week of astronomy that includes the construction of sundials, Alka-Seltzer rockets, or simulations of the Big Dipper. Part of the staff has even written a puppet play on Comet Halley.

5. Courses

We regularly teach astronomy to high-school teachers. (See article by Herrera elsewhere in these *Proceedings*).

POPULARIZATION OF ASTRONOMY IN SOUTH INDIA

R. Ramakarthikeyan
Department of Mathematics, RKM Vivekananda College, Madras, S. India, 600 004

I am presenting this paper only with reference to my city of Madras, the capital of Tamil Nadu, where more than 60 per cent of the population are illiterate.

I am an active member of the Madras Astronomical Association and I am the Vice-President of the Vivekananda Astronomy Club, both of which are popularizing astronomy among the masses.

Though we have been pleading for the necessity of a planetarium over the last two decades, we got one only in May 1988. But for the last seven years, we designed and set up a mini-planetarium at the Education Pavilion of the Trade Fair Exhibition conducted by the local government every year.

In this pavilion we showed the various kinds of eclipses and certain zodiacal constellations using an overhead projector. We did this to eradicate superstitions and to teach the public some basic concepts on astronomy. The response to these programs has been very encouraging. I have also given popular talks on astronomy in several schools in this city and helped them in starting astronomy clubs. We have given a number of television programs on the popularization of astronomy, broadcast statewide. I must mention here that astronomy is not in the curriculum for the state-government-aided schools.

For students and for the public we conduct regular observations at my college astronomy club. In addition to the regular observations, we have also shown the public and students the following celestial phenomena.

1. Comet Kohoutek in Winter 1975;

2. the grand spectacle of five planets arrayed within a longitude of 90° in March 1983;

3. several eclipses during the decade;

4. the transit of Mercury on November 13, 1986;

5. Comet Halley in March 1986; and

6. meteor showers.

We have also conducted seminars and exhibitions at regular intervals.

On February 16, 1980, hundreds of students and others attached to the astronomy clubs of various colleges and the members of the Madras Astronomical Association observed the total solar eclipse at Hubli, 500 miles from Madras. In this grand spectacle, we observed at 2:45 p.m.: 1. Baily's beads; 2. diamond rings; 3. the planets Mercury and Venus; 4. the star Sirius; and 5. shadow bands, and so on.

We now have plans to tap the talented school children who are interested in astronomy and give them training on various aspects of this subject. The Birla Planetarium attracts thousands of citizens to its shows on astronomy, and we hope that the day is not far off when every citizen of this part of India is aware of the basic concepts of astronomy.

A SERIES OF ASTRONOMY PROGRAMS FOR TELEVISION IN INDIA

Jayant V. Narlikar
Tata Institute of Fundamental Research, Bombay 400005, India
and
Inter-University Centre for Astronomy and Astrophysics, Pune 411007, India

1. Introduction

Astronomy, unlike most other sciences, arouses great curiosity amongst laypeople. It is a subject that can be described relatively easily in public lectures. Distinguished astronomers like James Jeans and Arthur Eddington in the past and many more in recent times have "stooped down" to the public level to share the excitement of astronomical discoveries. Today, the popularization program normally proceeds in four different ways — through popular articles, public lectures, planetarium shows, and radio – TV programs. However, this overwhelming public interest in astronomy brings its own difficulties. Not all of it is motivated by a scientific interest! Many persons read mystic significance into astronomical findings. Many more are guided by astrological interest. Many fail to perceive the scientific basis for astronomy, a subject whose laboratory is the whole cosmos with objects too remote to be subject to scientific experimentation.

Thus "stooping down" to the public level is not so easy! In fact, I consider it more difficult to prepare a public lecture on an astronomical topic than to give a technical talk on a research problem. After all, in the former case one has to surmount the above-mentioned difficulties that don't exist in the latter case. It is not surprising, therefore, that even today the number of really successful popularizers of the subject is small.

Yet, wherever successful, the rewards of such efforts have been great. The books and lectures of Jeans and Eddington thrilled an entire generation of educated public before the Second World War. I know several of my own generation who drew inspiration from Fred Hoyle's lectures on BBC radio that were subsequently published as the book *The Nature of the Universe*.

Today the most powerful medium for mass communication is recognized to be television. How does it fare in the program of science popularization, especially in astronomy?

2. The Four Modes of Popularization

Before coming to TV *per se*, I wish to compare it with three other modes of popularization that I have used in my own limited efforts: writing articles, delivering public lectures, and preparing planetarium shows — in India where the literacy level is much lower than in the West.

Because of the lower literacy level, the written article has very limited impact. Articles written in English are read by a small fraction of the population, mostly the college students of science, faculty members of universities and research institutes, and a small fraction of the intelligentsia. I have found that my articles written in Marathi or Hindi have generated a greater reader response than those in English.

The same language barrier shows up in public lectures. Although a public lecture is localized, it can produce a greater impact than a written article with larger circulation. In a series of public lectures I once delivered in a rural town in India, the size of the audience grew with each lecture until, on the final day, it was estimated at 10,000. By then the venue of the lecture had to be shifted from an auditorium to outdoor premises.

The feedback from a public lecture — in the form of questions from the audience — is also more direct. While quite a few questions tend to be trivial or ill-informed, there are a few mature ones that well justify the efforts put in by the speaker.

Planetariums have become very popular in India, largely because the stellar roof brings astronomy manifestly closer to the audience. Planetariums can be effective in debunking astrological superstitions, although social pressures have prevented this from happening as fast as one would like. Inevitably constellations with their imagined shapes take a major part of a typical program. However, the excitement of recent astronomical discoveries (pulsars, supernovae, quasars, gravitational lensing, the expanding universe, *etc.*) is finding more and more time in the programs of the planetarium in Bombay, with which I have close contact.

It is against this background that one has to view the contributions of television. Again, I wish to describe my personal experiences in this field.

In the early 1980's, the Bombay TV station staged a series of nine monthly programs of 45–50 minutes duration on Sunday mornings. These consisted of discussions among 3–4 persons with one moderator. Slides were shown to illustrate certain points. The viewer response was generally favorable even though the program was screened at prime time. There were requests that the program (in Marathi) be repeated in English or Hindi on the national network.

In 1985–86, Carl Sagan's *Cosmos* was telecast. Each program was preceded by a 3-minute introduction by me in Hindi, summarizing the highlights. Many viewers found the Hindi introduction helpful in following the English commentary. The program generated tremendous impression and illustrated the power of TV visuals in communicating astronomical concepts and facts.

3. A TV Series in Hindi

The *Cosmos* experience produced two negative reactions in the Indian context, however. Many viewers found the episodes too long to hold their attention. And, because it was in English, a vast majority of viewers could not benefit from its information content.

These reactions prompted me to propose a TV series in Hindi set against the Indian background, with a large number (about 30) of episodes of about 25-minute duration. Its aim was "education through entertainment" with promotion of the scientific temperament. Prime Minister Rajiv Gandhi immediately liked the idea and the Government of India asked the Films Division to provide all the facilities for the production of the series on 35-mm film. The National Council for Science and Technology Communication also advanced a handsome research grant for the research component of the project.

It has, however, taken longer than expected to arrive at a suitable concept for the series. My earlier attempt was to introduce two "gandharvas," space travelers from Indian mythology, through the dream of a girl of 13. The girl has read mythology through comics and also some popular astronomy. She is faced with a conflict as to what is reality. She asks the gandharvas questions on stars, supernovae, quasars, and galaxies. They are foxed by the jargon and approach scientists for answers when they find that traditional mythology is insufficient to answer the girl's questions.

This was a way of contrasting modern knowledge with mythology and superstition. However, trial episodes based on it created a conflict in the minds of lay viewers. Are the gandharvas real? If not, how do they meet real scientists? Do the scientists consider them real? Although the entire episode was in the girl's dream, the confusion was serious enough to prompt us to shelve the idea.

In the revised format, we have adopted a "down to earth" scenario. In the opening episode, a schoolboy on his way to watch a solar eclipse on a school expedition to the local planetarium is admonished by his grandmother, who believes that an eclipse denotes an evil. The contrast between modern astronomical facts and age-old superstitions is brought out through demonstrations of working models and discussions in the course of the school trip to the planetarium.

We proposed to link two or three episodes in a story that seeks to bring about such contrasts in an uncontrived way.

4. Conclusion

It is too early to judge the impact of such a format. Less dramatic than the earlier one involving the gandharvas, it may deliver a clearer message to the viewers. The linking of episodes in one story may keep their interest alive. The real challenge, however, lies in educating them without disturbing their sensitivities.

(Note: The first episode of this series was screened during the colloquium.)

Discussion

D. Brückner: *Do not mythology and astronomy form a complementary pair rather than a pair of opposites, and might not a complementary approach be more effective in presenting astronomy to the public?*

J.V. Narlikar: Yes, and in our present approach we are taking this into consideration, showing how both deal with the same subject in their individual ways, and so introduce the modern astronomical view.

THE MILLS PUBLIC OBSERVATORY IN DUNDEE

Fiona Vincent
Mills Observatory, Balgay Park, Dundee DD2 2UB, Scotland

The only full-time public observatory in Britain is the Mills Observatory in Dundee, Scotland. John Mills was a successful businessman in Victorian Dundee with an interest in astronomy. He bequeathed his money to build a public astronomical observatory in the city, for studying "the wonder and beauty of the works of God in creation." The Mills Observatory was eventually opened in 1935.

Dundee is situated on the northern shore of the River Tay, in central Scotland. The Observatory stands on a wooded hill (an extinct volcano) near the center of the city. The trees shelter it from virtually all direct city light, while leaving an uninterrupted view south across the Firth of Tay. There is some scattered light, but in clear weather the sky can be surprisingly dark, and the Milky Way distinctly visible.

The Observatory is run by the City of Dundee District Council. It stays open till 10 pm, Monday to Friday, throughout the winter; in summer it is open only in daytime (at latitude 56°.5 N it never gets dark in summer). There are just two members of staff, currently Fiona Vincent and Gary Hannan; between us we carry out all the work of the Observatory, from designing displays and giving public lectures to routine cleaning and maintenance.

The Observatory's dome is 25 feet in diameter, and is made of water-proofed pâpier-maché on a framework of steel ribs; it can easily be rotated by hand. The telescope inside is a 10-inch Cooke refractor, made in 1871, with the superb optics you would expect. It has its original clockwork drive, and it is still in very good condition; the telescope can be moved with one finger, and remains perfectly balanced in any position. With a focal ratio of f/15, it offers the high-magnification views of the moon and planets that are just what the public wants to see.

The Observatory also owns a 10-cm Merz refractor on a portable equatorial mount, which is used in the dome alongside the Cooke to give visitors a choice of views of a celestial object. These telescopes are always pointed and supervised by the staff, who give explanations and answer questions. Visitors are free to use the large first-floor balcony, where smaller (and more robust) instruments are available for unsupervised use. This is a useful position, too, for learning the constellations. By day there are sundials here to study.

The main exhibition area is on the ground floor, and uses posters, photographs and models to provide an introduction to astronomy and space science. There is a large wall-mounted planisphere that the visitor can adjust, and a "Helios" orrery operated by pressing a button. Other interactive exhibits are planned, when money and staff time permit.

Next door is a small lecture-room, seating up to 30. Booked parties first tour the building, and then come in here for a talk with slides — usually a basic introduction to astronomy, adapted to the audience's age. When the room is not in use, casual visitors can walk in, press a button and see a pre-recorded slide-show. These shows are all produced in-house, with slides from the Observatory's extensive collection.

A wire framed cotton dome sits in a corner of the lecture-room; hoisting it reveals a "StarLab" planetarium projector. (There is also an inflatable dome but this is rarely used.) Up to 20 seats can be squeezed in underneath, or a complete class of 30 children can be accommodated on cushions on the floor. A boxed-in section of the exhibition area houses an even smaller planetarium, hand-made by Harry Ford, the previous Curator of the Observatory. Only 10-12 people can get inside, but the highly-realistic star-images make this one ideal for teaching the constellations.

Visitor numbers are currently over 17,000 a year. Many of these visitors come in pre-booked groups; some, such as school classes, come by day, but most prefer an evening visit. All available evenings in the winter season are always booked, and many groups have to be refused. Clear nights bring plenty of casual visitors too. On average, about one night in two is "viewable."

Dundee Astronomical Society has for many years used the Observatory as a meeting-place. In return, some members of the Society act as unpaid helpers on busy evenings. In recent years the Observatory has also been used for evening classes in astronomy, organized by the University of Dundee's Centre for Continuing Education. This is an introductory course of 14 lectures, spread over the winter season. The size of the lecture-room limits the numbers to 30, and the course is usually fully-booked. The advantage of using the Observatory is that a large number of models and teaching aids are available, as well as the Observatory's library of reference-books and journals. And of course, if the sky is clear, the classroom work can always be abandoned in favor of some real observation.

Dundee's daily paper, the *Courier*, carries a monthly feature on the night sky, and the local radio stations, *Radio Tay* and neighboring *Radio Forth* (Edinburgh area), broadcast weekly talks — all contributed by the Observatory. A duplicated sheet, "The Sky This Month," is available free to visitors at the Observatory, and several copies are sent out each month to postal subscribers. Extra sheets are prepared as necessary for topical phenomena such as comets.

Other initiatives in recent years include summer workshops for children, where they have made such devices as sundials, planispheres or simple solar-system models. And there have been a couple of weekend astronomy courses for amateur astronomers.

This is a purely public observatory. The telescope is not for use by the staff, but by the public; and on a clear winter evening it is not uncommon for over a hundred people to look through it. The staff never get a chance! We content ourselves instead with watching the faces of visitors who are getting perhaps their first experience of that "wonder and beauty."

POPULARIZING ASTRONOMY AT PUBLIC OBSERVATORIES IN WEST GERMANY

Hans L. Neumann
Frankfurt Public Observatory and Friedrich-Dessauer-Gymnasium
Bismarkstrasse 68, D-6050 Offenbach, F.R. Germany

1. Introduction

Based on the pedagogical ideas of A. Diesterweg, a number of science associations have been founded since about 1880 under the name of URANIA; astronomy has been one of their subjects. In the 1920–1930's, the works of Bruno H. Bürgel and Robert Henseling initiated the founding of many more local and regional associations and of public observatories all over the country. But most of the currently active associations were founded to answer the sharp increase of general interest that followed the early successes and spectacular results of space science.

Aims of the associations always have been manyfold:

– to share a fine hobby with like-minded people;
– to participate theoretically or practically in scientific research as far as technical and local circumstances allow;
– to offer to the public means and advice for celestial observations, and to share the joy of deep-sky wonders with guests;
– to mediate the progress, and the results of astronomical research to the public.

Public observatories either have developed as a result of an association's activities in popularizing astronomy, or become the center for an association's work where both internal (amateur, scientific, social) and external (public) interests can be followed.

A list of 130 associations and public observatories has been compiled for this report from recent sources (3–6). 75 out of some 100 of them answered my questionnaire.

2. Distribution

The regional distribution of all the 130 associations and public observatories shows strong correlation to the general population distribution in the Federal Republic of Germany. As light pollution depends on population density too, the correlation generates severe restrictions on observational activities at most places.

3. Structure

Of the 75 associations that answered the questionnaire,

– 47 have the standing of a registered association,

– 11 are installations of communities, counties, or other corporations,

– 6 are private initiatives of single persons,

– 10 have a permanent full-time staff (including administration) of 1 to 8 people.

The sizes of the associations vary strongly from about 20 to more than 2600 members, and show a dependence on regional population as well as on the age of an association. The number of volunteers in the public programs ranges from very few to about 50 respectively.

4. Instruments

Though refractors of approximately 5 cm (2") and reflectors of approximately 10 cm (4") have been the starting instruments for many associations, by now the 20-cm (8") Schmidt-Cassegrain telescopes have become standard equipment. At public observatories, refractors up to 30 cm (12") and reflectors up to 75 cm (30") are installed. Small planetarium projectors are used regularly at least at 12 public observatories. Special educational installations (*e.g.*, exhibitions, demonstration apparatus) have been developed at various places.

5. General Programs

Programs for the public usually include

– observing sessions using the telescopes available at the observatory or on loan from association members,

– lectures giving information on general astronomy and on actual events or research progress,

– courses of introductory or practical astronomy or more specialized subjects (telescope making, astrophotography, *etc.*), often in cooperation with local institutions of adult education.

6. School Programs

In many places, a close cooperation between public observatories or their connected associations and local schools has been developed. Special programs for school classes, sometimes spanning the whole career from kindergarten to senior high school, have been created.

Introductory lectures may feature special items upon arrangement with the teacher to best fit regular lessons. Observing is very important, as only very few schools have telescopes of their own. Media available at public observatories and astronomical associations often are more numerous and elaborate than those in schools.

7. Attendance

Because of the size and standing of an association and other local circumstances, the numbers of public events and of participants varies widely. We cannot check for the ages or other structure of the attendance. Generally, young people form a good fraction.

In total, during 1987 more than 850 amateur astronomers of 60 associations introduced well over 181,000 people to astronomy, running over 5500 events in a wide variety of ways. Some 27,000 more people attended the 1300 events offered by the 9 communal astronomical institutions. The numbers provided give lower limits only, as statistical data are incomplete.

8. Financing

Public observatories generally are non-profit and self-preserving organizations. Public support in most cases is negligible compared with members' contributions and donations, though for a very small number of associations it provides up to 80 per cent of the budget.

References

1. Special inquiry for this report.
2. Personal information from several associations.
3. Manfroid, J. and Heck, A. *International Directory of Astronomical Associations.* Strasbourg: Centre de Données, 1988.
4. Vercoutter, Ph. A. J., *Directory of European Observatories.* Ieper: Astronomy Contact Group, 1988.
5. Koch B., Th. Jurriens, and J. Meeus *Sternführer 1988.* Düsseldorf: Treugesell, 1988.
6. Various reports in: *Sterne und Weltraum.* München, 1985–1988.

TEACHING AND POPULARIZING ASTRONOMY AND SPACE SCIENCES AT THE OBSERVATORY OF THE CITY OF BOCHUM

J.V. Feitzinger, M. Hünerbein, R. Kordecki, U. Lemmer, G. Monstadt, J. Prölß,
Observatory of the City of Bochum, Postfach 102148, D-4630 Bochum 1 , F.R.
Germany

The tasks and aims of the Bochum Observatory are popularization of astronomy and space sciences, and adult education. In general, as a cultural center for natural sciences we must translate scientific nomenclature into the language of the nonspecialist. Astronomy is ideal for presenting the basic facts of scientific methodology and reasoning to the public.

A planetarium is the most versatile instrument for teaching basic astronomy and space sciences to the general public as well as to school groups. We take great care to avoid a lecture-like style in the programs. Audience surveys have shown that most visitors don't want to get the feeling of being educated as in school. Nevertheless, we first have to motivate before we can educate. Instead, most visitors want merely to enjoy astronomy in the pleasant atmosphere of the dome. Consequently, our public planetarium shows contain elements of entertainment. We use many special effects, panoramas, and all-sky projections for a most precise simulation of astronomical phenomena. The audience should get the thrilling impression of witnessing things from close up. For example, they all become passengers on an imaginary spacecraft visiting the rugged terrain of Valles Marineris on Mars, the swirling clouds in Jupiter's atmosphere, or even the vicinity of a whirlpool-like accretion disc around a supermassive black hole in the core of an active galaxy. We use the potential of the planetarium as an "illusion factory" to increase the visitors' positive attitude towards astronomy and space travel. Special music, sound, and noise effects add to the impression.

In addition to astronomical presentations for the general public, we offer special programs devoted to more advanced audiences, such as amateur astronomers. Professional astrophysicists talk about their fields of research and the latest findings in astronomy and related sciences. We also have courses and lectures outside the planetarium dome for a thorough introduction to astronomy.

We should also mention that the planetarium is considered to be a cultural facility in the city of Bochum. Each year we have a couple of exhibitions with painting featuring all aspects of space art. And we present music in the dome together with planetarium projections. These light shows, entitled " Music-Startheater" have run regularly since 1984 with a tremendous impact on our visitor attendance. The three astronomical observing stations have telescopes of 12.5-cm, 32-cm, 40-cm, and 60-cm diameter.

To ensure a continuing high standard of the performances, the public sky shows in the Bochum Planetarium (20-m dome, 300 seats, Zeiss IV Projector) are taped and pre-programmed. Our automation system meets all industrial standards and allows the application of many special effects. Fig. 1 gives a gross overview of its configuration. Its most prominent characteristic is the enormous flexibility. The decentralized architecture allows a modular growth or easy re-design controlled with this automation system.

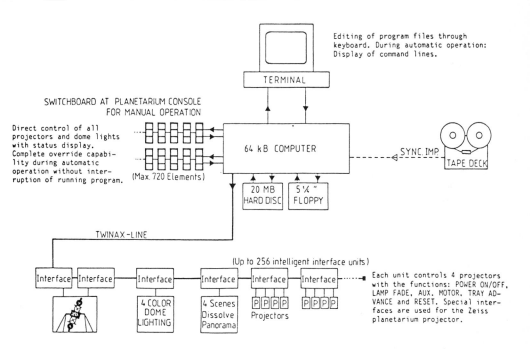

Fig. 1 The Bochum Planetarium Automation System

THE TEACHING OF ASTRONOMY AT
THE UNITÀ DI RICERCA ASIAGO

Laura Erculiani Abati
Unità di Ricerca Asiago, c/o Osservatorio Astronomico, Vicolo dell'Osservatorio 5,
35122 Padova, Italy

The Unità di Ricerca (U.d.R.) Asiago, of the National Astronomical Group of the Italian Research Council, consists of:
1) the staff of the Padua-Asiago Observatory
2) the staff of the Astronomy Department of Padua University
3) external members with interests in stellar, galactic, and extragalactic astronomy and in technology.

The teaching of astronomy at the U.d.R. Asiago has several aspects: popularization, educational visits, cooperation with amateur groups, promotional contacts with public institutions, participation in a commission on school programs, and participation in the educational activities of the Italian Astronomical Society.

1. Popularization

Almost all the members of the U.d.R. give public lectures and seminars in the schools, with the aim of providing a complete view of astronomy from its historical evolution to the more recent scientific discoveries with an emphasis on acquainting people with the latest research.

2. Educational Visits

Many schools and private groups ask to visit the Astronomical Observatory of Asiago, but, in spite of the staff's trying, not all requests can be satisfied. There is a standard program for the visit which, besides explaining the instruments in the dome, includes slides of celestial objects and describing the main fields of astronomical research and the future of astronomy. However, the program is flexible, according to the visitors' specific requests. These visits take a large amount of staff time, but it seems to us that lectures, however useful, cannot substitute for the direct contact with the instruments and with research.

3. Cooperation with Amateur Astronomers' Associations

In our region (the north-east of Italy) numerous amateur astronomers' groups, some of them extremely active, exist and interact profitably with the U.d.R. In the field of research, professional astronomers offer advice, suggesting programs for different instruments and different observers, provide materials at the larger telescopes, and write papers for amateur and professional newspapers. In the field of popularization, amateur astronomers are irreplaceable; for example, they guide the sky observations with the naked eye and with small instruments, organize local or traveling astronomical exhibitions, manage planetariums, and conceive new experiences particularly suitable for understanding celestial phenomena.

4. Promotional Contacts with Public Institutions

Some members of the U.d.R. cooperate with public institutions that are devoted to educating the population and training teachers. They propose astronomical subjects and give lessons, seminars, and courses. They act as reference points both for the institutions and for people who like to get deeper knowledge of specific subjects. It seems to us that this kind of cooperation is particularly important because it places the culture of the professional astronomers at citizens' and teachers' disposal.

5. Participation in School Programs Commission

A member of the U.d.R. belongs to the joint commission of the Italian Astronomical Society (S.A.It.) and the Physics Teaching Association; the commission's aim is to set the astronomical teaching program of the high school. So far there are new programs for elementary and the middle secondary school, but the programs for high school are old and generic. Furthermore, only a few of the teachers ever attended a regular astronomy course during their university studies. Thus astronomy teaching in the high schools is often superficial and inadequate. We hope to help in formulating suitable and specific programs, especially in view of the high school reform that has been long expected and is probably near at hand. We believe that setting an adequate program is important, since teacher training and textbooks are planned on the basis of the Ministry's programs.

6. Participation in the Educational Activities of the S.A.It.

The Italian Astronomical Society, through its educational commission, promotes initiatives for the teaching of astronomy at all levels. A member of the U.d.R. belongs to this Commission and many others are involved in the different activities. A particularly significant activity is the National Summer School for teachers; it has existed for three years in collaboration with teachers' associations and is supported by the Ministry of Public Education. It is a kind of residential school for small groups of teachers; it trains them in a specific astronomical subject, either in a field of science or on methodology; it includes lessons, practical exercises, and night observations. The U.d.R. helps in all phases: proposing the arguments, giving lectures, making practical exercises, coordinating activities during the school, and writing the final proceedings. U.d.R.'s work is important for improving teaching.

In conclusion: at the U.d.R. Asiago, we are trying to develop the teaching of astronomy in all its aspects. We believe that research, education, and popularization are different ways of promoting astronomy and that promoting astronomy helps humanity.

_____

TEACHING ASTRONOMY IN NEW ZEALAND: THE WARD OBSERVATORY

Christopher T. Harper[a]
Ward Observatory, Wanganui, New Zealand

[a]Present address: Phillips Exeter Academy, Exeter, New Hampshire 03833, U.S.A.

1. Introduction

New Zealand is a self-governed, independent nation, a member of the British Commonwealth, and a major center of Polynesian culture in the south Pacific. The country's two main islands lie between 34° and 47° south latitude, which places New Zealand south of many well known southern observing sites such as Sydney (Australia), Cape Town (South Africa), and Cordoba (Argentina).

The population of 3.5 million people inhabit a country slightly larger than the United Kingdom. The astronomical interests of this small population are nourished by no less than 25 local and regional astronomical societies. Amateur astronomers are active in New Zealand[1]. The Royal Astronomical Society of New Zealand coordinates national interests and hosts international activities[2].

2. Problems Associated with Growing Up Astronomically in New Zealand

Astronomy is taught in all New Zealand State and Private Schools in the 4th Form (the equivalent of the U.S. 9th grade) as part of the general science curriculum. However, New Zealand students are often faced with more than the usual problems when it comes to learning astronomy. Many of the astronomical publications they read are designed solely for the northern point of view. But, in contrast to an observer under northern skies, *everything seen by a New Zealand student is upside down.* The moon is upside down. Orion hangs by his feet, with his sword pointing upwards. Furthermore, a student in New Zealand facing the *south* celestial pole will see the stars rise on the *left* (left ascension?). All those convenient northern hemisphere clock drives have to be reversed if you are upside down!

There are other problems. The sun, moon, and planets are seen when New Zealand observers face north. Polaris is of course forever hidden. So is the Big Dipper. Deneb is on the northern horizon. Arcturus can be seen about 25° above the northwest horizon in *winter*. Orion, upside down, is visible 45° above the northern horizon in *summer*. Fomalhaut is seen near the zenith. New Zealanders find their celestial pole midway between Achernar and Beta Centauri. Alpha and Beta Centauri together form the "pointers" to Crux, the southern cross. In winter, New Zealanders looking up see right into the heart of the Milky Way galaxy. The star clouds of Sagittarius are directly overhead in July and August.

From a position of 40° south latitude, many of the classical constellations are lost. Even the stories of the ancient zodiac constellations, which link *northern* seasons with star patterns, make little sense to a young New Zealander who is brought up with reversed seasons. Indeed, much of the ancient lore about the sky, derived as it is from cultures in the northern hemisphere, makes no sense at all "down under."

Fortunately, the astronomical knowledge of the original settlers from the north, the Polynesian navigators, was quite extensive[3]. New Zealand teachers do have opportunities to replace Chaldean and Greek mythologies with Maori legends as far as these legends pertain to the stars. Other basic astronomical facts, particularly those dealing with position and orientation, have to be re-written for the southern observer.

3. Public Observatories in New Zealand

Observatories open to the public are run for the most part by devoted amateur astronomers who freely offer their time and knowledge for the benefit of all. Young people are generally introduced to these observatories by their teachers or parents.

A notable example of an active amateur observatory is the Auckland Observatory in the North Island — home of the Auckland Astronomical Society. The Auckland Society was formed in 1922 by an enthusiastic group of amateur astronomers. Funds accumulated from private donations and from local business communities enabled the society to open an observatory in 1967. The Auckland Observatory houses a 50 cm (20") Zeiss Cassegrain reflector built in Jena, East Germany. More recently, the addition of photo-electric equipment and computer facilities has helped make the Auckland Observatory one of the most respected amateur research establishments in the southern hemisphere.

The Auckland Observatory runs a weekly program of evening lectures. There is a beginning course held annually, and training courses in the use of telescopes. Teachers can arrange daytime visits for school classes throughout the school year. There are programs throughout the summer. The observatory has a well-stocked bookshop. A library is open to members of the society. The Auckland Observatory is operated by a charitable trust. It receives no Government or Local Body support, and relies on donations and admission fees to provide and maintain an astronomical facility for the Auckland region.

In Wellington, the Carter Observatory overlooking Wellington harbor, is New Zealand's National Observatory. Supported by government funds, the facility houses two refractors; a 15 cm (6") telescope donated by the late Peter Read, and a 23 cm (9"). In addition, the Observatory operates a 40.6 cm (16") reflector at a field station on Black Birch Mountain in the South Island, 20 km south of Blenheim. A U.S. Naval observatory is located at the same site.

Carter Observatory publishes an annual Astronomical Handbook for New Zealand observers[4]. This publication is similar in conception to the Observer's Handbook published by the Royal Astronomical Society of Canada. Carter Observatory is open to visitors during the day. A series of evening lectures is given each week from March through October.

Recent moves by the New Zealand government to close Carter Observatory on economic grounds were met by a storm of public protest. The government was persuaded to reverse its intentions.

4. The Ward Observatory

Ward Observatory was the first major astronomical observatory to be built in New Zealand. For many years Ward Observatory (originally known as the Wanganui Observatory) was *the* New Zealand Observatory. Opened in 1903, it is located in the city of Wanganui, (pop. 40,000) near the mouth of the Wanganui River on the west coast of the North Island of New Zealand, at latitude 40° south.

The observatory houses a 24 cm (9.5") refractor, built by Thomas Cooke in 1865. The mount for this Cooke refractor is a cast iron prototype of the new or modified English mount. After 85 years of continuous operation, this telescope is still the largest refractor in operation in New Zealand today.

As soon as the Cooke refractor was installed, Joseph Ward, founder and honorary director of the observatory, with the assistance of another Wanganui amateur, Thomas Allison, set to work to survey the southern sky for double stars. During the years 1903 to 1910, they discovered more than a hundred new pairs. Most are located between 50° and 80° south declination. This work was the first of many important contributions made by amateur astronomers in New Zealand. The Ward doubles found their way into the international catalogues under the designation "NZO" for New Zealand Observatory. The IDS lists a total of 88 NZO doubles[5].

Joseph Ward may not have foreseen how wise his choice of an instrument with a long 12 foot (367 cm) focal length really was. The refractor still provides excellent viewing despite the advent of modern street lights and an adjacent outdoor sports stadium. It now has a precision electric clock drive designed and built by members of the Wanganui Astronomical Society.

The successors to Joseph Ward continue both to maintain their own interest and development in astronomy while at the same time providing a public service in the middle of a busy city. Public viewing nights are held at the Ward Observatory on Monday nights, March through October. Meetings of the Wanganui Astronomical Society are held once a month, always with a lecture on a current topic of astronomical interest.

References

1. Blow, Graham L. "Astronomy in New Zealand." *Sky and Telescope* **63** (June, 1982), p. 555. See also *Sky and Telescope* **63**, p. 612.

2. The RASNZ publishes a quarterly journal *Southern Stars* containing reports and articles of interest to New Zealand astronomers.

3. Elsdon Best (1922), *Astronomical Knowledge of the Maori.* Dominion Museum publications, Wellington, New Zealand.

4. The *Carter Observatory Astronomical Handbook* for New Zealand observers is available from Carter Observatory, P.O. Box 2909, Wellington 1, New Zealand.

5. Jeffers, H.M. *et al.* (1963). *Index Catalogue of Visual Double Stars, 1961.0.* Lick Observatory.

Discussion

P.W. Hill: *Professional astronomers are extremely grateful for the work of New Zealand amateurs, particularly in the field of variable stars.*

SYDNEY OBSERVATORY GOES PUBLIC

N.R. Lomb and T. Wilson
Sydney Observatory, PO Box K346, Haymarket 2000, Australia

1. Sydney Observatory

In 1982, after a 124-year history of research, Sydney Observatory became a branch of a large local museum, the Museum of Applied Arts and Sciences. A four-year, million-dollar project was undertaken to restore the building and its grounds to their nineteenth century appearance. The services needed for a modern museum were also added. One of the larger areas became a modern lecture theater seating up to fifty people, with back projection video, film and slide projectors.

Exhibition space within the building is limited to eight rooms of approximately 200 m^2 total area. To overcome this lack of space, a proposal has been made for an extension to the rear of the building. An underground 100-seat planetarium is included in the proposal. There is a great need for this as there is no planetarium currently in Sydney.

2. Halley's Comet Exhibition

From January to May 1986 an exhibition on Halley's Comet was put on in a part of the Observatory building. In the first and main display area, visitors found themselves in a spaceship. This spaceship had spot lighting, synthesized music and a fluorescent background of stars and galaxies. Exhibits included:

- the written record from 1835 of the first (European) sighting of Halley's Comet in Australia and the telescope with which the observation was made;
- the telescope with which the Great Comet of 1861 was discovered;

- plates of the comet made at the Observatory in 1910, together with modern contrast enhanced prints from the plates;
- computers, interactive exhibits, and a 10-minute audiovisual program.

There were 47,000 visitors to the exhibition during the five months.

3. The 29-cm Telescope

Evening telescope viewings are held for the public on six nights a week. The remaining night is used for adult education courses in astronomy and an amateur astronomy group. Until recently the main telescope used for the public viewings was the historic 29-cm refractor. This was obtained by the Observatory to observe the 1874 transit of Venus. Its main limitation is that objects shown with it are limited to those that can be seen with the naked eye in a highly light-polluted sky. The newly installed computer-controlled C-14 telescope does not have the same problem and it is envisaged that with narrow-band filters and an image intensifier, regions of faint nebulosity will be visible.

4. Repeating Circle

The plans for the permanent exhibition include not only interactive exhibits, but also displays of antique astronomical instruments. The Museum has a large collection of nineteenth century items, most of which came with the Observatory. Possibly, the most important items are a collection of instruments from the first permanent observatory in Australia. This observatory was set up in 1821 at Parramatta, which is now an outer suburb of Sydney. Included in the collection are a Breguet regulator clock, a refracting telescope, a transit telescope and the repeating circle.

5. Hands-On Astronomy

The current exhibition at the Observatory is made up of interactive exhibits. They are prototypes that are being tried out for possible use in the planned permanent exhibition. Some of these are discussed below.

a) Weights on Other Worlds

This exhibit demonstrates the different pull of gravity on various solar system bodies: the sun, the Earth, the moon and Jupiter. Appropriate amounts of lead shot have been placed inside the empty cartons.

b) Colored Pictures

In this demonstration of image processing, a picture of the Horsehead Nebula is color coded according to intensity. Then, using a TV camera, the same process is carried out on real-time images of the visitor. The exhibit is based on an IBM PC and an image-processing card.

Other computer exhibits on display are based on Amiga computers. One interactive program tells the visitor about the solar system; with another the visitor can build a telescope on the screen and test it out.

c) Look for Galaxies

In this exhibit, visitors look with a magnifying glass at a back-lit film negative from the UK Schmidt Telescope. The photograph is of the Fornax Cluster and so contains a variety of types of galaxies.

d) Star Pictures

This is a constellation guessing game. Visitors can choose from four different constellations and try to guess the constellation before pushing the answer button. Optical fibers are used to display the star patterns. A similarly built exhibit demonstrates stellar scintillation.

e) Supernova

A zoetrope (an optical device in which figures on the inside of a revolving cylinder are viewed through slits in its circumference and appear like a single animated figure) is used to show the implosion of the core and the expansion of the outer layer in a supernova. Even more ambitiously, another zoetrope shows the "big bang" and the expansion of the universe.

f) Astronomical Picture Show

The videodisk based exhibit shows a pre-selected series of astronomical images. A title is displayed with each image. This visitor can choose to obtain extra information.

———————————————

ASTRONOMY COURSES FOR ADULTS IN SYDNEY, AUSTRALIA

J.W. O'Byrne[1]
The University of Sydney, Australia

1. Introduction

The University of Sydney has been associated with adult education courses for the general public for at least 70 years and astronomy has often been a part of this activity. Since 1976, these courses have been conducted by a succession of graduate students from the Astronomy and Astrophysics departments within the university. The courses were a part-time activity conducted with limited resources, but served as useful teaching experience. This arrangement continued until the end of 1987 when I left the university. I am pleased to report that this activity is considered sufficiently important to be continued by the permanent academic staff.

Here I seek to briefly report on these astronomy courses by describing those that I conducted at the university over the last five years. I should firstly acknowledge the work of Dr. Graeme White in particular, for his earlier courses and continuing interest. Also, this discussion should be placed in context by recognizing that other astronomy courses are run in Sydney by professional and amateur astronomers through local evening colleges and other organizations. These vary in format, with many of the amateur-run courses emphasizing the practical aspects of sky observing. The displays and facilities provided by the Sydney Observatory also play an important role. Each activity helps to meet the demand in the Sydney region for astronomical education and thereby undoubtedly serves the interests of Australian astronomy.

2. The Courses

The courses presented at the University of Sydney were all designed for the general public, assuming nothing other than an intelligent interest. On the other hand, the courses were "academic" in that they attempted to present a broad range of astronomy at a level typical of many college astronomy texts (*i.e.*, largely non-mathematical). The courses were relatively long, with two-hour night-time meetings once a week for up to 19 weeks, and cost up to A$80. A starting group of 60 was typical, with more than half attending regularly over a long course. Clearly these people form a highly motivated audience. Many different backgrounds were represented in a class, although the largest single group was usually formed by people in their 20's and 30's, some of them university graduates. Other than a

[1]Present Address: The Johns Hopkins University, Applied Physics Laboratory, Laurel, Maryland 20707, U.S.A.

small number of amateur astronomers who wished to reinforce their knowledge of the science of astronomy, few students had any practical experience of observing the night sky.

The first of the two major courses I presented each year was a conventional "survey" of astronomy given largely by a single lecturer. The course started with history and the night sky, proceeded through planetary, stellar and galactic astronomy, and ended with cosmology and SETI. The objective of the course was not to saturate the audience with information, but to impart some understanding that the students could use later in their own reading. There was no specified text and so handouts were used to summarize the lectures and provide diagrams to supplement the notes taken by many people. Slides were valuable tools, notably those provided by David Malin of the Anglo-Australian Observatory. They help to break down the "lecture hall" atmosphere and often prompt wide ranging discussion. *Sky & Telescope* "Laboratory Exercises" were of interest to a few, and simple demonstrations like handing out plastic diffraction gratings were always popular.

Questionnaires distributed at the end of the course proved to be a useful way to gauge the interests of the group and assisted in planning future courses. Despite the course being essentially one devoted to theory, its most popular single aspect was the field trips held outside the city. People willingly traveled over 100 km to view the sky through a variety of telescopes. Perhaps as a result, many people asked for more lecture material on the night sky. Australia's location and climate make the sky a year-round attraction. Also of interest was the description of local astronomical facilities and how astronomy is really done at these observatories. This often came through in discussion, but illustrating the description with the lecturer's own work was regarded as particularly interesting and personal. Other topics, like cosmology, are always popular, and many people requested more of this type of material at the expense of historical background.

Many students finished one course with their appetite whetted for another. However, to do justice to specialized topics requires considerable effort on the part of lecturers not working in the field. Thus shorter courses were organized, with professional astronomers each giving a lecture on a subject of their own interest. Fortunately, there are a large number of potential lecturers available at several astronomical organizations in the Sydney area. The level of presentation was not much deeper than the previous course, but more could be discussed when dealing with a specific topic and a relatively informed audience. Not surprisingly, the drop-out rate of students in this course was very low and the lecturers were often surprised by the interest shown and the intelligent questions asked. The topics and lecturers were changed from year to year and many students became regular participants.

There are always a few students who ask for more detailed argument and discussion in a course. This usually means more mathematics and physics, which presented an obstacle to most students who had little or no background in these areas. I presented a single 9-lecture course on stellar evolution in 1987 to try to meet this interest. It proved to be considerably more difficult for all concerned (including the lecturer!), but was well received by the 30 participants.

There are several major astronomical facilities within several hundred kilometers of Sydney and students often ask if they can visit one at night. Visits at night have been possible at a small university facility, but not at a large observatory. However, several daytime tours to major observatories were arranged and attracted 30 people each for a one-day trip using a chartered DC3 aircraft! The observatories provided a special guided tour and the trips were very successful.

3. Conclusion

The conclusion to be drawn from this experience is that there is significant public interest in astronomy in Australia, even at the level of demanding courses. Also, the interaction with enthusiastic practicing astronomers is a particular attraction to students. It is the responsibility of astronomers to be active in adult education, if only because in doing so they will help generate the public support that is vital to the future of astronomy.

Discussion

H.F. Haupt: *In Austria it is possible for all sorts of adults to enroll in university courses, which are free. Many of these people are active or retired teachers, doctors, engineers, etc., who have time and enthusiasm for doing astronomical work even if they do not aim at a degree. These people should be used to help spread astronomical knowledge.*

J.R. Percy: At the University of Toronto, those 65 years of age or older can enroll in courses free of charge. They are indeed enthusiastic — in contrast to many of the "regular" students — and represent one of the most receptive audiences for astronomy education.

INTRODUCTORY ASTRONOMY FOR MATURE STUDENTS

William H. Waller
Department of Astronomy, University of Washington, Seattle, Washington 98195, U.S.A.

Much of this meeting has been concerned with the teaching of introductory astronomy to children, teenagers, and young adults. Introductory astronomy for working and retired adults has been given short shrift, however. Because the mature population is significant in number and in its influence on governmental support for astronomy, I thought that some mention should be made. My own experience concerns the educational programs that are available to mature students in the United States.

Most working adults are restricted by their jobs to taking classes in the evening. Most large universities in the United States provide such classes. These are often administered by organizations that are completely autonomous from the "host" school. A typical arrangement is for the school to provide classroom space, access to audio-visual equipment (sometimes), and university credit for those who register for credit. The "adult education," "continuing education," or "university extension" organization handles the rest. It obtains the necessary faculty to teach the courses, advertises the classes, and administers the registration and grading. For the "credit" courses in astronomy, the most common teachers are graduate students and postdoctoral fellows looking for some extra cash and teaching experience. There is no real impetus for regular faculty members to teach these classes. This is an unfortunate circumstance that could be remedied by some private, state, or federal funding of endowed "chairs" in adult education.

The attitudes, expectations, and tolerances of mature students are very different from those of their younger counterparts. These differences must be recognized and addressed, if one wishes to teach introductory astronomy with any degree of success. Because the teacher is often faced with students who are older and more experienced in a general sense, he or she must carefully define the proper realm of authority. I have found that the students appreciate a well-run class complete with challenging homework and exams, just as long as they can deal with me as a fellow adult who will be responsive to their opinions. I have also found that, on-average, the mature students are more motivated and enthusiastic than their younger counterparts who are often taking astronomy to satisfy some institutional requirement.

One of the most exciting educational developments in the United States is the emergence of ElderHostel programs for students 60 years of age or older. These programs are typically one or two week workshops, where 10 to 30 students live together "dormitory-style" and participate in a coordinated program of classes and field trips. I had the pleasure of relaxing in a rustic lodge in the Berkshire Hills of Massachusetts while teaching astronomy in a program that also included cross-country ski lessons and lectures on famous authors of the Berkshires. A dizzying variety of ElderHostel packages are now available throughout the United States and in other countries as well. In 1987 over 140,000 hostelers enrolled in programs operating in all 50 states, all 10 Canadian provinces, and in 35 countries overseas. More than 1000 universities and colleges are involved as hosts of these programs. Perhaps most important, the programs are inexpensive compared to the lodging and eating expenses of equivalent vacations. Information on the ElderHostel program (and how one can get involved as an educator or host) can be obtained by writing to Judy Goggin, ElderHostel, 80 Boylston St., Suite 400, Boston, Massachusetts 02116, U.S.A.

Comment

Jay M. Pasachoff: I had the pleasure of teaching an ElderHostel course to 35 students here at Williams just two weeks ago. My five one-hour lectures were: Halley's Comet, Stellar Evolution, Supernova 1987A, Solar Eclipses, and Observing with a Field Guide. The students were attentive and devoted, and the feedback was very favorable.

Chapter 12

Planetariums

Discussion of planetariums was centered around a panel organized by Joseph Chamberlain, Director of the Adler Planetarium in Chicago. The presentations consider first general philosophy, then the role of the planetarium in schools and for teaching teachers, and next how to attract the public to planetariums first in general and then in developing countries. Further, independent papers dealt with planetariums in schools and colleges.

PANEL DISCUSSION: THE ROLE OF THE PLANETARIUM

Philosophy and Directions in Planetarium Programming

T.R. Clarke
McLaughlin Planetarium, Royal Ontario Museum, Toronto, Ontario M5S 2C6, Canada

1. Introduction

I am going to start with the premise that planetariums serve several roles in their communities:

1. They are popularizers of astronomy and space science
2. They support and enhance the teaching of astronomy and related subjects within the formal education system
3. They provide a community resource for astronomical information

Not all planetariums incorporate all of these roles or do so to the same degree.

Planetariums, as facilities, come in a variety of forms. At one extreme might be a space theater with its wide-format films. A major public facility would have a strong emphasis on public shows and school programming. A college or school facility would emphasize programming for the education system at some or all levels with some or no public programs. At the other extreme would be the small portable planetariums with 100 per cent use for school activities.

2. Popularizers of Astronomy

Others on the panel will speak more directly to school or course issues. I will direct the balance of my remarks to public programming, which is the primary means by which planetariums act as "popularizers of astronomy."

In many major facilities that emphasize public programs, the trend is one of greater sophistication in presentation. This sophistication involves greater use of projection technology but more importantly a better knowledge of the audience and its needs, more effective development of show concepts, a good writing style, and high production values.

Such planetarium programming responds to what we perceive the audience is looking for or expecting. Good communications (and, as we have heard in this colloquium, good education as well) starts with knowing your audience. I submit that the audiences in many cases want to hear about and see the results of new discoveries, new theories, and current speculations. They respond less favorably to historical programs and sky lore. I suggest that we should not be surprised at this trend for what we are seeing is part of the cultural dimension of astronomy. Astronomy is perceived as new, futuristic, high-tech, *etc.*, to such an extent that commercial advertising readily uses astronomy to sell products. Likewise, film and television raise in our audiences high expectations in visual presentation.

In short, if you are big, visible, and invite an audience in at a significant fee, you had better be good at what you offer. In terms of astronomy content, you should not expect a lot of hard in-depth science. A public show in a planetarium is not a classroom nor does it speak to a homogeneous audience.

To become more effective popularizers of astronomy, planetariums need some or all of the following:

1. to have access to good, up-to-date information and materials;

2. to have good communication skills to interpret that information;

3. to know the needs and expectations of their audience;

4. to exploit and to use the knowledge of learning processes and new technologies of presentation;

5. a staff that combines a good knowledge of astronomy, including concepts and processes, and good communications skills. In these requirements planetariums share a common need with science museums and schools, and one wonders if the 80 per cent of astronomy majors who, it has been said, do not pursue a career in astronomy research or formal teaching, might be better served by an education that gives them the other skills with which to pursue these other careers.

Process and Practice

Jeanne E. Bishop
Westlake Schools Planetarium, 24525 Hilliard Road, Westlake, Ohio 44145, U.S.A.

I would like to discuss strategies for participation within planetarium programs. The planetarium is not only a theater in which passive audiences can hear lectures and see demonstrations. It is also a classroom in which people of all ages can benefit from involvement. Participatory programs have been advocated in the US since the early 1970's, and the idea has grown and spread.

One participatory method involves drawing. For example, an elementary school group can draw a planet group for two or three different times. The psychological value of learning by drawing is well known. And drawing positions of stars and planets with appropriate labels produces take-home sky charts. Families can and do use them to learn astronomy and enjoy the sky together.

Drawing can be combined effectively with hypothesis-testing (predicting what one thinks will happen). In a lesson about the daily path of the sun at the beginnings of equinoxes and solstices, students can draw sunrise, noon, and sunset positions for each date, always being asked to predict from their previous observations (and drawings) where the sun will be next. A lesson on positions, shapes, and orientations of lunar phases is very effective with drawing and predictions.

With interactive questioning and hypothesis testing, the instructor quickly learns helpful learning characteristics about the audience: misconceptions, level of abstractive ability, and problems with visual perception ability. (An example of a perceptual problem is a student saying that the moon "looks like the capital letter 'D'" when, in fact, it is the *last* quarter moon presented in the sky.)

Use of model materials is also useful in the planetarium. With flashlights and small spheres, pairs of students can recreate the reasons for the phases of the moon at each step of change of phase in the planetarium sky. The topics of seasons, planet motions, and ideas of celestial sphere are also enhanced by student-manipulated models. Where it is impractical for students to all have model materials, the instructor can move or have students move one demonstration model. Dynamic human models can also be used (See my paper, "Dynamic Human Models," elsewhere in these proceedings.)

Other types of participation activities include role-playing and singing. Role-playing means that students assume the names or designations of particular characters from history or made-up roles in plays or contrived situations. Examples are navigators in historical voyages, Galileo discovering with his small telescope, and a government body discussing pros and cons (a cost-benefit analysis) of a costly space project.

Very young children love to sing. They remember ideas well when they are attached to music. Instructor-written words that match well-known tunes can be learned quickly — songs about directions, star patterns with mythology, changing moon phases, and day-and-night are both fun and educational. Commercially-produced records of children's songs about the sky are also available.

There probably are as many participatory styles for planetarium teaching as there are styles in a normal classroom, and then a few more. The latest addition to participation techniques — initiated by one of our panelists, Terence Murtagh — is interactive decisions on topics to explore in a planetarium program. Measuring, using data to calculate answers to mathematical problems, creative writing under stars with a musical or natural-sounds background, and fantasy journeys in which audience members close their eyes and use imaginations to extend what the planetarium can demonstrate — perhaps a mind trip into a black hole, complete with references to relativity, aching feelings as one is stretched, and reflections on the predicament — all can be successfully used in the planetarium classroom and theater.

Project STARWALK

Bob Riddle
Lakeview Museum Planetarium, 1125 W. Lake Ave., Peoria, Illinois 81614, U.S.A.

Project STARWALK is an Earth/space science program, developed for elementary students in grades 3 and 5 or 4 and 6. The program is designed around the teaching of some basic earth science concepts (Earth's rotation, revolution, and axial tilt), and their consequences. Classroom lessons are designed to both prepare and follow-up student visits to a planetarium facility. The planetarium is used as an instrument to display models to help the students understand the concepts. There are three visits to the planetarium (fall, winter, and spring).

The role of the teacher is to prepare the students for a laboratory experience at the planetarium. This is done by way of the materials provided, and whatever other strategies the teacher deems necessary. There are a variety of activities designed for both pre- and post-reinforcement of the concepts.

An integral part of Project STARWALK is the teacher preparation. This is done by way of a two-day inservice workshop, one day for each grade involved, and is conducted at a planetarium facility. Administrators and other grade-level teachers are welcome to attend as well. The participants are introduced to the teacher materials and the concepts and objectives of the program. They will be involved with the same activities as their students. Emphasis will be placed on understanding basic astronomy and earth-science concepts as they relate to the objectives of the program. The use of models, visual aids, and computers as aids in enhancing student comprehension are also part of the teacher workshop.

Project STARWALK is one of many programs within a nation-wide network known as the National Diffusion Network. It is funded by a dissemination grant from the US Department of Education. Within the National Diffusion Network are state contact people who help in coordinating efforts between projects like STARWALK and interested educators.

Project STARWALK is currently in use by elementary schools in fifteen states. Some of the schools use the services of the staff at fixed-base planetariums for conducting the student planetarium lessons. Other schools, in more rural areas, have purchased a portable planetarium system for use with STARWALK lessons. In these situations, the teachers have been trained to teach the planetarium lessons as well as the classroom lessons.

Teaching Astronomy at School in a Planetarium

M.F. Duval and D. Bardin
Observatoire de Marseille, 2 Place Le Verrier, 13248 Marseille Cedex 4, France

1. Introduction

In some countries, such as the U.S.A., there are quite a lot of major planetariums and an enormous number of small fixed-operating planetariums. In France, the situation is not as good. There are only three planetariums with domes in excess of 15 meters across, six with diameters between six and nine meters, and ten transportable ones. For these last — five EX3s, three Starlabs and two of personal design — the inflatable domes are between four and five meters in diameter, providing a dedicated astronomy teaching facility significantly large to accommodate a typical class.

The principal advantages of these transportable planetariums are:

- easy transportation
- installation inside the school
- quickly erected in less than 20 square meters of space
- capacity of six to eight sessions per day
- low cost: less than $30 per class and often free
- good contact between the instructor and the students, who sit on the ground or on small stools in a confined space.

Balancing these great advantages, the main deficiency is the limited program: only stars, planets, star clusters, galaxies, and the fundamental circles can be projected. Another difficulty is the manual control of the apparent motions of the planets. In any case, the facilities are not a place for entertainment; they are a place for pedagogic experiences.

Ten years ago, almost no astronomy was taught in French schools. A small group of teachers decided to begin extensive training of their colleagues. Thanks to their enormous efforts, different programs in astronomy are now included in both primary and secondary curricula.

2. Fundamental Aims of Planetariums

A. Discovery of Constellations

Children living in cities are unable to observe the night sky and do not usually know the locations of the cardinal points. So, allowing for the ages of the children, it is essential to show them how to find these through the motion of the sun and the direction of Polaris. Next we begin a description of the main constellations and how to recognize them. At the very start of the session, they should be shown for the current season, so that the students can easily find them in the sky. The session could then develop into a discussion of the names and meanings of the zodiacal constellations.

Pointing out the relative brightness of stars provides the opportunity to tell students about stellar distances and about the nature and origin of starlight.

B. Diurnal Movement

A complete rotation of the Earth can take place in three to twelve minutes, depending on the planetarium mode. In either case, the speed is sufficient to illustrate the phenomenon. Most students have not noticed this nighttime movement of stars and associated it with the diurnal motion of the sun; however, we regard this observation as one of the most important to explain.

We place special emphasis on the locations of rising stars, the duration of their visibility, and on their culmination. It is possible to compare star trails near the celestial pole with those at the celestial equator, and in the southern hemisphere.

C. Annual Solar Movement

In most planetariums, annual solar movement along the ecliptic is made manually. Doing so is not especially easy or convenient, but the ecliptic and equatorial circles aid the demonstrations, and with secondary-school students, it is possible to explain the seasons and the variations in the night sky throughout the year. By changing the apparent latitude of the observer, the reason for the "polar night" and the "midnight sun" can easily be demonstrated.

D. Planetary Movements

As previously stated, the small commercial planetariums are not well equipped to exhibit the annual motion of planets among the stars. However, students are most interested in information about the nature of the planets, and it is a pleasure to encourage them to talk about it. At this moment in the session, the proximity of the planets to Earth compared with the stars can be pointed out, and their relative brightnesses and physical properties can be explained.

Venus is a good (and easily observed) example of the apparent motion of a planet around the sun and illustrates the phenomenon of the phases of illumination.

E. Lunar Motion

The motion of the moon is a topic of particular interest to students. The planetarium helps present a clear and understandable explanation of lunar phases, and of the moon's rapid monthly motion among the stars. These phenomena, emphasized by the visible phases of the moon at night, are easily comprehensible even by 10-year olds.

3. Secondary Aims of the Planetarium

Planetariums give also an opportunity to describe stellar distributions in clusters, such as M44 (Praesepe), and h and χ Persei.

Some galaxies can be shown too, such as M31 and the Magellanic Clouds. The location of the Milky Way should be clearly shown, although presenting it with a diffuser is only fair.

4. How Transportable Planetariums are Utilized in Schools

These facilities generally spend from a day to a week in each location. During this time, the whole student population or only selected classes may receive instruction.

Because of the limited space and ventilation, a standard session in the dome is no longer than 30–45 minutes. Still, transportable planetariums are perfect for children who visit a planetarium for the first time and whose curiosity is great. We allocate a short period in the dome and a longer period outside to answer the questions that inevitably arise from this curiosity.

The planetarium session is usually a part of a larger activity including exhibits, video or films, and a description of astronomical devices (celestial globes, sundials, astrolabes...). Whenever possible, we organize an outside session to observe the night sky.

We must point out that the person conducting this planetarium show can be a volunteer teacher with specific training, an amateur astronomer, or an enthusiastic astronomy student, but in all cases this activity is normally unpaid at the present time. We operate these programs with inflatable planetariums 25 to 30 weeks a year. They serve about 10,000 students a year.

5. A Unique Experience in Nice

At Valeri's College in Nice, children from 12 to 15 years of age decided to build their own planetarium in the schoolyard. Sixty square meters were allocated, and in 1985, a facility was completed that included a planetarium with a four-meter dome, an observatory, and an astronomical laboratory.

The students, who are very proud of their accomplishment, were greatly motivated to produce exhibits and models for display in the surrounding schools and general community. This project is a remarkable example of the student's enthusiasm for astronomy.

6. Concluding Remarks

Most large cities in France do not have a planetarium. This lack is a source of frustration to teachers who are aware of the great interest that their students have in astronomy. The inflatable planetariums are an answer to this need.

Their number has been increasing in these last three years, but as no resources are yet available from official agencies, encouraging this activity is a most difficult problem and necessitates the use of highly motivated volunteers. In any case, we feel that increased planetarium use must develop in France and in many other countries in the context of a comprehensive plan for the better teaching of science.

The Pre-College Planetarium

Brendan Curran
Bronx High School of Science, 75 West 205 St., Bronx, New York, New York, U.S.A.

The Bronx High School of Science is a public school in the New York City school system. It is a specialized school whose students are selected on the basis of a competitive exam. Our planetarium projector was installed out his spring, so many of these suggested activities are still in development stage.

In the high school setting, the planetarium has a dual role as a teaching tool and as a motivational tool. The planetarium is used in the Astronomy and Astrophysics course, an advanced elective. This use permits a detailed study of the sky, which is otherwise difficult in New York City. Students who are familiar with thinking in two dimensions often have trouble making the transition to the three dimensions required in astronomy. The planetarium is an invaluable tool to develop students' spatial orientation. Further, the planetarium generates tremendous enthusiasm.

The planetarium will also find a place in the physics curriculum, maximizing its exposure to students since every student must take the physics course. Students will be given a planetarium experience when they study geocentric *versus* heliocentric models of the solar system and Kepler's laws of planetary motion. The planetarium will provide reinforcement, making the lessons memorable.

An additional benefit will be to generate interest in astronomy. The result will be a supply of students for the course as well as encouraging amateur interest.

We hope to develop additional programs, particularly some linking the humanities and astronomy. When English classes learn mythology, a visit to the planetarium would highlight the connection between culture and the sky. Planetarium observation would be appropriate for art-history classes. In social studies, a discussion of various calendars used by different human societies and their relation to celestial motion would provide enrichment.

We hope that these programs will maximize the use of the planetarium and increase students' awareness of astronomy.

Teaching Teachers in the Planetarium

Roland Szostak
Institut für Didaktik der Physik, Universität Münster, Wilhelm-Klemm-Strasse 10,
4400 Münster, Federal Republic of Germany

Many people do not know what a planetarium is. Since this disappointing fact is even true for well-educated people, including students and teachers, many pupils will never hear about planetariums. In order to spread knowledge about planetariums, universities can play a major role. One of the best ways is to visit a planetarium with students, especially those who will become teachers. Such a visit is a very efficient multiplying factor.

But in order to make a planetarium visit efficient, one should not go unprepared. We therefore go with education students who are enrolled in an astronomy course. Before their visit, we inform them about what a planetarium is and about how it is used. We teach them, for example, specific terms of the rotating sky, its coordinates, and the significance of the ecliptic and of retrograde loops. And of course, we teach the students how the sky appears from different latitudes, including the equator and the southern hemisphere. We also teach the technical aspects of how the projector works.

Students visits take place by special arrangement, without any other audience, so all technical details can be discussed individually. The operator can run the machine to answer specific questions that arise. And some of the students can even put their own hands onto the buttons of the control panel to run the machine. After the exclusive technical visit, we usually attend a normal public performance. We see teachers and their classes, so we can observe how children behave and react to the phenomena projected onto the dome.

Students are very motivated by their visit to the planetarium. As a result, they come back with their own classes later on. Such cooperation between university and planetarium is highly recommended. It will help in getting more knowledge about astronomy into schools, promoting a more profound understanding of nature.

Astronomy for the Visually Handicapped

Pierre Lacombe
Planétarium Dow, 1000 rue Saint-Jacques ouest, Montréal, Québec H3C 1G7, Canada

It is important to add that in addition to the general public and all the students and teachers involved in school systems, the planetarium community in general tries also to reach some special groups of the public. The groups could be those with hearing problems, those who have never seen any stars (the visually handicapped), amateur astronomy groups, and even gifted students.

As you can imagine, these special groups represent a very small fraction of our attendance at the planetarium but they need special treatment, and for that reason many planeteria around the world have at one time or another prepared a unique program in the star theater for them, built new and special exhibits, or given technical tours of their installations.

Earlier this year, the Dow Planetarium was involved in an unusual experience. At the planetarium, it is relatively easy to initiate the general public in astronomy. With shows in the star theater, observing sessions, and spectacular exhibits, the public can make direct relations between images and the real celestial objects visible in the night sky. But how can we initiate people who have never seen any stars — people without sight?

The Dow Planetarium in collaboration with the Botanical Garden, the Aquarium, and the Zoological Garden of the City of Montreal was involved in an exhibition to initiate these special people to the world of science. Our exhibit was a scale model of the orbits of the nine planets of our solar system. To permit exploring the exhibit, each orbit was drawn in relief so as to be easily located with the fingers and each planet was represented by a unique symbol. Experiments done with the visually handicapped have shown us that this concept was ideal for the exhibit. We also indicated in braille the names of the planets and the sun. But as only a small fraction of these people is able to read braille, an audio-guide describing each planet and the sun was available and highly recommended.

The exhibition was a great success. Imagine, for the first time, a complete scientific exhibition designed especially for the sightless person! At the Planetarium, we have learned a lot about this neglected clientele and the work we have done sets new standards for future exhibits (height, angle, and so on...).

Planetarium Activities in the Federal Republic of Germany

H.-U. Keller
Planetarium Stuttgart, Neckarstrasse 47, D-7000 Stuttgart 1, West Germany

The planetarium was invented by the German engineer Walther Bauersfeld of the Carl Zeiss Company in 1919, and the first projection-planetarium in the world was installed in the Deutsches Museum in München (Munich) 1923. Most of the German planetariums were destroyed during World War II. Today, nine major planetariums with dome diameters greater than 15m are in operation in the following cities in F.R. Germany; the numbers in brackets are the year of opening, the dome size and the seating capacity: West Berlin (1965; 20 m; 320), Bochum (1964; 20 m; 300), Hamburg (1933; 20.6 m; 270), Mannheim (1984; 20 m; 287), München (1925; 15 m; 156), Münster (1981; 20 m; 280), Nürnberg (1961; 18 m; 255), Stuttgart (1977; 20 m; 277), and Wolfsburg (1983; 15 m; 148).

The total seating capacity of all the major German planetariums is 293, and the total attendance is 1,130,000 at 16,300 performances and star shows. The main goal of the German planetarium shows is to teach astronomy with an element of entertainment, but not entertainment only or primarily.

Besides the major planetariums, there are 14 smaller ones, the most active being located in Bremen, Freiburg, Kiel, Osnabrück, and Recklinghausen. In the next few years planetariums will be put in operation in: Augsburg, Laupheim, Kassel, Ulm, and Wiesbaden.

At the opening of Stuttgart Planetarium in 1977, the Arbeitsgemeinschaft Deutschsprachiger Planetarien (Working Group of German-Speaking Planetariums) was founded (for the planetariums of Austria, F.R.G., G.D.R., and Switzerland). Since 1987, the Rat Bundesdeutscher Planetarien (Council of Federal German Planetariums) represents the planetariums in the F.R. Germany.

Bochum Planetarium: The Bochum Planetarium is a part of the Public Observatory Bochum. The tasks and aims of the Bochum observatory are popularization of astronomy and space sciences, and adult education. About half of the planetarium presentations in Bochum are public sky shows. A well-chosen proportion of astronomical information and entertainment has successfully proven to yield a maximum impact on a broad range of visitors. Regular light shows with a multitude of musical styles enhance the planetarium's importance as a local cultural center.

To ensure a continuing high standard of performance, the public sky shows are taped and pre-programmed. Our automation system meets all industrial standards and allows the application of many special effects.

Hamburg Planetarium: Located in the impressive Municipal Park Water Tower, the planetarium's Zeiss-Mark VI projector is the third instrument in a long history of the Planetarium. Hundreds of auxiliary devices enrich the performances with planet representations, galaxies, pulsars, panoramas, northern lights, and cloud formations. The dome of the Hamburg planetarium seats 270 visitors, to follow the monthly-changing programs or one of the special performances. The program is rounded out by exhibitions and telescopic guided tours.

Münster Planetarium: Münster Planetarium is an integrated part of the Westphalian Museum of Natural History. As in most museum-planetariums, we offer several taped programs a day for a general audience. But the scope of our many different programs (taped and live) includes general shows, shows on special astronomical topics, shows for children and advanced courses. Our presentations emphasize information rather than show elements.

Nürnberg Planetarium: The Nicolaus Copernicus Planetarium Nürnberg was inaugurated in 1961 as the first new planetarium built in Germany after the war. For 15 years, it was operated with a modernized version of the projector that had belonged to the old Nürnberg Planetarium (in operation 1928 to 1934). In 1976, a new Zeiss-projector model V was installed. More than 20 special-effect-projectors were bought during the last 20 years.

Public shows are carried out by tape; 10 to 12 special lectures per year are given by foreign astronomers. Most demonstrations for pupils are live. Occasionally we have music performances.

Stuttgart Planetarium: Stuttgart, the capital of the German Federal State Baden-Württemberg, the Swabian metropole, was one of the first cities to have a major planetarium. It was opened in 1928 and destroyed in 1943. In April 1977, the new Stuttgart Planetarium was opened. The Carl Zeiss Company donated a Mark IV projector and additional financial support toward the cost of the building. It was the first planetarium in Germany with a fully automatic control system. The Zeiss-projector can be lowered under the floor with an elevator. In the first ten years, more than 2 million visitors came to the star performances. Besides shows for the general public, we give live presentations for school and special groups. About 50 per cent of the visitors are students and scholars. The full-time staff of eleven persons is supported by 21 part-time workers. The main tasks of our planetarium are to teach astronomy for the general public and for advanced students, too, and to demonstrate how small our blue planet is — a tiny island of life in the giant, deep, boundless universe — and that the fate of mankind depends on skilled behavior of all human beings for living in peace and freedom.

Marketing for Planetariums

J.V. Feitzinger
Observatory of the City of Bochum and Astronomical Institute,
Ruhr-University, Postfach 10 2148, D-4630 Bochum 1, Federal Republic of Germany

The sales promotion of the planetarium is a marketing problem. Program arrangements must take this into account. Marketing is not exclusively a tool to obtain a maximum profit: marketing methods can control the exchange problems between the public and the nonprofit planetarium. The planetarium produces a service. This service must be distributed to the public. The public must know that the planetarium distributes a special service.

The public visits the planetarium during leisure time. (One exception — the school classes — is not considered here.) Therefore one marketing problem of the planetarium is to sell its service as recreational value.

The leisure time activities of the public can be divided into 10 categories. These categories contain the basic patterns and needs of the public. The content of the public planetarium performances must try to cover the leisure-time categories. The communication windows of the public can only be reached by simplifying astronomical facts.

Table 1. The planetarium as a leisure time activity

Leisure time categories met by	Planetarium visits
1. Physical movement; sporting — playful activities	No correspondence (sporting activities)
2. Calmness, loneliness, self-realization	All planetarium performances also pure music and star presentations
3. Social life and social communication	Planetarium visits by groups and the resulting conversations
4. Social presentation	Technical and scientific interests; status thinking
5. Amusement and dissemination	Planetarium performances are ideal; not much own initiative, diversion
6. Intellectual and political interests, discussion, education	Planetarium performances are well suited
7. Mobility, search for new environment attractions	In the course of planetarium performances: Presentation of realistic planetary or science-fiction environments
8. Participations at competitions, achievement, rivalry	Astroquiz
9. Playful pastimes, not determined by its function	Similar to point 5 (the boundaries are flowing)
10. Well-being, sensual impressions	Similar to point 5 (especially music performances)

The planetarium is a versatile institution that fits well into the leisure-time categories of the public. There should be no exchange difficulties, if these categories are taken into account in the program.

Innovative Astronomy Education Programs for Developing Countries

B.G. Sidharth, B.M. Birla Planetarium, Adarsh Nagar,
Hyderabad — 500 463 India

1. Introduction

It is desirable that planetariums in developing countries should make the maximum and most efficient use of the planetarium infrastructure and facilities to cover as much ground as possible in the popularization and dissemination of astronomy. After all, the number of planetariums in developing countries necessarily has to be small, and so specialization in specific disciplines or fields becomes a luxury. In India, for example, there are about ten planetariums, and another five or six will come into operation in the next few years. But these planetariums have to cater to a large population. In the U.S.A., which has a fraction of India's population, on the other hand, there are hundreds of planetariums. The following suggestions are based on successfully implemented projects at the B.M. Birla Planetarium, Hyderabad.

A golden rule for planetarium programs anywhere, and certainly in developing countries, is to start a planetarium sky show or activity with a local flair. For example, the local names of stars and constellations, local myths, local astronomers or, more specifically, topics like the history of astronomy in the region should be highlighted.

2. Planetarium Sky Shows

It is important for planetariums in developing countries in particular to choose the levels of their astronomical presentations (sky shows) carefully. These could be broadly classified as follows:

a) Popular programs for the lay public. Such sky shows should have a minimum of subtle inputs. The lay audiences in developing countries definitely prefer a spectacle and a dramatic experience to a pedagogic presentation.

b) Sky shows, preferably live and participatory in nature, for school groups. It would be advisable to prepare them in consultation with the relevant teachers. In any case a dialogue with the teaching community is essential.

c) More specialized, again preferably live, programs for amateurs and astronomy students.

3. Educational and Amateur Activities

Educational activities should be spun around planetariums. It would be helpful if these planetariums also acquire mini/portable planetariums, and, of course, telescopes for such activities. At the B.M. Birla Planetarium, for example, a project entitled *Astro School* has caught the imagination of educators. In this program, a set of school children is given a one-day exposure to all the excitement of astronomy through sky shows, exposure to a mini planetarium, illustrated lectures, computer graphics, astronomical games, science-fiction films, and, whenever possible, sky observation sessions.

Planetariums in developing countries would also do well to organize informal astronomy courses for lay persons. At Hyderabad, the course is of a multi-media nature and lasts for three months, with two evening lectures per week. The course has generated so much enthusiasm that rejection of applicants is a major problem. The teaching community in schools, particularly in the rural areas, is in general scientifically impoverished and ill-equipped to teach astronomy or science itself while preserving the sense of excitement and discovery. So planetariums would render a service to the community by conducting two or three-day camps for science teachers from schools, particularly the rural schools. An important extension of this idea is camps of shorter duration, for example on selected Sundays, for the lay public. Such one-day camps using portable planetariums, for school students/teachers in the rural areas, would be a great boon for this less fortunate segment of society in the rural areas.

Regular lectures by interesting speakers, workshops — for example, on telescope making — and exhibition of astronomical and scientific films and videos should be a part of the culture of every planetarium. Such events are logically linked with the amateur astronomical activities that can build up around a planetarium.

It is interesting that at Hyderabad, where one of the only astronomy university departments in India has been functioning for nearly three decades, the first amateur astronomical association came into being shortly after the inauguration of the planetarium. The association was formed by a group of lay persons who attended a three-month multi-media astronomy course offered by the planetarium.

4. Research Activities

Seminars on catchy topics organized by the planetariums not only give a boost to the interest in astronomy but also earn for the institution a lot of publicity. This has been the case in Hyderabad, where the B.M. Birla Planetarium has organized major seminars on Halley's Comet, Indian Astronomy, and finally Ancient Astronomies. In fact, one of these seminars coincided with an equally important international seminar on a topic in theoretical physics. While the former got wide coverage in the press, the latter was almost totally ignored.

5. Information Centers

A planetarium would render its community a great service by organizing a research center for the history of astronomy of the region. Even the planetarium-projector facility itself could be directly used, for example, in the dating of historical events by using relevant astronomical allusions.

Planetariums in developing countries should act as centers for dissemination of astronomy information. At Hyderabad, this is done through posters, press releases and educational films which are telecast all over the country. In this connection, an acute problem faced by planetariums in developing countries is the lack of immediate access to important astronomical-event information. People look to planetariums, and not university departments, for information. And it is of vital importance that there should be an agency on the lines of the IAU telegrams that can transmit — at a price, if necessary — such information, preferably in layman's language, to planetariums.

The planetarium at Hyderabad for example receives even long-distance calls from people wanting to know where and when a comet can be sighted and so on. So the importance of such an international information facility cannot be overemphasized.

Another sad problem faced by planetariums in developing countries is the lack of resources to meet the need for take-away literatures, audio tapes, video cassettes, planispheres, books, souvenirs, and so on. The problem is even more acute in a country like India with its import regulations.

6. Leadership

For developing countries, it is important that the established and successful planetariums should provide leadership for the new and also smaller planetariums. An acute problem in this context is the lack of trained personnel, for the simple reason that a planetarium culture is either non-existent or nearly so. With exactly this in mind, the B.M. Birla Planetarium, Hyderabad, has introduced a university-recognized post-graduate diploma in Planetarium Techniques and Management. This will, we hope, build up a cadre of planetarium directors and educators.

7. Conclusion

Lastly, it should be pointed out that the success of a planetarium depends to a great extent on its showmanship and marketing ability. In fact, this task is easier in the developing countries, where media coverage is much more accessible than in the advanced countries. An author once advised novices that they should break their necks to get into print. A good dictum for a planetarium would be that it should break its neck or whatever to remain in public view.

Follow-up remarks by Joseph M. Chamberlain

My colleagues on the panel have presented an excellent cross-section of activities and opportunities in the teaching of astronomy through the planetarium medium. Since our time allotment is brief, I do not want to duplicate, but I would like to summarize two very attractive programs that are offered in several planetarium settings, and to invite the cooperation of this group of astronomers involved in the more formal aspects of astronomy, and of others who may read what we are here presenting orally.

The first type of program that I have in mind is a series of lectures, demonstrations, and discussions for high-ability high-school students. One such program, known as the Astro-Science Workshop, has been offered in my own institution, The Adler Planetarium in Chicago, for twenty-seven years. I know of similar programs at the American Museum — Hayden Planetarium in New York and others offered on college campuses rather than in a planetarium. Typically, the students are juniors and seniors in high school, carefully selected and recommended by the head of the school science department. The program might be conducted on a daily basis in the summer or on Saturdays through the school year. The planetarium staff organizes the program and often presents many of the lectures. Professors from nearby colleges are invited to discuss their research interests. Sometimes grades are assigned; sometimes scholarships to colleges are awarded. Frequently, the planetarium program is accepted for advanced standing in the typical introductory astronomy course during the first year of college.

I encourage all of you here, and your colleagues elsewhere, to be aware of such programs, to participate in them if asked, and even to volunteer your services. Experience tells us that such programs have been effective in producing research-oriented astronomers as well as science-oriented professionals in other disciplines.

The other type of program that I wish to mention often comes under the heading of continuing education. A planetarium can offer a series of lectures or experiences ranging from elementary-level descriptive astronomy to advanced astrophysics, and even to special topics in contemporary astronomy. The students usually are planetarium members or astronomy aficionados who are participating simply because they are interested in the subject. At Adler, professors from the University of Chicago, Northwestern University, and elsewhere supplement the Planetarium staff as course lecturers. Similar patterns of instruction prevail at other planetariums. In recent years, I have found that even the busiest research astronomers are willing to share the excitement of their involvement in their astronomical specialty.

These continuing education courses are sometimes offered for credit through a college or university, but more often are set up on an informal basis that does no more than satisfy the curiosity of the participant. Some students are so diligent, however, that they continue to take such courses and become amateurs with advanced standing.

Discussion

C. Harper: *Planetariums in the northern hemisphere should be urged to present to their audiences the southern sky — that is to say, the sky as seen from the southern hemisphere. Opportunities for hypothesis testing and an awareness of our global unity are important educational goals.*

E.V. Sprouls: *Some effort has to be made to keep members of the planetarium staffs in touch with astronomical events, because the planetarium is often the first place that people turn to find out more information.*

Suggestion: Planetariums can ask to be placed on the press-release mailing lists of astronomical institutions and associations.

B.G. Sidharth: *The planetarium is looked up to by people and media for giving the latest information on astronomical events. It would be good if there were an agency to which planetariums could subscribe for "Astronomical Telegrams" — but at the layman level.*

O. Gingerich: *But if an astronomer is announcing a new discovery, how does he inform the hundreds of planetariums? He can't telephone them all!*

A. Fraknoi: *To help U.S. and Canadian planetarium and other educators to learn about new developments in astronomy research, the A.S.P. has set up an astronomical hot-line with weekly recorded messages on new discoveries and theories. The number is (415) 337-1244. [Ed. Note: From other countries, after getting your international access line, dial 1-415-337-1244.]*

C.R. Chambliss: *I teach astronomy in the state of Pennsylvania with only 5 per cent of the population of the U.S.A. but with more than one-third of the nation's planetariums. As noted in my talk, we make very heavy use of our planetarium (10.5-meter dome) at Kutztown University in astronomy courses and school-group presentations. Two of my good friends are high school planetarium directors. Their duties include both teaching astronomy at the senior-high-school level and giving a wide variety of shows for kindergarten through 12th grade. Both use planetariums comparable in size to the one in Kutztown University. Regrettably, there are also in Pennsylvania many planetariums which are under-used or not used at all. Our state suffers more from a dearth of qualified planetarium instructors than from a deficiency in planetariums.*

L. Gougenheim: *1992 is planned to be the International Space Year and is recognized as so by the International Council of Scientific Unions. Much attention should be paid to educational activities and general public understanding. Funding is expected from the space agencies. One of the topics that has been selected is precisely the development of educational activities around planetariums. This International Space Year would provide a good opportunity to develop planetariums, especially in developing countries.*

KNOWLEDGE AND WONDER IN THE PLANETARIUM

J. Lawrence Dunlap
Flandrau Planetarium, University of Arizona, Tucson, Arizona 85721, U.S.A.

Planetariums around the world attract millions of adults and children each year to public programs popularizing astronomy and related topics. For many urban school children, the planetarium experience is a unique opportunity to observe and to wonder about the night sky and objects in space. In particular, the planetarium provides children with a needed observational basis for increasing scientific literacy and for understanding astronomical concepts.

This paper explores the role of the planetarium in the astronomy education of school children, grades Kindergarten–12, in Tucson, Arizona, over the past 12 years.

1. Background and Scope

The Grace H. Flandrau Planetarium, located on the University of Arizona campus, is Arizona's primary resource for astronomy education for the general public. The mission of the Planetarium is to enhance the public appreciation and understanding of science and astronomy by communicating the beauty, knowledge, and wonders of the universe to persons of all ages. The planetarium produces a visual simulation of sky phenomena. Thus it is a teaching machine that creates a direct visual learning environment that is especially "friendly" to children and laypersons.

In 1974, a cooperative agreement between the University and the Tucson Unified School District established a means to provide instructional programs for precollege students. In 12 years of operation, the school programs have provided instructional activities in optics, astronomy, and space science to over 260,000 students and teachers from Arizona.

2. Goals and Objectives

The principal goal of the school planetarium programs is to teach selected facts and concepts about the night sky and to motivate students to seek a deeper appreciation of astronomy and the universe.

The general objectives of all the programs are to encourage students to use and develop those intellectual skills described as "the processes of science." Students participate in the learning process by responding to questions in both the theater and exhibit situations. Acceptable responses indicate that students are able to:

— make visual observations

— communicate information accurately

— classify objects by similarities and differences

— use numbers to make quantitative comparisons

– make predictions
– make inferences.

3. The Planetarium and Concept Learning

We acquire concepts by "formation" and by "assimilation." Concept formation is primarily inductive, as children learn by observing, imitating and trial and error behavior. Childrens' intuitive world concepts are built from everyday experiences that reinforce their observations. The child's world is flat, unmoving, and in the center of the universe. Forces are needed to keep things in motion. Rocks fall faster than leaves and feathers. Heavy things are harder to move than light things.

Concept assimilation is primarily deductive. We learn a general rule and then try specific examples. Without a concrete experience, students learn to say or to write the words (the "keyed response") to answer questions, but they may not develop an intuitive understanding of the concept. The planetarium can provide the experience they need to give physical meaning to the words. Observations and questions can help students build more accurate physical concepts.

4. Typical Observing Activities

Sample observations and grades	A	B	C	D[a]
1. find directions by sun and stars.	X	X	X	X
2. locate star patterns.	X	X	X	X
3. daily motion of sun and stars:				
– for 24 hrs.	X	X	X	X
– for 1 year.			X	X
4. motion of stars at other latitudes.		X	X	X
5. appearance and daily motion of moon.		X	X	X
6. conditions for lunar or solar eclipse.		X	X	X
7. apparent motion of planets relative to stars.		X	X	X
8. synodic periods of planets.			X	X
9. computing relative distances to planets.				X
10. light curve of a variable star.				X
11. precession of the equinoxes.				X
12. landscapes of moons and planets.	X	X	X	X

[a]Key: A = Primary, B = Intermediate, C = Middle, D = High School

5. Asking Questions

Questions can be used to acquire feedback about what students have learned, to foster concept formation, or to encourage wonder. Content feedback questions ask for information. Concept formation questions ask for inferences or predictions that require the student to test their conceptual models. A correct response reinforces the model and an incorrect response can lead to new insights and a new model.

Questions to encourage wonder are based on observations, but no explanations are given. Using a mixture of all kinds of questions encourages students to review their knowledge, to test and improve their concepts, and to consider some new mysteries.

6. Some Questions to Wonder About

Primary level:

1. Where is the sun after sunset?
2. Where will the moon be a week from now? Will it be more full, less full, or about the same as it is now?
3. What makes part of the moon look dark?
4. Does the moon have a night and day?
5. What would you see from the moon in the daytime?

Intermediate level:

1. What makes the twilight?
2. What makes starlight twinkle more than planet light?
3. (When one or more planets are in conjunction), where are the missing planets?
4. What makes the ocean appear blue?
5. What would you see in the skies of other planets?
6. Are there planets orbiting other stars?

7. Comparing Live and Recorded School Programs

Live school programs focus on visual observations, concept development and information. Recorded programs permit more extensive audio-visual program content and attract a larger audience. Our most popular school programs have been a 50/50 combination of a live current sky interpretation "talk" and a short recorded program showing landscapes on other worlds, imaginary voyages to other parts of the universe, or highly visual treatments of interesting topics. Many students are motivated to ask questions, to read astronomy books, or to do special projects after seeing an excellent program. Perhaps this is the most important role of the planetarium — to promote and sustain an interest in astronomy and our place in the universe among children everywhere.

USING TELECONFERENCING FOR A PLANETARIUM LESSON

Elizabeth S. Wasiluk
Berkeley County Planetarium, Hedgesville, West Virginia, U.S.A., and
800 West Addition Street, Martinsburg, West Virginia 25401, U.S.A.

Before coming to West Virginia, I worked for Delaware-Chenango BOCES as their telelearning coordinator. BOCES stands for "Board of Cooperative Educational Services." Such organizations exist throughout New York State for the purpose of pooling resources of school systems and providing services and equipment they might be unable to afford on their own. Examples of the services they provide include distance learning, vocational education, computer services, *etc.* Our BOCES owned a portable Starlab planetarium for loan to member school districts.

Carol Kwiencinski, from Downsville Central School in Downsville, New York, used the portable planetarium with her students. Her students asked her many questions she could not answer, particularly about black holes. As an add-on to her planetarium session, she approached me to do a teleconference with a planetarium director, Mitch Luman of the Koch Planetarium and Science Center, Evansville, Indiana. To set up the conference, we used a speakerphone, which is basically a set of amplifiers and microphones that allow students to speak and listen to a telephone conversation as a group.

Prior to the conference students read information on black holes and devised a list of questions. The day of the conference, students addressed their questions to someone many miles away who had become familiar to them. A tape made of the conference sounded as though the students and Mr. Luman were in the same room.

The key to success of a teleconference is selecting the appropriate equipment, preparing both ends prior to the teleconference, and choosing a dynamic speaker who can make the conversation a "dialog" rather than a lecture.

Teleconferencing could be of great use to a small school or planetarium incapable of having a speaker travel to their facility. Slides and/or videos could be used before, after, or during the teleconference.

With the use of AT&T's Alliance Service, it is possible to link more than one location, enabling several groups or individuals to speak at the same time from many locations. This system is useful in holding committee meetings.

Perhaps teleconferencing can be used to bring more astronomers into classrooms that couldn't afford to have access to them before.

Ed. Note: As of 1989, picture-phone systems capable of transmitting one image every 5 seconds are available at a cost of $400 per instrument.

———————————

A PLANETARIUM ORIENTED SEQUENCE OF EXERCISES

Carlson R. Chambliss
Kutztown University, Kutztown, Pennsylvania 19530, U.S.A.

The introductory course in astronomy intended for non-majors is taken by about 150 students per semester at Kutztown University. This institution is fortunate in having a planetarium, which is used extensively both in course work in astronomy and with visiting school groups. The projector is a Spitz model A3P, the dome is of perforated aluminum and has a diameter of 10.5 meters, and the seating capacity of the planetarium is about 120. The introductory astronomy course at Kutztown consists of two hours of lectures and a two-hour laboratory session per week. There are five or six laboratory sections, each of which contains about 25-30 students.

The planetarium has proven to be extremely useful in demonstrating numerous concepts of the celestial sphere. At Kutztown University, a three-part sequence of laboratory exercises on the celestial sphere is required near the beginning of the introductory course. In the first week of this sequence the students are introduced to the altazimuth system and to such terms as cardinal points, celestial poles, meridian, equator, and great circles. All of these can be demonstrated with the illuminated circles and markings that are provided in the planetarium. The relationship between the altitude of the north celestial pole and of its compliment (the inclination of the celestial equator to the horizon) to the latitude of the observer is stressed.

In the following unit, such concepts as hour angle and circumpolar stars are introduced together with the annual motion of the sun and the equatorial coordinate system. The appearance of the night sky is shown at a wide variety of latitudes, and the role of astronomical observations in determining one's position on the Earth's surface is discussed. The relationship between the altazimuth and equatorial coordinate systems is examined, although the trigonometric equations that allow for the transformation of positions from one system to the other are not used in this introductory course. Various stars and constellations are pointed out, and later on in the course students are given a quiz in which they are required to identify various bright stars and prominent constellations.

In the third and final part of the exercises on the celestial sphere the seasons are discussed together with ecliptic coordinates, precession of the equinoxes, and a brief introduction to climatology. Throughout the sequence, students are required to make estimates of various parameters from observations that they make on the planetarium dome.

The conceptualization of the celestial sphere involves many terms that are initially unfamiliar to the non-science student, but by breaking up this study into three units, usually spread over a period of three weeks, we find that nearly all students can successfully grasp the terminology and the interrelations among the various parameters of the celestial sphere.

A question arises as to how much material of this sort should be included in one-semester course for non-science majors. Formerly the equatorial coordinates of the north and south poles of the ecliptic were derived from diagrams, but we have concluded that this is superfluous to the syllabus of this course. Galactic coordinates are also omitted, as they are not used in any of the earlier portions of the course. Some references to sidereal time and to mean and apparent solar times are made, but students are not required to be able to convert from one of these to the others in their exercises.

Although we are able to make full use of the planetarium that is available at Kutztown University, these exercises can also be done using a celestial globe. For instance, the celestial globes manufactured by the Denoyer-Geppert Co. (5235 Ravenswood Ave., Chicago, Illinois 60640, U.S.A.) can satisfactorily produce all of the scenarios concerning celestial sphere concepts that we demonstrate in the planetarium. Thus these exercises would also prove most useful at institutions that do not have access to a planetarium.

Chapter 13

Developing Countries

With the widespread international representation at the conference, much attention was paid to astronomy in developing countries — a particular concern of the International Astronomical Union. The papers that follow discuss first case studies about India, Egypt, and Thailand. (Some related papers appear in the Curriculum section.) A paper from Uruguay relates the teaching of astronomy to economic development. The existing International Schools for Young Astronomers and the Visiting Lecturers Program of the International Astronomical Union, the experience with the latter program in Peru, and the Vatican Observatory's summer schools for students from a wide variety of countries are next discussed. Finally, two papers deal with new projects to contribute surplus texts and journals and to provide a well-equipped small telescope for developing countries.

A panel discussion organized by Silvia Torres-Peimbert considered the astronomical needs of developing countries from the points of view of astronomers from Nigeria, Malaysia, Mexico, and Paraguay. A paper about textbooks in developing countries is included in chapter 6. Interest in textbooks for developing countries was so high that a special impromptu session was organized on the topic, of which a summary is included here.

TEACHING OF ASTRONOMY IN INDIA

H.S. Gurm
Department of Astronomy & Space Sciences, Punjabi University,
Patiala, 147002, India

1. Historical Perspective

Studies of the skies have dominated intellectual activities since ancient man. In this respect, India has a very long tradition of such recorded activity, covering the observations of celestial bodies both as a science and as mythology (Gurm, 1980). The first half of the Christian era witnessed the evolution of spherical astronomy as a part of the study of mathematics (algebra and trigonometry) and its application to astrology. The evolution of spherical astronomy culminated in the concrete manifestation in the northern parts of India in the form of Jantar-Mantars by Raja Jai Singh (Mayer, 1979) in the early eighteenth century. Interestingly, spherical astronomy remained one of the most important activities in the study of astronomy during the British period too. Some of the older treatises on this subject during the nineteenth century were written in the Offices of the Survey of India.

Even today spherical astronomy is taught all over the country in colleges and in some universities as a part of the mathematics curriculum. Somehow, it does not generate a feel of present day astronomy. It reminds one more of spherical trigonometry.

A meterological *cum* astronomical observatory was started in Madras in 1786 and is known as "Madras Observatory" in records (Kochar, 1985). It was shifted to Kodaikanal in 1899, where the Evershed effect was first observed in 1909. It is now a part of the Indian Institute of Astrophysics. A few research centers were formed during the post independence period. Practically all astronomy is covered (Daniel, 1983; Bhattacharyya, 1986). The Astronomical Society of India (established in 1978) has about 300 members.

For any evaluation of the teaching of astronomy in India, we need to have a look at the system of education. It fits the "traditional pattern" of Wentzel (1989). Five years of primary schooling, five years (three and two) of secondary schooling, two years of intermediate level, two years of undergraduate, and two years of masters course is the teaching pattern. One year is being added to the undergraduate course as part of a new education policy (Ministry of Education, 1985). Doctorate work can extend from two to any number of years. The education in India is supposed to cater to a population of 800 million. Presently there are about 150 full-fledged universities with 6000 colleges affiliated to these universities where the bulk of undergraduate teaching is done. There are 40,000 high schools and at least five times as many primary schools.

2. Teaching of Astronomy

Positional astronomy based on texts like Todhunter (1952) and Smart (1931) as a part of the mathematics curriculum has been taught for the last hundred years and is still being taught in a large number of colleges and universities. A large number of small telescopes were purchased by some of the Maharajas in the 1930's and donated to some of the colleges. These are under the control of the departments of mathematics. Over the years, interest has faded in their use for the teaching of astronomy, particularly with the introduction of new mathematics in the 1950's.

Astronomy in small capsules is being taught in about 20 universities as a part of physics at the masters level. It is taught as an independent subject at the undergraduate level at only two places: Lucknow University and Shivalik College at Naya Nangal, affiliated to the Punjabi University. The number of students so involved is less than one hundred.

Two independent departments of astronomy exist, at Osmania University and at Punjabi University, where two-year masters-degree courses in astronomy are taught. The total number of students spread over two years is about 40. Delhi University taught such a course as a one-time exception in 1960-62.

There is a lot of activity at the amateur level, with about twenty planetariums and with coverage in newspapers and on TV as a consequence to the space program. Recently, primary-school books have a nice introduction of the solar system and related topics. But beyond this, there is a complete blackout up to the first degree. A few centers with about 100 students, in an estimated population of 800 million, is all that India has in colleges and universities.

3. Recent Efforts

Representatives of various research establishments and from the universities evaluated the manpower requirements at Osmania University in a Round Table Discussion on Training Requirements of Astronomers in India in 1977 (Abhyankar and Sanwal, 1978). Research centers prefer physics graduates and, as such, no independent teaching of astronomy. There was a difference of opinion over these estimates and over the contents of courses to be taught.

The University Grants Commission (UGC), Government of India, under the guidance of two of its successive vice-chairmen (B.R. Rao and Rais Ahmed) set up two working groups in 1982 and 1985, respectively, to evolve teaching and research in astronomy in the Indian universities. The first group got involved with the day-to-day affairs of the Advanced Centre of Astronomy in Osmania University. However, the second did formulate wide-ranging recommendations to initiate teaching of astronomy in the Indian universities. The recommendations were accepted by the UGC (1987) for implementation. Some of the recommendations are as follows:

- A few independent departments of astronomy to be established in the universities during the coming 5 to 10 years.
- Five colleges of the Punjabi University to teach astronomy as an independent subject at the undergraduate level.
- The teaching of astronomy to be supported as part of physics at the undergraduate and graduate level, depending upon the availability of staff.
- Some of the existing astronomy infrastructure in the universities to be supported.
- A center of research to be established around the 122-cm telescope of Osmania University for the use of astronomers in the universities.

The Indian Space Research Organization (ISRO) initiated and funded the manufacture of 7.5-cm and 40-cm telescopes in the country. This was based upon the recommendations of the ISRO-UGC Panel for Telescopes in 1981. One thousand 7.5-cm telescopes were manufactured and distributed among the universities and schools all over the country for Comet Halley. The prototype of a 40-cm telescope is proposed, with some focal-plane instrumentation. New departments of astronomy may develop their teaching program around these telescopes. The colleges and universities have about a hundred telescopes larger than the 7.5-cm of the ISRO. These were mostly purchased in the 1930's. There are about 20 of 15-cm Cassegrain systems of Carl Zeiss and of Celestron. Two medium-sized telescopes are in Osmania and Punjabi Universities.

A joint astronomy program leading to a Ph.D degree in astronomy has been initiated by the Indian Institute of Science, Bangalore, in collaboration with the Indian Institute of Astrophysics, the Raman Research Institute, the Tata Institute of Fundamental Research, and the Physical Research Laboratory. The Inter-University Centre for Astronomy and Astrophysics (IUCAA) has been set up at Pune very recently (July 1988). It is supposed to cater to the needs of the astronomers in the Indian universities and is funded by the UGC.

There are a very few astronomers in the Indian universities, and this presents one of the serious problems in the teaching of astronomy. It leads to a vicious circle: no teaching, no teachers. At one stage, we at Patiala were asked to close down the teaching of astronomy, as teachers of astronomy were not available in other universities who could act as external examiners! Further, locally manufactured small- or medium-sized telescopes with focal plane instrumentation are not available. The import of such instruments is a long and cumbersome process. Similar problems regarding reference material, textbooks, and journals exist.

A concerted effort is required to make the teaching of astronomy popular. For India, the UGC working group recommendations of 1987 is the best way to start.

Recognizing the teaching of astronomy as an independent discipline and having professional astronomers to work in it is the issue facing the astronomy community.

4. Astronomy Teaching at Punjabi University

The Department of Astronomy and Space Sciences was established in 1978. We run a two-year M.Sc. course in astronomy and space physics, with ten students admitted every year. The Ph.D. program has led to six doctorates and five are in progress. Astronomy is also taught as an independent subject in one of the colleges and is to be introduced in five more colleges to be supported by the UGC. The department provides a basic infrastructure for training and research. One 60-cm Cassegrain is operational. A CCD system and a spectrograph are being added.

The two-year course leading to the M.Sc. degree is for those candidates who have a B.Sc. degree with mathematics, physics, and chemistry. The courses have a 40 per cent content of mathematics and physics. The practical training is oriented around basic electronics, spectroscopy, numerical analysis, data handling, and observational programs on stellar and solar physics. Some exercises are based on the plates of solar and stellar spectra acquired from other observatories. Experiments to evaluate site selection and seeing conditions have also been set up. Seminars and projects are an essential part of training. Laboratory training is along the same lines as at the University College London (Dworetsky, 1989; Gurm, 1983). There is a powerful support of meteorological observatories at the campus for weather parameters.

The department has organized two summer schools of six weeks each for college teachers of physics. It has generated short programs for local school teachers and TV programs in the local language (Punjabi). It caters to various enquiries on astronomy, meteorology, and allied fields. A number of amateurs have used the optics workshop of the department to make small telescopes.

References

Abhyankar, K.D., and Sanwal, N.B. (eds.), 1978, "Proceedings of the Round Table Discussion on Training Requirements of Astronomers in India," Osmania University Press, Hyderabad.

Bhattacharyya, J.C., 1986, "New Telescopes in India," *Astrophys. & Space Science,* **118**, 45.

Daniel, R.R., 1983, "Astronomy and Astrophysics in India — A Profile for the 1980's," Indian Space Research Organization, Bangalore.

Daniel, R.R., 1985, " Career in Astronomy," *Science Today,* **26**, No. 9, 29.

Dworetsky, M.M., 1989, "Classification of Stellar Spectra," these proceedings.

Gurm, H.S., 1980, "Indian Astronomy Through Ages," *Akashwani,* **45**, No. 49, 6.

Gurm, H.S., 1983, "Department of Astronomy & Space Science, Punjabi University," *Bull. Astr. Soc. India,* **11**, 246.

Kochar, R.K., 1985, "Madras Observatory: The Beginning," *Bull. Astr. Soc. India,* **13**, 162.

Mayer, B., 1979, "Jai Singh Observatories," *Sky & Telescope,* **50**, 6.

Ministry of Education, Government of India, 1985. "Challenge of education — A policy perspective," New Delhi.

Smart, W.M., 1931, "Text Book on Spherical Astronomy," Cambridge University Press.

University Grants Commission of India, 1987, "Report of the Working Group on Teaching and Research in the Indian Universities."

Wentzel, D.G., 1989, "Diverse Structures, Same Science," these proceedings.

Todhunter, I., 1952 (revised by G. Prasad), "Spherical Trigonometry," Pothishala Private Ltd., Allahabad.

TEACHING OF ASTRONOMY IN INDIA: With Special Reference to Teaching of Astronomy at Lucknow University

P.P. Saxena

Department of Mathematics & Astronomy, Lucknow University, Lucknow, India

1. The Teaching of Astronomy in India

a) Introduction of modern astronomy in India

Modern astronomy started in India when an astronomical observatory was founded in Madras as early as 1786 by the East India Company and to which the Indian Institute of Astrophysics traces its origin. There are, however, records of astronomical observations taken through a telescope from Pondicherry that elucidate the double-star nature of Alpha-Centauri as early as in 1689. Since then many more research centers in astronomy have been established. Today, institutions like the Indian Institute of Astrophysics (Bangalore), the Raman Research Institute (Bangalore), the Center of Advanced Study in Astronomy (Hyderabad), the Tata Institute of Fundamental Research (Bombay), and the Physical Research Laboratory (Ahmedabad) are engaged in pioneering work in theoretical and observational branches of astronomy and astrophysics.

b) Present teaching status of astronomy in India

At the pre-university level (grades 9 to 12): The Central Board of Secondary Education has prescribed a course of astronomy/astrophysics at the school level that is good enough to familiarize the student with the subject.

At the university level: Formal teaching of astronomy at the university level is imparted in more than two dozen universities as a part of the physics/mathematics syllabus. Facilities for imparting the teaching of astronomy as an elective subject exist in three universities, *viz*, for undergraduate (B.Sc.) teaching — Lucknow University, Lucknow; for post-graduate (M.Sc.) teaching — Osmania University, Hyderabad; and for both undergraduate and postgraduate teaching — Punjabi University, Patiala. Lately, Poona[1] has also been chosen for graduate teaching and research in astronomy. There exist, however, a variety of syllabi of astronomy at the various universities.

c) Need for a "core" syllabus of astronomy at the university level

Efforts must be made for uniformity in the syllabi of astronomy in mathematics and in physics and as a full elective subject at the B.Sc. and M.Sc. levels. This would have the following advantages:

[1]The name of the city of Poona has now been officially changed to Pune.

1. It will help in maintaining a uniform standard of astronomy education.

2. In the initial stages it will be easier to train teachers in astronomy (presently there is an acute scarcity of teachers of astronomy).

3. Books may be written on this "core" syllabus to the benefit of all students of astronomy.

2. Teaching Astronomy at Lucknow University

a) As an elective subject at the B.Sc. level

Teaching astronomy as an elective subject at the undergraduate level started at Lucknow University as early as 1950 at the initiative taken by the then Head of the department, Professor A.N. Singh. The university offers a course in astronomy to the students who opt for astronomy, physics, and mathematics as subjects of their study at the B.Sc. level. In the third (final) year of B.Sc. studies, the number of subjects is reduced to two so that the subjects may be taught more extensively.

The number of places available is 10. The number of students actually admitted, however, is around 60 for the last two years, partly because of the students' increased liking for the subject and partly because of the fact that Lucknow University is the only university in the state of Uttar Pradesh where astronomy is taught as an elective subject at the undergraduate level. The six-fold increase in the number of admitted students creates difficulties in laboratory classes, but it is hoped that the situation will become normal soon when we shall have more facilities available to us.

The syllabus and courses in astronomy in the three-year degree program are shown in Table 1 for each of the three years. Copies of the syllabus are available from me.

Table 1. Courses in astronomy at the B.Sc. level
(Lucknow University)

	Theory			Practical
	Paper I	Paper II	Paper III	Paper IV
B.Sc. I	Spherical Trigonometry Spherical Astronomy	General Astronomy	General Astronomy	Experiments & Problems
B.Sc. II	Spherical Astronomy	Stellar Astronomy	Stellar Astronomy	Experiments & Problems
B.Sc. III	Astrophysics	Astrophysics	Astrophysics	Experiments & Problems

In framing the syllabus, due consideration is given to the fact that the student learns about the celestial bodies and the observed phenomena gradually. During the first year, the students' knowledge of astronomy is confined to the solar system. In the second year, this knowledge is extended to stars and stellar associations. In the third year, stress is on the physics of the observed phenomena relating to celestial bodies. The teaching hours per week are tabulated below (Table 2):

Table 2. Teaching hours per week

Class		Astronomy		Total hours/week
		Theory	Practical	
B.Sc.	I	6	12	18
	II	6	12	18
	III	9	12	21

b) Teaching astronomy as part of the mathematics syllabus at the M.Sc. level at Lucknow University

There is an optional paper of "Astrophysics" for the M.Sc. final-year students of mathematics. Copies of the syllabus are available from me. An average number of students opting for this paper is ten and the number of teaching hours per week is four.

c) Facilities at Lucknow University

Astronomical equipment includes a 15-cm (6-inch) f/10 Watson refractor, a 10-cm (4-inch) f/15 Unitron telescope on an equatorial mount with a drive, four each of theodolites and sextants, a transparent star globe and a celestial globe, and a number of binoculars. The university acquired a planetarium (Spitz model) in 1951 and astronomical activities at the Department of Mathematics and Astronomy include, besides photography of the moon (which constitutes a practical at the B.Sc. level), planetarium shows, open-night shows, and astronomical film show on a regular basis. Educational tours to an observatory once every two years provide the students with an opportunity to familiarize themselves with the observational techniques used in astronomical observations.

3. A Few Suggestions for the Development of Teaching of Astronomy in India

a) In India, modern astronomy started as a discipline for research much earlier than as a discipline for teaching. At the university level, the number of departments of astronomy are, therefore, extremely low at present. There is an imperative need to increase this number at the earliest possibility.

b) There are very limited teaching facilities in astronomy in view of the acute scarcity of qualified astronomy teachers and also of equipment, which is quite expensive. A more liberal attitude for supplementing is needed.

c) Presently, astronomy seems to have limited job potential in India. The postgraduates in astronomy have no avenues for entering the teaching profession, as neither the subject nor the post of astronomy teacher exists in most of the colleges and universities. Unless they are created, there is little hope for availability of qualified staff in astronomy.

d) Astronomy books are not easily available. They should be made available at low cost.

e) Usually astronomy is mistaken for astrology. Ignorance of the subject and its importance among the general public and those who matter could be removed by educating them through the media.

TEACHING ELEMENTARY ASTRONOMY IN VILLAGE PRIMARY SCHOOLS IN INDIA

Dr. M.F. Ingham, Institute of Astronomy, University of Cambridge, U.K., reported on the work of Professor V.G. Kulkarni and his colleagues at the Homi Bhabha Centre for Science Education, none of whom could attend this meeting. They are preparing some excellent material for teaching elementary astronomy in village primary schools in India. The text of Dr. Ingham's paper was not provided to us, but we have included, below, some of the discussion which followed this interesting paper. Further information can be obtained from Professor V.G. Kulkarni, Homi Bhabha Centre for Science Education, Tata Institute for Fundamental Research, Homi Bhabha Road, Bombay 400 005, India — Editors.

Discussion

J.V. Narlikar: *The Homi Bhabha Centre for Science Education gets financial support from the Sir Dorab Tata (charitable) Trust. It also gets considerable infrastructure support from the Department of Atomic Energy of the Government of India, which supports the Tata Institute for Fundamental Research.*

J.L. Dunlap: *Did the materials suggest the use of common objects to model the concepts presented to the children?*

M.F. Ingham: The drawings, for example, for day/night, show the use of a ball and a lamp. This showed a method by which the ideas could be confirmed.

C. Iwaniszewska: *How would the language (local language) problem be dealt with when producing the cheap booklets for villagers in India?*

M.F. Ingham: Although the captions to the pictures were written in English, the language of the text was very simple and could easily be translated into any other language or dialect.

TEACHING AND POPULARIZING ASTRONOMY IN EGYPT AND OTHER ARAB COUNTRIES

A. Aiad
Astronomy Department, Faculty of Science, Cairo University, Egypt

The recent astronomy of the Arab countries began by the last decade of the 19th century in Alger and the first decade of the 20th century in Helwan. Two Arab countries have been members of the IAU, namely Egypt (since 1925) and Iraq (since 1976). Saudi Arabia, Algeria, and Morocco became members in 1988. We restrict ourselves here to the teaching and popularizing of astronomy.

1. Egypt

In Egypt there is a single department of astronomy; since 1937 it has belonged to the Faculty of Science of Cairo University. A B.Sc. in astronomy requires two years' study of mathematics and physics followed by two more years devoted mainly to astronomy, including a small project. Elementary courses for other sciences, such as geography and geophysics, are also taught at Cairo University and the American University in Cairo.

Alazhar (Islamic) University has approved an astronomy department. The first staff candidate has recently been sent in a mission to University College London for his Ph.D.

Daily astronomical shows and specials shows for schools have been presented in the planetarium of Cairo since 1967.

The Astronomical Society of Egypt was founded in 1976 and has published a quarterly Arabic magazine, "World of Astronomy and Space," since 1983. A one or two-week school for 30–60 physics and mathematics teachers has been held annually by the society since 1983. The teachers attend lectures, receive practical training, and visit the planetarium and observatory. They have been more interested in lectures illustrated with slides, video, and movies than with normal lectures. They were most impressed when they saw the heavenly bodies and stellar spectra through the telescope.

After each school the teachers become interested in astronomy, apply to be members of the society, and, of course, popularize astronomy more among their students. Some of them discover an interest and start to study astronomy. There is a one-year qualification program if they have their B.Sc. in physics or mathematics.

The result of such schools showed in the 1987/1988 academic year, when some short units on astronomical concepts were added to the physics curriculum of the secondary schools.

2. Other Arab Countries

The activities of the other Arab countries could be summarized as follows:

2.1 Teaching of Astronomy

In Ryadh and Jeddah, astronomy departments teach astronomy for all sciences. The university college of Bahrain, The Benighazi University (Libya), and some Iraqi universities are introducing astronomy for physics students.

2.2 Popularization of Astronomy

Kuwait has a planetarium and a science club with a nucleus of astronomy that publishes a quarterly Arabic magazine: "The Galaxy." Saudi Arabia, Jordan, Morocco, and Syria each has an astronomical society. Sudan has an astronomical club.

To sum up: The Arab area has about 50 astronomers (among about 150 million inhabitants), 3 departments, 2 planetariums, and 7 astronomical societies and clubs.

Generally each of those institutions works alone. They suffer from a deficiency of educational materials. Limited funds or memberships or both do not allow attractive programs or publishing of books and magazines.

In order to improve teaching and popularizing astronomy in the Arab area, one has to seek some sort of regional astronomical group that would sponsor meetings, cooperation in publishing books and magazines, exchange of educational materials, and cooperation with other organizations.

Discussion

S. Torres-Peimbert: *It is of great interest that we find out about the astronomical situation in Arab countries. We think it would be particularly interesting to identify the universities where astronomy is being taught and perhaps establish programs with other universities to obtain information like periodicals and books.*

———————————

THE PAST AND PRESENT STATE OF ASTRONOMY EDUCATION IN THAILAND

Yupa Vanichai
Srinakharinwirot University, Bangsaen, Thailand

1. Ancient Thai Astronomy

About 300 years ago in Lopburi and Ayudhaya, many astronomical observatories were designed, constructed, and supported by the Roman Catholic missionaries from Europe and by King Narai the Great of Siam (Thailand). The interesting recorded history of Thai astronomy was found in the national museum of France a few years ago. Afterwards, the ruined observatories were searched for and found. A tercentennial commemorative ceremony of Thai astronomy was held on April 30, 1988.

The French emissary and the Roman Catholic missionaries in the reign of King Louis XIV first visited Thailand in 1685. Besides intending to spread the Catholic religion, they carried out research on surveying local and celestial positions. A Thai royal astrologer calculated and predicted a total lunar eclipse on December 11, 1685. King Narai, together with the missionaries, observed the eclipse at Lopburi through a telescope having magnification of 30 to 72. A partial solar eclipse was also observed near the same place on April 30, 1688. Sanpaolo, a Catholic church on the outskirts of Lopburi, was the site of the first astronomical observatory in Thailand, built in 1685. Another observatory in Lopburi, built in the house of a Persian emissary, later became a Thai temple. Other astronomical observatories were supposedly built in Ayudhaya, the capital of Thailand in King Narai's period, and should be worthy of search and restoration. However, the historical events on record have not yet been clearly found out.

About 120 years ago, King Rama IV in the Charkri dynasty, greatly interested in astronomy, precisely calculated and predicted a total solar eclipse on August 18, 1864. The king, together with Thai, French, and English people, were eyewitnesses to the eclipse at Wakor in Prachuabkhirikhan. Until now, no one clearly understands what method he successfully used for this masterpiece. An astronomical observatory was built at Khaowang in Petchaburi in 1860. The king measured the height of the sun and pursued his calculation in astronomy. He set 100° East as the Thai Prime Longitude. A clock tower built at this longitude started Bangkok Mean Time, earlier than Greenwich Mean Time. His palace was furnished with telescopes, microscopes, barometers, and other scientific instruments. He was declared "The Father of Thai Science" not long ago. August 18, the date of his predicted eclipse, was selected to be "the Thai Science Day," in which scientific exhibitions, demonstrations and/or other activities yearly take place in schools and universities throughout the country.

2. Planetarium and Popularization

The Bangkok Planetarium, situated beside the national science museum, was established in 1964. The planetarium, with a dome 20 meters in diameter, has a seating capacity of 500. Within the planetarium precincts, an astronomy exhibition is well planned. It contains various photos, posters, and models, but lacks computers.

The Thai Astronomical Society was formed in 1980, the year in which the Astronomical Amateur Society closed. The society office shares the rooms in the planetarium building. This society has about 1,000 members, mostly amateur astronomers. Their activities are various. Astronomical training, seminars, and colloquia are occasionally held with astronomy lecturers. Competitions in astronomy tests, photos, and drawing pictures are periodically held and/or alternatively held on special occasions such as the Thai Science Day. The society issues books, slides, pictures, constellation maps, and a quarterly journal in astronomy. New aims are to possess bigger telescopes and to restore the ancient observatories in Lopburi and Ayudhaya.

3. Publications

Television, newspapers, and popular journals are rarely helpful in regularly distributing articles on astronomy except for news about launching of spacecraft and well-known celestial objects such as Halley's Comet and Supernova 1987A. At present, there are two important monthly scientific journals, which are great favorites among students. The journals regularly publish articles on astronomy. Popular pocket books on astronomy are usually not beyond amateur comprehension. Textbooks written by university lecturers are published merely to benefit their teaching, and are seldom sold in bookstores. Advanced textbooks from abroad are insufficient.

4. Primary and Secondary Schools

At the primary-school level of education, there are many self-taught books to practice reading and comprehension of the physical environment. Five of these books are on elements of astronomy: the sun, the moon, stars, rockets, and a journey to the moon. At the secondary-school level (grades 7 to 9), no astronomy book is available. Lessons concerning astronomy appear in a few chapters of science textbooks. Only liberal arts students, not science students, in all high schools (grades 10 to 12 can take optional subjects in physical science. One of the books, entitled "The Earth and Stars," largely contains concepts about motion of the Earth, the moon, and planets. However, it is widely said that this book hardly arouses students' interest in astronomy and should be improved. The fault likely lies in the fact that the editor is not an astronomy teacher.

5. University Level

In each education college, students majoring in general science are required to take an astronomy course. For example, the course in astronomy and space is compulsory in Surin Education College. The contents of this course are mainly electromagnetic waves, astronomical instruments, celestial objects, and space exploration. This course in general astronomy is neither profound nor complex; it is appropriate for students who are going to be teachers in secondary schools.

In some universities, an elementary course on astronomy or partially on astronomy is open to most undergraduates. The contents of the *Earth and Space Science* course in Srinakarinwirot University mainly includes elementary climatology, geology, and astronomy.

An Introduction to Astronomy is a compulsory course for general science students in Srinakharinwirot University. This follows a textbook written by Baker, *Astronomy*. A similar course is elective for all science students in Khonkaen University but elective for physics students in Chiangmai University. The other elective course for physics students in Chiangmai University is *Astrophysics*, largely concerning the foundation of astrodynamics. Physics students in Chulalongkorn and Khonkaen Universities are offered *Astrophysics* and *Spherical Astronomy* as elective courses. If the astronomy section of the physics department has only one astronomer, only one astronomy course can be offered[1]. A seminar and/or a project in astronomy can also be supplementally selected by physics students. A few laboratory exercises in astronomy are usually included in physics laboratories for physics students.

6. Research Programs and M.Sc. Level

Only Chulalongkorn and Chiangmai Universities offer the M.Sc. in astrophysics. Astrophysics students mostly take the same compulsory physics courses as other physics students; they also take 3 to 4 selective astrophysics courses, seminars, and a thesis on astrophysical research. In Chulalongkorn University, theoretical stellar research, or observational solar or comet research can be chosen as a thesis topic for a master's degree in science. Variable and binary stars can be selected as a research project in Chiangmai University, which has equipment for photoelectric photometry and a 40-cm (16-inch) telescope. An analysis of eclipses is carried out by an astronomer in Khonkaen University. Silpakorn University is likely to begin solar observation with a filter passing the K-line of ionized calcium. Without technicians or advanced technology, some of these research projects are managed successfully but with difficulties; sometimes they are not much too different from work in laboratories or by amateurs.

7. Problems and How to Solve Them

Astronomy in Thailand is improving at a slow pace, and sometimes, oscillates

---

[1]In some universities, astronomy is taught by a physicist with no training in astronomy, so only a basic astronomy course can be taught.

up and down. Difficult conditions exist more in advanced levels than in primary and secondary-school levels. For example, the best observatory admits the public but hardly allows cooperation in research with other universities. The government ordinarily pays little attention to astronomical support since such support is not in the country's development plans. Many astronomers, lacking sufficient equipment and encouragement, often find it so difficult to carry out their research that they turn their attention to other fields. Moreover, advanced journals and textbooks are insufficient. All these cause gaps in astronomy education.

These problems may be eradicated by cooperation with astronomers from well-developed countries in astronomy. It would be good to establish well-planned observatories in Thailand, managed by an efficient staff of foreign astronomers who could make justified allowances to enable enthusiastic Thai astronomers to participate in research. Starting from such a step, Thai astronomy would be likely to be on the threshold of prosperity.

THE TEACHING OF ASTRONOMY AND SOCIAL AND ECONOMIC DEVELOPMENT

Gonzalo Vicino
A.N.E.P., Consejo de Educación Secundaria, Uruguay

Sometimes we may ask ourselves how can astronomy, a science with apparently few points of contact with technology or industrial production, help the economic development of people?

My paper is an account of an experiment undertaken in Uruguay on the teaching of astronomy in secondary schools. We seek to show how astronomy may contribute to the development of nations that, like Uruguay, need to create their own technologies to overcome their economic backwardness.

Though the introduction of astronomy in the secondary education curriculum in Uruguay dates back to 1889, in 1986 the Educational Supervisory Office made a radically different proposal for the methodological and programmatic orientation of astronomy teaching. The proposed goal is clear: the teaching of astronomy in secondary education should be geared not to train technicians but to create a scientific-minded youth. It is not important if later, in their university studies, these young people go into medicine, engineering, or economics; the important thing is a reevaluation of sciences in the eyes of adolescents, who often slide along the comfortable slope of certain humanistic areas, in which they wind up by the process of elimination. (One often hears the argument, "I'm studying law because it doesn't have mathematics.") Further, above all those considerations, the most important thing is to create thinking minds, the minds of free people.

We firmly believe that astronomy is the integrating science *par excellence*, in which adolescent students can find clear bonds with almost all the branches of knowledge, including exact and natural sciences as well as social sciences, philosophy, and the arts. It is this integrating vision of astronomy that makes it the door leading youth along the way to the sciences.

But how can a subject be made attractive enough to draw young minds towards the sciences? We propose the concept of "active learning" in place of the classical "passive learning," in which the pupils receive the knowledge from their teachers. In our concept, teachers must not "show" the students the facts: they must show them the way to discover facts by themselves. Education cannot be imposed: nobody educates anyone else, but only himself; adolescents must be the protagonists of their own discovery adventure. Likewise, we must recover the cheerful sense of study, of discovery achieved through a healthy and vital curiosity. Moreover, the study must have a close relation with daily life; the abstraction capacity of young people, to a great extent reduced by the so-called "image culture" in which we live, cannot be developed overnight, or by mere imposition. Therefore, as a starting point for science studies, we must inculcate the habit and pleasure of the observation of nature.

To be able to adequately develop this observational stage, before analyzing and discussing phenomena, the students must learn to build their own elementary instruments, for instance an "astronomical crossbow" to measure angular distances, or a simple spectroscope with which to observe light dispersion. This activity thus stimulates the adolescent's manual ability and creativity, and at the same time involves many other students, perhaps intellectually less able but manually skilled, who otherwise must be left out in an eminently theoretical course.

From the very beginning of the course, we seek to have our pupils learn to observe (to measure) several phenomena with which data they will have to reach some conclusions some 5 or 6 months later; these observations must be ordered and recorded in such a way that students can learn to use them in a statistical way, and then deduce the regularities that lead to identification of natural laws. These systematic observations can be, for example, about the sun's visibility, the movements of the moon and planets, and so on.

But while they record these data, in the classroom the course is developed with the treatment and discussion of such subjects as the magnitude scale, the origin and denomination of constellations, the chemical composition of stars (presenting a wide view about spectral analysis and matter's atomic structure), the stars' surface temperatures (including experiments about radiation laws), always seeking to relate each discovery in a deductive way with the subsequent ones.

I am not going to describe all the program, but I can mention the remaining subjects: "The Universe," "Life in the Universe" (which implies a study of the solar system), and "The Steps of Astronomy" (a historical vision involving the discovery of the causes of seasons, the moon's movements, the geocentric and heliocentric systems, and so on). The historical view of the last of these leads us through Greek and Arabic astronomy, up to the Renaissance through the great developments made by Kepler, Galileo, and Newton, and through the development of modern astronomy.

For many years, our country has suffered from a strong propensity in university studies toward law and humanities over all other disciplines: on the average, 30 per cent of students go to law school, and together with humanities students the total is near to 50 per cent; meanwhile, we find some 5 to 10 per cent in engineering, 3 to 4 per cent in chemistry, 3 to 5 per cent in architecture, and only 2 to 3 per cent in agronomy.

It is still too early, only two years after the implementation of this new astronomy methodology, to show statistical data to prove the influence of this renewed approach on vocational orientation. Perhaps it may be necessary to wait 10 to 15 years to see the evidence of that influence. But there are some very positive indications that let us affirm we are in the right path: during the first two years with the new astronomy curriculum, many astronomy clubs have sprung up around the whole country. We cannot be absolutely certain about the future of these young people, but we have evidence of their interest, their yearning for scientific knowledge. They are our hope; they are the ones who will be able to change the ratio of students in scientific vs. humanistic disciplines. And this is probably where the future economic and social development of our country comes into play.

PROGRAMS OF I.A.U. COMMISSION 46 FOR DEVELOPING COUNTRIES

Donat G. Wentzel
University of Maryland, College Park, Maryland 20742, U.S.A.

1. International Schools for Young Astronomers (ISYA)

ISYA typically last about three weeks and attract roughly twenty students at the Master of Science level. About half the students are from the host country, the others from neighboring countries. ISYA are initiated by astronomically developing countries and have been held in Venezuela, Nigeria, Egypt, Portugal, Yugoslavia, Indonesia, China, and others. Two or three central topics are chosen. For each topic, a suitable faculty member is invited to stay for the entire school. Other topics appear for enrichment and are chosen according to local expertise and availability of other visitors. The IAU provides the prestige of an international venture and seeks to garner local support not only for the ISYA but also for the longer-term increase in local astronomical capabilities. Financial support is provided both locally and by the IAU, when possible ultimately from UNESCO. Perhaps the greatest benefits of an ISYA are the resulting lasting acquaintances and scientific cooperations.

2. Visiting Lecturers Program (VLP)

The VLP is designed to lead to a significant improvement in the astronomical capabilities of the host country. Primarily, lecturers offer a series of courses. The VLP in Peru has operated for three years, the VLP in Paraguay has just started, and a VLP in Nigeria has been postponed for technical reasons. J. Sahade identified the lecturers for Peru and Paraguay; it was considered essential that lectures be in Spanish. The IAU supports travel costs; living expenses are provided locally. A contract between the IAU and the host institution seeks to assure the continued support of astronomy in the host country, including the employment of astronomers trained during the VLP.

THE VISITING LECTURERS PROGRAM AT SAN MARCOS UNIVERSITY

María Luisa Aguilar H.
Av. Arica 830, Lima 5, Peru

Professors and students of San Marcos University (USM) have found that the Visiting Lecturers Program (VLP) of the IAU has helped to improve the study of astronomy in Peru. The Peruvian people and I are profoundly grateful.

The VLP and the International Schools for Young Astronomers (ISYA) advised us, pointing our perspectives and capacity for work in the correct direction. I feel that both the VLP and the ISYA are noble expressions of the universal message of scientific collaboration, and are the voices of the highest altruism of the astronomers who participate in them.

With reference to the VLP, the state of astronomy education can be described in three parts:

1. Before VLP

In Peru, there were a few introductory courses, and there was a little research in celestial mechanics in the 1950's. There was and still is a solar observatory, but one that does not carry out research. There was some popularization of astronomy, but in spite of it, the local population was not familiar with astronomy. Under these circumstances, our most overwhelming task was to identify new methods and strategies. So we studied the local conditions and characteristics, and chose the most appropriate institution to begin teaching astronomy. Later, we would progress to the rest of the country.

ISYA involvement provided communication with top-level astronomers. In 1982, we formed a group: *Seminario de Astronomía y Astrofísica* (SAA) which consolidated a group of young students. Conferences, courses and contacts with all people and institutions related in some way with astronomy were worked out.

2. VLP is Running

The first IAU-USM contract for a VLP was signed in January 1984, and the courses were begun in August. The university encouraged science (astronomy) education based on academic tradition (it is the oldest university in the Americas) and the development of our country.

There were 8 courses: *General Astrophysics* (Jorge Sahade), *Galactic Structure* (J.C. Cersosimo), *Astronomical Optics and Techniques in Astronomy* (Horacio Dottori), *Stellar Spectroscopy* (Roberto Mendez), *Stellar Atmospheres* (Jean Zorec), *Extragalactic Astronomy* (Luis Sersic), *Solar Physics* (Josip Kleczek) and *Internal Structure and Stellar Evolution* (Stella Malaroda). All these courses gave credit, and were options in our physics curriculum. The visiting lecturers kindly accepted an extra task: to be advisors of our students for their thesis work for the Licenciate degree. The visiting lecturers and we worked in perfect understanding and collaboration, discussing different aspects and problems.

Many people wanted to show the link of such a basic science as astronomy with technology and production.

A crucial problem was, and still is, that the VLP courses were theoretical only, because we do not have any telescope. Problems and experiments in class are not sufficient.

In February 1988, the IAU-USM contract was renewed for 3 more years, with 6 visiting lecturers on special topics: spectroscopy, photometry, solar physics, and celestial mechanics. The goal will be to improve research.

To date, 18 students have shown a definite interest in astronomy: 4 got Licenciate degrees and are working in astrophysics and celestial mechanics; 5 physics students will finish their pre-graduate studies and will travel to do their Licenciature theses on astronomical topics.

SAA is very active: it permanently organizes seminars, courses, and conferences for university and high school students, teachers, and the general public. It has "Astronomical Friday." In August 1986, SAA organized the first international astronomical seminar in the Facultad de Ciencias Fisicas, USM, with J.L. Sersic (Argentina), J. Kleczek (Czechoslovakia), P. Pittluga (U.S.A.) and M.L. Aguilar (Peru). This year, it will begin popularization of astronomy in different cities of Peru. SAA has 53 active members.

3. Present and Perspectives

What is needed?

a) To consolidate and enhance professional astronomy at USM and in Peru.

b) To avoid academic spread because of a lack of funds.

c) To have efficient international co-operation, and a bigger budget.

d) To co-ordinate the use of isolated instruments existing in various institutions.

e) To improve astronomy education through: education of high school teachers, development of high school curricula, conferences, seminars, radio and TV programs *etc.*

Finally, the proposed relocation of one of the Vatican Observatory telescopes in Peru will be a happy opportunity to form a Peruvian astronomical center.

THE SUMMER SCHOOLS OF THE VATICAN OBSERVATORY

Martin McCarthy
Vatican Observatory, Vatican City State

Starting on the 10th of June, 1986, the staid and quiet halls and courtyard and corridors, including the giant circular staircase designed by Bernini for the little donkeys that carried Popes to their quarters in the summer palace, echoed to the swift patter of lightfooted students and the buzz of their conversations as 17 young men and 8 young women met at the first Vatican Observatory Summer School in Astronomy and Astrophysics at Castel Gandolfo.

These 25 scholars had been chosen from a list of 135 candidates at university and graduate school campuses all over this planet. More than 30 candidates were rejected for "excellence"; they were judged to be too far advanced for admission to the classes on *Galaxies and Dark Matter* and to those on *Spectral Classification* and on *Instrumentation for Photometry and Image Processing.* The classes were aimed at students just beginning or planning immediate entrance into graduate level classes; the School was not planned for "new Ph.D.s," for "post-docs," or for those already well into thesis work. Criteria for admission included academic grades, recommendations of university professors, plus personal statements from the candidates on reasons why they felt they wanted to attend the sessions of the school. Applications were studied by faculty and staff members and results were announced in January, giving students some four months to arrange their travel and home commitments so they could be free to respond to the school bells on June 10.

It was decided early on to assign a majority of places to young scholars from the developing nations on our planet. Fourteen came from homes and schools there, while eleven voyaged to Castel Gandolfo from industrialized nations. Neither race nor creed nor sex were discriminants.

Three students each came from India, Argentina, Italy and the United States; and two each from Greece, New Zealand, Brazil, and Argentina. China, Austria, Korea, Bangladesh, Yugoslavia, Holland, Belgium, Poland, and Nigeria had one representative each. English was the required language for admission to the school.

Conversations through the days and nights were conducted in many diverse tongues and on the blackboard used to welcome the members of the Vatican Press Corps, who had asked to meet the students, greetings were inscribed in 23 languages.

Two features of the school merit mention here. The generosity and concern of the Holy Father made our orientation towards the Third World both possible and practical. For all students, free tuition and free registration were offered and there were no fees assigned for laboratory use, for the libraries, for stationery supplies, or for computer programming. In addition, coffee and croissants (the latter quite a rarity in the papal palace) were offered each schoolday and a "free lunch," Italian style, was served to all students and faculty on the lovely palace terrace above the volcanic crater lake called Albano and looking towards the 900-m (3000-foot) peak of Monte Cavo. For the students from the developing nations (14 of 25), the Holy See paid for air transport plus 70 per cent of board and room. Travel and lodging could not be offered to students from the industrialized countries, but the Local Organizing Committee headed by the director of the Vatican Observatory, Dr. G.V. Coyne S.J., was able to find and arrange for inexpensive housing facilities for these also.

The second feature of this Summer School in 1986 was noted by the Holy Father in the address he gave to the students on 30 June, the eve of his departure for Colombia in Latin America; he declared that the school was possible in large part because of the significant gift of the professors' lectures, which were offered without salary. The Vatican covered travel expenses for the professors and their spouses and hosted them during the period of the school.

The professor who lectured most frequently in the school was Dr. Vera Rubin of the Carnegie Institution of Washington, whose lectures on *Galaxies and Dark Matter in the Universe* ran through the entire month. The Dean of the School, Dr. Martin McCarthy S.J. of the Vatican Observatory Staff, and Prof. David Latham of the Harvard-Smithsonian Center for Astrophysics divided the lecture set that ran parallel to Dr. Rubin's. McCarthy lectured on *Spectral Classification of Stars* with application to problems of galactic structure; Latham described new observational techniques as used in current problems in astrophysics. Supplementary lectures were given by Dr. Coyne and by Dr. V. Piirola, visiting research professor from Helsinki University; they described for the students some of their exciting research in polarimetry of the interstellar medium; Dr. C. Corbally of the Vatican Observatory Staff gave two lectures on *Stellar Evolution*. These main lectures were given each morning in the classroom, which had once contained the large grating and prism spectrographs used by the Vatican team of Gatterer, Junkes, and Salpeter in making the world-famed *Vatican Spectral Atlases* found in physics labs and in observatories the world over. The first class ran from 9:00 am to 10:30 am, the second from 11:00 am to 12:30 pm. Labs and special lectures by visiting scientists plus laboratory research and library consultation for assigned projects occupied the post-prandial period, while the late afternoon and early evenings were kept free for recreational activities (swimming, tennis, jogging, cycling, *etc.*).

Evaluation is a long process and continued throughout the ensuing months.

The students seemed to have profited by their working and studying and playing together; a common complaint was that the school term was too short. All of us in planning the school realized that this brevity was necessary, since the summertime for students and professors from northern and southern latitudes could not allow more extensive time away from home institutions. Further, here at the Vatican the coming of the Holy Father for his annual and much needed vacation meant that the scholarly (and sometimes noisy) activities of young scholars had to be completed before the palace became the center for the organizational, diplomatic, and spiritual-cultural events that always mark the papal visits to Castel Gandolfo.

On 6 June 1988, bells rang again at Castel Gandolfo for the second session of the Vatican Observatory's Summer School. Yes, it was the same and yes, there were differences between the two sessions. Same was the number of students, same was the emphasis on the international composition of the student body, and similarly the solid encouragement for young scholars from developing nations. The appendix gives the enrollment statistics for the 1986 and 1988 Summer Schools. The differences? This year there were two Deans: Prof. Charles Lada of the Steward Observatory of the University of Arizona and Dr. Christopher Corbally S.J. of the Vatican Observatory Research Group, also located in Tucson, Arizona. The co-deans soon enlisted the professional aid of Dr. Jean Keppel of the Kitt Peak National Observatory and Prof. Frank Shu of the University of California in Berkeley to be their colleagues in the principal teaching roles at the school. These four, with consultation from Dr. George Coyne S.J., who was also residing in Tucson in connection with his regular teaching role at the University of Arizona, were able together to decide on the composition of the student body and to initiate communication with the young scholars selected for Summer School '88. The topics for the lectures in 1988 differed from those offered two years previously. Only *Galaxies* remained a common feature for the two schools. *Star Formation and the Interstellar Medium* replaced *Spectral Classification and Stellar Photometry* as lecture topics.

Another difference concerned the activities outside the two formal morning lectures from 9:00 am to 12:30 pm. In 1986, activities had included the laboratory, library, and computer exercises concerned chiefly with practical progress in classifying and analyzing galaxies and their stellar composition and in spectral classification from the Anglo-Australian 1.2-m (48") Schmidt photographs. In contrast, the scholars of 1988 were invited during the opening days of the school to report formally on the research work that they had been engaged in at their home institutions; this consisted of 10-minute reports (with warnings at 8 minutes) made to their fellow students under the moderation of Dean Corbally and Dean Lada. Then, at the close of the 1988 school days, the students, now in pairs (with no two scholars from the same country allowed to collaborate), presented 20-minute talks before the class, the faculty, and observatory staff members. Topics had been assigned by the professors, and the results were presented orally by each student in turn with slides and transparencies; each talk was followed by questions from the audience. The present writer, as former Dean, was very impressed with the high level of student reporting: they seemed to have mastered the art of literature searching and demonstrated clear writing skills and presentation techniques plus a quite surprising maturity in their evaluation of scientific projects. Of course, many of the faults that beset scientific presentations even at a much higher level of expertise could be (and were) noted (and pointed out to the speakers). Some showed a tendency to speak too rapidly, some too softly; some transparencies appeared too crowded and data-dense. Castel Gandolfo seemed to be a very good place to start and we hope the students will look back on these first steps in public presentation of science results with joyous recollections they move on to higher things in the future: reports on their own research.

Another difference noted between the 1986 and 1988 versions of the Vatican Observatory's Summer Schools was an increased mental and verbal familiarity that the students of 1988 showed at the end of their courses with the fundamental notions of stellar evolution and star formation. The interactions of observations and theory had been well presented to them and they seemed ready for this and were not "snowed" or "overawed" by the formulations of the transfer equations or by the laws of radiation and absorption.

The length of the school, 30 calendar days, was for the Director, Deans, and Staff a non-variable that cannot be altered for the reasons stated above. Hence also formal examinations, written or oral, and grading of students were not part of the plan. Opportunities for student-teacher interaction were many and easy to arrange; other exercises in visual observations were offered and accepted, with Dr. E. Carreira S.J. using the 40-cm visual Zeiss refractor on the palace roof. Alas, the telescopes, including a 60-cm reflector, and a 40-cm four-lens refractor, the Vatican 63/93/250-cm Schmidt, and the ancient Carte Du Ciel camera, located in the papal gardens of the Villa Barberini, have long been "light-struck" by the growing industrial complex on the Roman plain some 450-m (1500 feet) below the observatory and are no longer usable for modern research problems. We note here that the courses offered were lecture-based and not observing-based; in large part this method was used because of the brevity of our school season and in part because of the deterioration of the Roman skies.

One worthwhile feature common to both the 1986 and 1988 summer schools that bids fair to be continued and extended in future sessions is the scheduling of historical talks and visits. Invited talks were given by visitors and staff members in both the 1986 and 1988 sessions; topics ranged from archeoastronomy to relativistic astrophysics; a strong emphasis was accorded to the life story and problems of Galileo Galilei, as seemed most fitting for a school in Italy and in the palace built by Pope Urban VIII, who so greatly admired Galileo and later imposed so harsh a penalty on him. Together, faculty, staff, and students visited sites of historical scientific interest in Florence, Arcetri, Padua, and Venice on "break-days" during one weekend of each session. It seemed to the Faculty of the School to be most fitting to offer some introduction to what we hoped might be an early and continuing historical interest among our young scholars. In 1986, as guests of the Vatican Observatory, students of the School went together with the faculty by buses to Florence and Arcetri, where we saw at close range the "roba di Galileo"; in 1988 two other "Galilean" cities — Padua and Venice — were visited, this time via overnight train-couchettes plus a one-day stay at an excellent and central Casa del Pellegrino in Padua. At the Galilean sites in Arcetri, Florence, and Padua, we were guided and helped by our astronomical colleagues from the observatories of Padova-Asiago and Arcetri.

There were also several visits to historical scientific sites in Rome. Through these we hope that our young scholars can begin early to reflect on the beginnings and early developments in science. Perhaps in the future one could think of "break day" visits to the north of Italy and someday even to the lands of Pythagoras and his fellows in Calabria and Sicily. No, indeed we cannot even imagine transporting the whole Summer School troop for a short and glorious visit to Greece, even though all of the young scholars from Greece both in 1986 and 1988 have been most enthusiastic in suggesting this.

Now we will see how our young scholars mature and grow and develop in their chosen scientific work. We shall look to see how they in their turn will help younger brothers and sisters along the rugged pathways that lead to the stars. We hope that they will remember the last words that the Holy Father spoke to the students in 1986, the same ones that are written in marble on the walls of the Observatory at Castel Gandolfo: *Deum Creatorem Venite Adoremus: Come let us adore God our Creator and Lord.*

Appendix

Nations sending scholars to Vatican Observatory Summer Schools in 1986 and 1988:

Argentina, Austria, Bangladesh, Belgium, Brazil, Canada, China, Finland, Great Britain, Greece, Holland, Hungary, India, Indonesia, Italy, Korea, Mexico, New Zealand, Nigeria, Poland, Sri Lanka, Turkey, United States, Venezuela, Yugoslavia.

Total: 25 nations, 14 developing nations, 11 industrialized nations.

Scholars at Vatican Observatory Summer Schools: 1986 and 1988

Scholars	In 1986	In 1988	Totals
	25	25	50
From Developing Nations	14	16	30
From Industrialized Nations	11	9	20
Women	8	8	16
Men	17	17	34

Discussion

D.G. Wentzel: *What happens to the Schmidt and other telscopes?*

M. McCarthy: The research telescopes at Castel Gandolfo (see *Annual Report of Vatican Observatory 1963*, D.J.K. O'Connell) are "lightstruck" by the growth of suburban Rome with its large industrial parks in the Roman plain. The Vatican Observatory Research Group moved to Tucson, Arizona, U.S.A., five years ago, where the new 1.8-m telescope will be erected (squirrels permitting) on Mt. Graham, Arizona. Dr. George Coyne, Director of the Observatory is directing the disposal and transfer of available telescopes. At present he is working with colleagues from Peru concerning the moving of the Zeiss doublet south of the equator.

Ed. note: *There is uncertainty whether Mt. Graham will be allowed to be used as a telescope site for environmental reasons, including whether a certain type of rare squirrel there will be affected.*

SURPLUS TEXTS AND JOURNALS FOR ABROAD
— A CANADIAN PROGRAM

Alan H. Batten
Dominion Astrophysical Observatory, Victoria, British Columbia, Canada

Dieter W. Brückner
Department of Astronomy, University of Toronto, Toronto, Ontario, Canada

We wish to report briefly on a program for sending donated astronomical books and journals to overseas institutions that have not been able to acquire them in other ways. The program is being sponsored for a trial year by the Canadian Astronomical Society (CASCA), in co-operation with the Canadian Organization for Development through Education (CODE), which has generously offered to pay for the shipments to some countries, mostly in Africa and the Caribbean.

We have been soliciting donations from among the membership of CASCA for materials surplus to their needs, and have offered these through the Newsletter of IAU Commission 46, the IAU Bulletin, and Colloquium 105. Our efforts have been successful, and at the time of writing, about half the journal runs and texts, and almost all of the monographs available at the time of the Colloquium have found recipients[1]. We are gratified by this result, as we had been unsure of the usefulness of what we were doing. Discussions at the Colloquium have given us additional

[1] in Bulgaria, Poland, India (2), Egypt, and Nigeria.

reassurance in this regard.

At the time of writing we are in the process of arranging shipping for the requests we have received. The program will then probably come under review by CASCA's Education Committee, which will make recommendations to CASCA's Board. Most of the donor activity has been centered around Toronto and Victoria, and so an expansion of activity to other Canadian centers seems to be a plausible development.

As we anticipate the program's continuation, our invitation for requests remains open. We recapitulate the conditions presently applying:

> When the recipient institution is in a country currently receiving CODE support (please see appendix), shipment of the requested materials will be arranged and paid for by that organization. If the recipient institution does not fall into that category, it would have to assume the shipping costs itself, as we are not able to pay for these. In that case, however, we are willing to explore creative alternatives to help keep costs to a minimum. Soliciting assistance from national airlines or embassies may be one of the ways of doing this.

> CODE have asked us to bind as many as we can of the journal sets that they ship for us. CASCA has set aside a modest budget for this purpose. Institutions making requests are asked to let us know if they are able to look after the binding themselves, so that we can stretch these funds as far as possible. We will probably not be able to assume the binding costs for journals sent to non-CODE countries, but hope that in most cases this additional support will not be seriously missed.

> Generally we will honor requests in the order that we receive them, but we reserve the right to direct requested materials where we feel they will bring the maximum benefit. Our operation is not centralized and has limited manpower — the filling of requests is thus expected to require time, particularly when binding arrangements must first be made.

Requests for materials may be made to either of the following:

Alan H. Batten	Dieter Brückner
Dominion Astrophysical Observatory	Department of Astronomy
5071 West Saanich Road	University of Toronto
Victoria, British Columbia	Toronto, Ontario
Canada	Canada
V8X 4M6	M5S 1A7

We wish to acknowledge the support given to the initial organization of this project by the late Michael Ovenden of the University of British Columbia. His participation added a distinctly humane touch. We also gratefully acknowledge the support offered to the program by CODE.

Appendix: CODE Program Countries

Please note that this list is subject to change. Readers of our previous communications will notice its shorter length, the result of restructuring at CODE. Eligibility for CODE support will be confirmed at the time a request is received.

AFRICA	Botswana	The Gambia	Mali	Togo
	Burkina Faso	Ghana	Mozambique	Uganda
	Burundi	Guinea	Rwanda	Zambia
	Cameroon	Kenya	Senegal	Zaire
	Chad	Lesotho	Sierra Leone	Zimbabwe
	Comoros	Liberia	Swaziland	
	Ethiopia	Malawi	Tanzania	
CARIBBEAN	Antigua	Grenada	St. Kitts, Nevis, Anguilla	
	Belize	Guyana	St. Lucia	
	Dominica	Haiti	St. Vincent	
OCEANIA	Papua New Guinea	Solomon Islands		

THE TRAVELLING TELESCOPE

Dieter W. Brückner and John R. Percy
Department of Astronomy, University of Toronto, Toronto, Canada

1. Introduction

Astronomy is deeply rooted in the culture of almost every society, including those in the so-called "developing countries." By virtue of its fundamental applications and implications, it has had a profound influence on philosophy and on religious ritual and belief. Astronomy is vigorously pursued in the industrialized countries where, with multi-million-dollar facilities on the ground and in space, exciting discoveries in astronomy are constantly being made. In the developing countries, astronomical research and teaching activity ranges from little to none. There is tremendous potential for growth, however, especially if the industrialized countries can be persuaded to help. At any time, there are several countries which are in the process of raising their astronomy to a level at which they can become full members of the international professional astronomical community. The IAU strongly supports this process.

It would seem inappropriate to encourage developing countries to make large expenditures on astronomical research at this time, but there are many benefits to be gained from making more modest investments in astronomical research and education. Astronomy, with its wide popular appeal, has the potential to raise the level of scientific literacy in a society and to attract young people to a study of the sciences. In addition, and no less important, is the potential of its insights to touch those deepest human chords that transcend national boundaries and are shared by all. To paraphrase the words of Professor Mazlan Othman of Malaysia: "Astronomy is the stuff of dreams and of youthful fascination, and when you in the developed countries achieve your dreams, we hope not to be too far behind you."

2. The History of the Travelling Telescope Project

The concept of the travelling telescope was developed in 1984 by Derek McNally and Richard West. They were aware that, in the developing countries, astronomers had little opportunity to gain experience with modern astronomical instrumentation. Without such experience, however, they could not hope to formulate credible proposals for modern astronomical facilities in their countries. Extended visits abroad could help to alleviate the problem, but were seldom possible. If astronomers from the developing countries could not visit a telescope, then perhaps it could visit them!

One of us (JRP) became aware of the idea of the travelling telescope as a result of articles published by McNally and West in the IAU Newsletter on the Teaching of Astronomy and in the daily newspaper of the 1985 IAU General Assembly. An application was submitted on behalf of the IAU to a new grants program of the Canadian Commission for UNESCO and the Canadian International Development Agency. The application was successful; indeed, at C$15,000, it was the largest of the 40 grants awarded, out of more than 400 applications.

After considerable thought and consultation, we purchased the following equipment: Celestron Powerstar 0.2-m (8") telescope, Meade heavy-duty tripod and various accessories, Canon F-1 camera (old style, with mechanical shutter), Optec SSP-3 solid-state photometer with BVRI filters, Optomechanics Research Model 3C grating spectrograph, objective grating, deep-cycle 12-volt battery with Hammond regulated power supply as charger, and several sets of star charts and books. We have constructed a set of four heavy-duty shipping cases, specially designed for the travelling telescope kit. We are also preparing instruction manuals and outlines of some possible research projects which can be carried out with the telescope. The travelling telescope was displayed at IAU Colloquium #105 and at the 1988 IAU General Assembly, where it attracted much attention and useful advice. It will be ready for its first assignment in the summer of 1989.

3. How the Telescope Will Be Used

The travelling telescope will be used primarily in conjunction with two existing programs of the IAU:

International Schools for Young Astronomers. At these intensive, three-week schools, the participants will be able to carry out useful astronomical observations with the telescope, under the guidance of the instructors. The planning, execution, reduction, and interpretation of these observations can be incorporated into the daytime schedule of the school. Short projects can also be carried out.

Visiting Lecturers Program. The travelling telescope will provide the host country with observational facilities over an interval of several months. Local astronomers can obtain hands-on experience with various astronomical techniques and can carry out research projects, perhaps in collaboration with one of the visiting lecturers.

By mutual agreement of the Canadian Commission for UNESCO, the University of Toronto, and the IAU, ownership of the telescope will be vested in the IAU, because of its administrative structure and its long-term commitment to improving the state of astronomy around the world. The specific assignments and itinerary of the telescope will be determined by IAU Commission 46: The Teaching of Astronomy. Countries or institutions wishing to use the travelling telescope should apply to the President of IAU Commission 46[1], who will consult with the Organizing Committee of the Commission and with the co-ordinator of the travelling telescope project. If the application is approved, detailed arrangements for the shipping of the telescope will be made by the co-ordinator, in consultation with the applicant and the IAU Secretariat in Paris.

4. Acknowledgments

We are grateful to the UCAP Program of the Canadian Commission for UNESCO and of the Canadian International Development Agency for funding the travelling telescope project, and to the many organizations and individuals who have contributed significantly to the project in so many ways.

ASTRONOMY EDUCATION: THE NEEDS OF DEVELOPING COUNTRIES

A Panel Discussion

S.E. Okoye
Department of Physics & Astronomy, University of Nigeria, Nsukka, Nigeria

It is perhaps somewhat ironical that although astronomy is regarded as the oldest of the sciences, yet the study of astronomy is still to be accepted. In many cases, astronomy appears to be the last of the basic sciences to be accepted as a teaching subject in the educational curricula of most of the developing countries. It is not too difficult for one to fathom the reason for this state of affairs. Taking Nigeria as a country that might be considered typical of the developing countries, one finds

---

[1] During 1986–1991 the President of IAU Commission 46 is Professor Aage Sandqvist, Stockholm Observatory, Saltsjöbaden S-13300, Sweden.

that the official policy towards science is oriented towards its utility. Thus science teaching is seen as an important prerequisite for the development and acquisition of technical skills needed for technological and industrial development. In this regard, the teaching of such basic science subjects like mathematics, physics, chemistry, and biology in the primary and secondary levels of the educational system is seen as being a prerequisite for future careers in engineering, medicine, agriculture, architecture, *etc.* On the other hand, the relevance of astronomy to the development of a modern national economy is less than apparent and many enlightened officials as well as politicians would in fact regard astronomy as an esoteric subject that is worthy neither of the investment of the lean resources of the developing countries nor of the tuition or time of their pupils.

It would then appear that one of the major needs of the developing countries is for astronomy education itself to be justified, for without this justification it would be next to impossible to secure government support and funding for any meaningful activity in astronomy education. The problem is, in fact, an aspect of the spreading of a scientific culture across the various strata of the citizenry. The name of the game is popularization, of which a lot will, I am sure, be said in other sessions of this colloquium.

For now, it is necessary to make a broad distinction between those developing countries with significant astronomical activity and those that are astronomically deficient or undeveloped. Their needs are obviously not the same. In the one case, the major need is to develop and consolidate existing structures and resources for astronomy education, while in the other, the need is for the initiation of astronomy education programs.

The problem of those developing countries with some astronomical activity is perhaps more straightforward. This revolves around the broad issues of the lack of human and material resources for the prosecution of the laid-down programs in astronomy. These are already addressed directly by such programs as the Visiting Lecturers Program, International Schools for Young Astronomers, and the Traveling Telescope Program; there is a lot of scope here for international activity and assistance. One may add to this list the possibility of establishing a regional or inter-regional center or institute for astronomy and astrophysics located in a third-world country, funded with international money (*e.g.*, from UNESCO, UNDP, *etc.*) and associated with the UN University or a family of first-class universities around the globe. Such a center would help the nucleation of astronomy in the developing countries; provide M.Sc. and Ph.D. programs and refresher courses; conduct basic research as well as advanced research workshops and schools; and run Fellowship, Associateship, Visitor, and Guest Observer programs. Perhaps the newly-established Inter-University Centre for Astronomy and Astrophysics in Poona, India, could be expanded to fulfill these objectives, initially for the Afro-Asian area.

We now turn to the more difficult problem of initiation of astronomy education where none exists in a developing country. There is no easy strategy, but the following strategy appears to have worked in Nigeria. Basically, the strategy involves identifying and developing potential resource persons in astronomy. In the case of Nigeria, the potential resource person was a fresh physics graduate who was persuaded about twenty-seven years ago by an expatriate astronomer on the staff of the local physics department to do a doctorate degree in radio astronomy at the University of Cambridge, U.K. Three years later, this resource person had returned to Nigeria, equipped with a Ph.D. in radio astronomy, and joined the physics department of his alma mater. It was this same resource person, who some years later had seen to the injection of a good quantity of astronomy topics in the national science curricula at both the primary and secondary levels; the initiation of astronomy programs at the University of Nigeria, Nsukka, with an astronomy option in the B.Sc. (physics) degree, as well as M.Sc. and Ph.D. programs in astronomy and astrophysics; the local production to date of 4 astronomy/astrophysics Ph.D.'s (with another 2 in the pipeline) as well as 10 M.Sc.'s in astronomy/astrophysics; the transformation of the former department of physics into the present department of physics and astronomy; and the recent establishment of a Space Research Centre at the University of Nigeria with a 10-meter parabolic antenna being equipped with a VLBI terminal to provide a long north-south baseline for the European VLBI Network. Other highlights include the formation of a National Committee on Astronomy and Nigeria's joining of the IAU in 1985.

It is proposed that potential resource persons be self-identified through competition for international scholarships and fellowships for doctoral studies abroad. It is also proposed that a package of support incentives be made available to these resource persons on the successful completion of their doctoral programs. These initiatives perhaps may not fit into the purview of IAU Commission 46 on the Teaching of Astronomy. Perhaps the time is ripe to consider upgrading Commission 46 or merging Commissions 38 (The Exchange of Astronomers) and 46 into a new Commission on "The Training of Astronomers," or perhaps an entirely new commission on the training of astronomers should be established. Alternatively, instead of having a Commission just on the Teaching of Astronomy, we might upgrade it to a Commission on Astronomy Education, which is a more comprehensive title.

Mazlan Othman
Department of Physics, Universiti Kebangsaan Malaysia
43600 UKM, Bangi, Selangor D.E., Malaysia

The needs of the developing countries pertaining to teaching resource materials — namely, books, audiovisual aids, *etc.* — and understanding of concepts have been dealt with already.

One problem that has not been covered in sufficient detail is the training of school teachers. Most teachers have very little exposure to astronomy and themselves have problems grappling with basic astronomical concepts. While this conundrum also exists in the developed countries, the situation is more severe in the third world, where the media coverage of astronomy is non-existent or at most scant.

These teachers are many in number, they are scattered throughout the country, and have no resources at hand. The training process thus becomes a mammoth undertaking for what is usually only a handful of dedicated astronomers.

This takes us to the question of whether a developing country needs a greater number of qualified astronomers (as opposed to whether it needs more, for instance, qualified engineers). It is usually appreciated that in order to advance research, a critical number of people is required to carry out and sustain the research effort. The problem highlighted in the previous paragraph, however, indicates to us that if the science education in a country is to progress it must have qualified scientists. Therefore, developing countries need a greater number of qualified astronomers, not only so that research is advanced (which is not normally regarded as a justifiable reason in the developing countries) but especially so that the education base may be broadened and so that a high standard of education is ensured.

Another dire need of the astronomy teacher in the developing nations is to be in contact with the outside world. Isolation leads to stagnation and boredom, which do not make for enhanced teaching standards. Links between astronomy teachers in a region should be established in the form of cheaply produced newsletters, *etc.*

L.F. Rodríguez
Instituto de Astronomía, UNAM, Apdo. Postal 70-264, 04510 México, D.F., México

The astronomy education needs of the developing countries are so many and so varied, that I decided to restrict myself to the two that appear to me as most important.

While listening to the talks of the U.S. participants, one gets the impression that in the U.S., supply and demand are more or less balanced in what refers to astronomy education. If anything, supply may exceed demand. This is, of course, not the case in México where we have about 40 professional astronomers for a population that is above 80 million people. This gives one half an astronomer per million people. In contrast, there are 10 to 20 times more astronomers per million people in the developed countries. This situation is further compounded by the fact that in México a large fraction of the population is young and demanding education.

Then, while in the developed countries the concern is in improving the quality of education, we face a more basic problem of quantity. Any teaching or popularization activity made by one of these very few Mexican astronomers is valuable and is absorbed like a drop of water in the desert. What can we do to increase the number of professional astronomers in México? It is quite interesting to summarize the growth in the number of astronomers in México during the last few decades. In 1960, a few pioneers started sending students abroad to get doctoral degrees in astronomy. By 1970, the number of professional astronomers had grown to 9. An explosive increase marked the period 1970–1980: at the end of this decade we were about 30. The reason for this increase was, I believe, a result of the oil boom of those years. There was a lot of money around and some was put to good use. Research and education developed in México as never before. It appeared that it was a matter of a few decades to catch up with the developed countries. But the mirage did not last long. Oil prices dropped again and now, very close to 1990, we have gained only a few more astronomers with respect to 1980 and are nowhere near 100, the expected number had the trend of the seventies continued. It turns out that in México the salaries in education and research are tied to the nation's economy and seem to reach reasonable levels only in epochs of bonanza. Some countries have had the good fortune of having visionary leaders who valued and supported education and research even in the worst of circumstances, knowing that this is the only way to build a strong country. We have not shared this fortune.

Nevertheless, even in our limited situation, a lot is being done and the second major need that I see can be satisfied at no cost (that is, money). By the peculiar structure of our universities (research and education are considered as totally separated activities, with institutes and centers doing the first and schools doing the second), teaching is considered as a less "intelligent" activity than research. Popularization does even worse, and can be valued as a negative component in one's *curriculum vitae*. It is true that teaching and popularization can be done disastrously, but the same is true for research. A manifestation of this scale of values was the creation by the government of the program of "scholarships" (actually, a supplement to our meager salaries) that is based mainly on our research publications. We need our scientific community as a whole to recognize education activities as important, as long as they are made seriously and professionally.

Silvia Torres-Peimbert
Instituto de Astronomía, Universidad Nacional Autónoma de México, Apdo. Postal 70-264, 04510 Mexico City, Mexico

A healthy astronomical education system can exist only if there are active research groups in a given country. I will discuss some of the cultural and economic problems that astronomical research groups face in developing countries.

In every country, in general, the higher the level of education, the smaller the group that benefits from it. This pyramid is immensely steeper in poor countries, where elementary education is not even available to all children, and where the higher levels of education are reached by only a very small fraction of the population. In atmospheres where frequently there is a general lack of culture and scant official interest in science, and where the mass media continuously advertise fantasy and escape from reality, it is difficult to promote and supervise science education at any level (including astronomical education). Even in those developing countries with a rich historical background, modern scientific research and science education are only incipient activities.

In underdeveloped countries, most of the scientific research and scientific education are carried out in state-supported universities. In the last 5 or 10 years, with the increased world economic imbalance against the poor countries, many governments are demanding that the universities become self-supporting or assist directly in the economic development of the country, extending their demands to scientific research groups. For a basic research discipline such as astronomy, the possibility of becoming self-supporting is out of the question since, by nature, astronomical research has no direct application to the country's economic welfare.

I am convinced that, in order to establish satisfactory astronomical education at all levels, there has to exist, within the country, some activity in astronomical research. Although, in principle, the research group does not need to interact directly with all levels of astronomical education, it can act as a catalyst and promote sound scientific material at schools (and to a lesser extent supervise the material that is presented on radio and TV programs); it can also create an interest in the topic among the population. Of course, the scientific structure at all levels is self-propagating, because how can an active group be established if the proper bases have not been laid out previously?

For a successful education program in astronomy at university level there are several basic resources that should be available: human resources, observatories, laboratories, libraries, and computing facilities. Each one of these items is important in its own right. To train astronomers, a group of active scientists must carry out the training programs. To have active scientists, they must have available some research facilities that include data-gathering equipment (observatories), laboratories for data measurement, data handling systems (computing facilities), updated library facilities, and accessible publication means.

Training of astronomers can also be carried out through scholarships to study in advanced countries. In some cases it has worked very favorably, and it has helped establish astronomical communities with a modern outlook. This process, although slow, is probably the best investment that poor countries can make; however, it is not necessarily foolproof, since there are many instances that, once the student has been trained, he does not want to return to face the unfavorable conditions of his own country.

Regarding observing facilities and equipment, it should be pointed out that a few years ago the working hypothesis was that, through hard work and shrewd planning, poor countries would eventually reach the same levels of facilities as the developed ones (at least in a few specialized areas). In the meantime, developed countries have improved their astronomical equipment at a very fast rate through the application of modern technology, and the use of this equipment has increased our knowledge of the universe and has enlarged the scope of astronomical problems. Alternatively, the astronomical future for developing countries has become uncertain. It is very expensive to obtain modern equipment; it requires a technological background not always available, even to buy the most convenient piece of equipment and to put it into operation. And in general, it is my personal feeling that the opportunity of giving significant contributions to the general knowledge of astronomy by developing countries is becoming more and more difficult. Thus the previous working hypothesis does not seem to apply any longer. The technological gap, instead of disappearing, is widening at an ever-increasing rate. Even trying to maintain a few well-stocked libraries with modern journals, books, and astronomical catalogues can be beyond the resources that most developing countries have allocated for this purpose. A meeting such as this — IAU Colloquium 105 — can only address the question of improving astronomical education in the countries, mainly through interaction with the professional astronomers, who themselves can have an indirect influence in their society. Given the enormous needs of the developing countries, that touch all aspects of society and that cannot be met by a small community such as that of professional astronomers in the world, perhaps it would be simpler to attempt to establish a one-to-one relationship between an astronomical institution in a developing country and another one in a developed country. From it, a personal relationship can be established to further astronomical training, selection of equipment to be acquired, and assistance to obtain astronomical materials to advance a specialized field of research of interest to both institutions. Commission 46 of the IAU could act as a clearing house to help establish contacts among institutions. Young astronomers should be encouraged to advance their astronomical training; the exchange of astronomers of different institutions should be extended to astronomers of developing countries.

In spite of the drawbacks described above, I consider it very important for astronomy to be advanced in all countries since: a) it is an important part of human knowledge and culture, b) astronomy is one of the basic sciences and its advance is related to that of other basic and applied sciences and to the advancement of technology that can lead to economic development, c) citizens of all countries have the right to do research in all areas and contribute to science development; this activity in turn can help them establish their national identity.

Efforts are being made and we should continue to pursue astronomical research to the best of our abilities. At the same time, public awareness of modern astronomy, communication of the fundamental ideas of astronomy to school children, and the strengthening of higher education programs have to be carried out.

Conclusions

Astronomy education cannot exist if there are no active astronomical research groups in a given country. The difficulties of carrying out modern astronomical research are enormous and are growing. However, groups should be encouraged to continue their efforts to promote astronomical research.

The Relevance of the Teaching of Astronomy in the Developing Countries; a Case: Paraguay

Alexis E. Troche-Boggino

Instituto de Ciencias Basicas, Universidad Nacional de Asuncion, Asuncion, Paraguay

Last year, I enjoyed having dinner with an American astronomer who was visiting my country. He mentioned that astronomy is an active science that is an expensive and luxury enterprise, and one for which it is difficult to receive government support because other needs have higher priority. He works in a state like my country, where cattle and agriculture are the most important products. I answered him as follows: Scientific research — particularly astronomy, of course — requires much effort, skill, and a great deal of economic support, but there are many young, bright students in my country, as everywhere, who have the potential to develop fully, provided that challenges are offered to them.

I would like to remind you about some famous astronomers who became IAU Presidents: the late Dr. Bappu from India and the current President, Dr. Sahade from Argentina; also, one of the present Vice-Presidents, Dr. Peimbert, is from México. These astronomers have come from developing countries. Other first-rate astronomers from these countries have helped educate young astronomers and build astronomical institutions of research. It is worthwhile to mention here Dr. Guillermo Haro — a Mexican astronomer who passed away just a few months ago — who made several contributions to astronomical research, especially the discovery of Herbig-Haro objects. Also, astronomers from countries where astronomy is fully developed have done much for astronomy education, observatories and research institutes in third world countries.

Further, I would like to mention two IAU programs that are helping a lot in some countries. These are: a) The International Schools for Young Astronomers (ISYA), under the secretariat of Dr. Kleczek. I am particularly grateful to this program, being myself twice a student of it and having been helped to become an IAU member, initially as a Commission 46 consulting member. b) The Visiting Lecturers Program (IAU Commissions 38 & 46). Dr. Donat Wentzel, then president of IAU Commission 46, proposed this program, which is doing a lot for astronomy in Peru, Nigeria, and since last April in Paraguay. In this program 20 students from the two Paraguayan universities, the Universidad Nacional de Asunción and the Universidad Católica N.S. de la Asunción became associated in the VLP. I hope that this program will enable our young scientists to work professionally in the field of astronomy with opportunities of collaboration with colleagues from other countries in the future. There are plans to cooperate with the Argentinians in very-long-baseline radio interferometric observations. The economic support to continue with the VLP program for the next term is being sought and your advice about the sources for building a radiotelescope are welcome.

I would like to quote here the words of Dr. Richard West, past IAU General Secretary, during the last General Assembly at Delhi in 1985 about future directions for the IAU: "There are two kinds of activities, perhaps in every scientific union, which are important. The IAU has put much emphasis on one of these all the time...The first is the question of top science, frontline science, having scientific meetings, symposia, colloquia, regional meetings, and by having General Assemblies you promote top science. But on the question of science in the developing countries, if I may put it that way, or science for the young people, perhaps there has been a little less emphasis in the IAU in the early days. We now see that this comes more into the picture."

I am now thinking of another answer to the question of the relevance of teaching astronomy in developing countries. This question is about the need for astronomy for the sake of the people; or, in different words, the human side of astronomy. Most of us agree about the contributions provided through astronomy education besides answering people's curiosity about space and the universe. These contributions are: a) learning about the scientific method, in which astronomy serves so well; b) displaying the interdisciplinary relationships in which astronomy is doing a beautiful job, combining the efforts of physicists, chemists, biologists, historians, and others. It is possible to provide teachers from these different fields with good examples in which astronomy is involved. This possibility has been pointed out several times by a leader in the teaching of astronomy in Commission 46, Dr. Josip Kleczek. c) Another important point that IAU President Dr. Sahade has pointed out and which should be kept in mind is the importance of applied science and the transfer of technology, two major topics of great interest to developing countries. Dr. Sahade has said, "It is a mistake to think that the application of science and the transfer of technology *alone* can solve the problems of underdevelopment. The transfer of technology, to be efficient, needs a medium which provides scientific and technological potential, which should be able to take advantage of and use as a starting point to develop its own resources, for the future. Otherwise underdevelopment and dependence will remain and worsen." The progress of astronomy is an indicator of improvement.

I would like to stress other subtle points concerning the interaction between the decision-making countries and countries that follow the decision makers. I shall point out only a few aspects, the most relevant ones:

a) National security (meaning defense of democracy, territorial defense, and maintaining cultural and social values).

b) Human rights (as declared by the U.N.).

c) Environmental problems and the saving of natural resources and energy.

These policies being applied equally, by both types of countries should help them grow harmoniously both internally and externally, and thereby to end wars, poverty, and other human misery. But unfortunately these aims are seldom achieved. The enemies are ignorance, vanity, and selfishness. It is, of course, utopian to hope to cure these illnesses rapidly. Now astronomy might do a lot for our young people, to help them to have decent ideals, and to build up more sensibility towards human beings and nature. There is a great need for sensitive people from the two types of countries to cooperate in looking for solutions to the problems, and building up a more humane society. Remember that nations' attitudes are just the same as the attitudes of its individual members. I am sure that most of us present here have a positive attitude. So, let us make known our efforts and our science, to improve humanity. I feel strongly impelled to do so.

Educational Material in Developing Countries

The panel discussion on textbooks stimulated a great deal of interest and discussion, especially about the problem of educational material in and for the developing countries. About 50 participants in the Colloquium therefore reconvened one afternoon for an impromptu open discussion of this problem. the following notes were compiled by John R. Percy; they are based on the comments and suggestions of many participants.

1. Acquiring Current Material from Abroad

Current monographs and journals are essential for graduate teaching and research, and for the "professional development" of instructors, but there are major problems resulting from currency shortages or restrictions. Institutions must therefore be particularly selective in choosing material. IAU Commission 46 produces a triennial listing of astronomy education material in all major languages to help to deal with this problem. Institutions in developing countries can obtain some periodicals through exchange which institutions elsewhere, but problems of language, content, and shipping costs make this difficult. Another approach might be to "twin" institutions in the developing and the developed countries.

2. Acquiring Less-Current Material from Abroad

Several organizations are now engaged in collecting unused books and journals, and sending them to developing countries[3]. Publishers might be also induced to donate material, such as slightly-outdated editions of textbooks. In North America, new editions of introductory astronomy textbooks are introduced approximately every three years!

3. Producing Translations of Important Material

Translations of important and popular material could be made in the language of the developing country (and then shared by other countries with the same language). This makes good use of good material, but requires the cooperation of the author and publisher of the original edition. It also requires a good translator, who is willing to do it as a "labor of love": there is not much profit to be made in translating the average astronomy book! In the case of elementary and secondary school

---

[3]See the article by Batten and Brückner elsewhere in these proceedings.

textbooks, there may be a larger market, but there are also additional problems. The book may have to be translated into several local languages (the example of India was quoted). Most school textbooks are written for a specific curriculum, which is far from standard from country to country.

4. Reprinting Material Locally

With the cooperation of the author and publisher of the original edition, certain books and journals could be reprinted locally, to take advantage of lower labor and shipping costs. These low-cost editions could then be used in other developing countries. Some projects of this kind have been undertaken by organizations such as UNESCO.

5. Reprinting Material Locally, with Modifications

One problem with acquiring or reprinting "foreign" material is that such material may not reflect the cultural and scientific environment of the developing country. This is particularly true for popularizations and for school textbooks. With present developments in electronic or "desktop" publishing, however, it may be possible for an author or publisher to provide the text and illustrations for a book in electronic form. These could then be modified to fit the local environment, and printed locally in order to take advantage of lower labor and shipping costs.

6. Local Authorship and Publication

As a developing country evolves into a developed country, the ultimate goal might be to produce all educational material locally. During the transitional phase, this is probably feasible only for standard textbooks. The few potential authors of such books are usually overloaded with the task of developing the local astronomy education system, and local publishers are reluctant to produce books for a limited market.

Discussion

C. Harper: *Modern astronomy is derived from the astronomy that developed in western Europe. It will be a great loss if the alternate astronomies that arose elsewhere round the globe disappear.*

The developing and the developed countries could do much more to recover and preserve the original native astronomies that emerged in their own regions. For instance, teachers of astronomy in New Zealand have opportunities to draw upon the rich heritage of the astronomical knowledge of the Maori. The cultural needs of individual countries can be served by incorporating regional native astronomies in introductory astronomy courses.

P.W. Hill: *Visiting lectures might be obtained at no cost by countries where stopovers are permitted on major air routes. For instance, British astronomers using the Anglo-Australian Telescope might be persuaded to visit Kuala Lumpar.*

J. Fierro: *It would be highly desirable for Paraguayan astronomy textbooks (at all levels) to circulate throughout Latin America.*

P. Sakimoto: *There is room for a two-way street here. Not only can the developed nations provide technical assistance for developing astronomy in other nations around the world, but also astronomers in those nations can assist us in gathering information about astronomy around the world. Many colleges and universities in the U.S. are struggling to diversify the cultural and intellectual content of their curricula beyond that of the Western world. Perhaps some body — the IAU or the Astronomical Society of the Pacific — could collect and produce an information packet on astronomy (past and present) around the world. This would be quite valuable to American educators.*

B.G. Sidharth: *The idea of centers of excellence in developing countries, catering to a wider group of neighboring countries, if necessary with international funding, makes good sense, a) logistically and b) because of the fact that these countries have comparable levels of expertise/training and similar problems. An example of this could be the Inter-University Centre at Poona (Pune), India.*

M. Gerbaldi: *Every 3 years, IAU Commission 46 publishes since 1970 the* Astronomy Educational Material Part C, *a list of most of the educational material (books, visual aids, etc.) published at any level, in all languages except the English language and the Slavic languages. In this booklet, published as a special issue of the Commission 46 Newsletter, you can find several books published in developing countries.*

G. Vicino: *I propose that the IAU should recommend to its member countries that astronomy should be taught in secondary school, and that the IAU should establish an Education Office on an experimental basis. Uruguay would be willing to host such an office.*

B.W. Jones: *Open-type[4] learning institutions exist in several developing countries (India, Pakistan, Iran, Thailand, etc.) and are already teaching science subjects in a cost-effective way. I don't think that any are teaching astronomy. However, if astronomers in such countries could contact their local open-type institutions, then perhaps astronomy courses could be offered, in a cost-effective way. Contact me (Dr. Barrie W. Jones, Physics Department, The Open University, Walton Hall, Milton Keynes, MK7 6AA, United Kingdom) for details of your local open-type institution if you do not know whether such an institution exists and if it would be difficult for you to find out.*

J.V. Narlikar: *With reference to the comments of Dr. Okoye and Dr. Sidharth, the*

---

[4]Open-type: distance teaching; open entry; part-time; can be subdegree level; tends to be for adults.

Centre in Pune, India, does hope to play a regional role in Afro-Asian countries by promoting interchanges of teachers and students, conducting workshops and schools, organizing joint experimental projects, etc. These aims will be more focused in two to three years as the Centre develops its infrastructure. In the meantime, I am exploring the possibility of assistance from IAU, UNESCO, UNDP, etc., for such programs. The model of the International Centre for Theoretical Physics at Trieste comes to my mind in this context.

D. Brückner: *What I have to say touches on things said by A. Troche-Boggino (from Paraguay) and C. Harper (from New Zealand).*

A while ago, my wife took me out and showed me the Milky Way, which I hadn't seen in a while, and pointed out where its center lay, the direction the sun is traveling in, and the direction away from the center. I was powerfully touched by this experience, and feel it is the type of experience we seldom communicate when we teach. I'd like to suggest that by paying attention to the communication of this aspect of astronomy, especially in the developed world, we may in a simple way lay seeds for profound shifts in the awareness of what it means to be a vulnerable human being on the face of our Earth, a situation we all share, no matter how well-off we are materially. I believe that facilitating such shifts in awareness may be a most effective way open to us for influencing the way human beings relate to their fellow human beings elsewhere, something that lies at the root of the problems we are discussing here.

M.L. Aguilar: *There are two problems:*

1. *The international institutions have to learn how to teach isolated astronomers (in developing countries) to study their local problems and how they can organize and plan. There are always limitations in resources — money and instruments — but the situation of the low organizational capacity is also severe.*

2. *Scientists from the developing countries sometimes obtain the highest level degrees (Ph.D., etc.) with funds from their countries, but then travel to the major countries and forget their origins. I think the international scientific institutions have to stress to them that they should help their own countries develop.*

C. Iwaniszewska: *There exists a group for a part of the world, the Working Group for Astronomy Education for South-East Asia. S. Isobe is the chairman, M. Othman is a member, and there are also delegates from Australia, Indonesia, etc. Would it not be a good idea to have such groups for the other parts of the world?*

S. Torres-Peimbert: *The needs of the developing countries are so great! Perhaps we could start in a small way by "twinning" institutions in the developing countries with willing institutions in more developed countries.*

Silver tankard, by Rebeccah Emes and Edward Barnard I. Clark Art Institute, Williamstown. Further information on page 433.

URANIA OBSERVED

Rafael Fernandez[1]
Jay M. Pasachoff[2]
Robert Volz[3]

[1]Clark Art Institute, Williamstown, Massachusetts 01267, U.S.A.
[2]Hopkins Observatory, Williams College, Williamstown, Massachusetts 01267,
U.S.A.
[3]Chapin Library of Rare Books, Williams College, Williamstown, Massachusetts
01267, U.S.A.

The public interest in astronomy, so often cited in this Colloquium, is demonstrated by the many works of art over the centuries with astronomical content. We mounted an exhibition, named *Urania Observed* after the muse of astronomy, at the Clark Art Institute to coincide with IAU Colloquium #105 on the Teaching of Astronomy and with the Sesquicentennial of the Hopkins Observatory. We thank Wayne Hammond of the Chapin Library for his expert assistance with this exhibition.

1. **François Denis Née** (French, c. 1735-1818)
   After a design by Charles Nicholas Cochin fils, French, 1715-1790
   ALLEGORY OF ASTRONOMY (Line engraving)
   Lent by Jay M. Pasachoff   **Frontispiece**

   Née was much in demand as an engraver of designs by illustrators such as Eisen, Gravelot, Marillier, Moreau le Jeune, and Cochin. Cochin produced large numbers of drawings to be turned by engravers into book illustrations. Although the print on view is signed as if designed by Cochin fils, it was his father who published a book entitled *Iconology in Figures*, with the assistance of Hubert Gravelot. Astronomy is presented as part of what was then defined as Mixed Mathematics, a discipline that taught how to know the celestial bodies, their sizes, movements, distances, eclipses, *etc.* There is a sphere, following the Copernican system, a telescope, and on a roll of paper appear four tracings of ellipses of several comets. The print on view, dated 1773, is identical to the one in the book except that it lacks a full inscription.

2. **Honoré Daumier** (French, 1808-1879)
   FOUR CARTOONS FROM A SERIES COMPRISING 10 PIECES ENTITLED:
   THE 1857 COMET (Lithographs)
   Lent by Jay M. Pasachoff

   Daumier's comic lithographs, executed to be published in a newspaper, are magnificent testimonies to his genius. With witty captions supplementing his eloquent crayon, he captured the very essence of his times and dramatized the appearance of his contemporaries in memorable fashion. Daumier has chosen in his captions to the 1858 Comet series to ridicule superstitions about comets prevalent in his day. In No. 2, a man rejects a raffle ticket because it is dated two days after the expected appearance of the comet. Two neighbors in No. 3 discuss the comet but it turns out that one of them has mistaken the smoke of the chimney for it, so they reassure each other and continue to feel safe. No. 4 shows a servant giving notice because she does not want to be overtaken by the end of the world at her workplace. No. 8, perhaps the most ironic, has an old bourgeoise brokenhearted in expectation of the world's end and the inevitable demise of her beloved little dog Azor, as she has no fear of death for herself.

3. **Albrecht Dürer** (German, 1471-1528)
   MELANCOLIA I (Engraving)
   Clark Art Institute Acc. No. 1968.89

Dürer's *Melancholy* is perhaps his best known print. It is recorded that he used to present it to friends with his engraving showing St. Jerome in his study. The complexity of allusion in *Melancolia I* has been studied in great detail by Professor Panofsky in his unsurpassed book on Dürer. Panofsky contrasts Dürer's vision of Jerome, secluded in his warm and sunlit study, with Melancholy, placed in a chilly spot "dimly illuminated by the light of the moon — as can be inferred from the cast shadow of the hourglass on the wall — and by the lurid gleam of a comet which is encircled by a lunar rainbow."

4. **Etienne Léopold Trouvelot** (French, active in the United States, 1827-1895)
   THE GREAT COMET OF 1881 (Chromolithograph)
   Lent by Sawyer Library, Williams College   p. 188

Trouvelot's magnificent chromolithographs were issued in 1882 by Charles Scribner's Sons. Trouvelot was a professor at Harvard, an expert in astronomical drawings, and the author of a manual on the subject. Although Trouvelot considered himself a scientist and his name does not figure in art historical works, his drawings have artistic significance and importance as independent works of art. Photography rendered obsolete Trouvelot's remarkable power for recording his observations, which had gained him the admiration of colleagues and contemporaries. His objectivity and accuracy were greatly appreciated.

5. **Etienne Léopold Trouvelot** (French, active in the United States, 1827-1895)
   SOLAR PROTUBERANCES (Chromolithograph)
   Lent by Sawyer Library, Williams College

Trouvelot entitled this print *Solar Protuberances*, which in our day are generally referred to as solar prominences. A solar prominence is a glowing gas at a temperature of about 15,000 degrees Celsius, suspended over the sun by the solar magnetic field.

6. **Nathaniel C. Sanborn** (American, 1831-1886) and **John L. Lovell** (American, 1825-1903)
   TWO VIEWS OF THE HOPKINS OBSERVATORY ABOUT 1870 (Albumen prints)
   Lent by the Williamsiana Collection, Williams College

The Hopkins Observatory, the oldest extant observatory in the United States, was erected in 1836-38 after the design of Professor Albert Hopkins, who built it with the help of his students. The handsome building echoes the Federal style adopted for buildings of the time in the college.

Sanborn and Lovell were 19th-century Massachusetts photographers, active in Lowell and in Amherst, respectively. Lovell was especially noted for scientific photography.

7. **Anonymous** (19th-century American photographers)
   STEREOSCOPIC VIEW OF THE HOPKINS OBSERVATORY (Albumen prints)
   Lent by the Williamsiana Collection, Williams College   p. 322

Stereoscopic views became very popular after the advent of photography in 1839. The first used daguerreotypes, and special cases were developed to see them. After the development of paper prints, a stereoscopic viewer was invented in which views could be changed easily. Mounted on cardboard, stereoscopic views became a standard feature of American living rooms and continued to be commercially produced until as late as the 1920s.

8. **Anonymous** (19th-century English artist)
   DONATI'S COMET SEEN OVER A LANDSCAPE, 1858 (Watercolor)
   Lent by Roberta Olson and Alexander Johnson

9. **Workshop of Wolgemut and Pleydenwurff** (German, active last decade of 15th century)
COMET OF 813 ON A PAGE FROM SCHEDEL'S *LIBER CHRONICARUM*
(Woodcut)
Lent by Jay M. Pasachoff

The text of the Latin *Chronicle*, published in Nuremberg in 1493, was written by Hartman Schedel (1440-1514), a prominent humanist of Nuremberg as well as the town physician. Michael Wolgemut (1434-1519) was a painter and printmaker who married the widow of the artist Hans Pleydenwurff (died 1472). Associated with his stepson Wilhelm, Wolgemut operated a workshop where woodcut illustrations for books were produced. For a time, among the young apprentices in this workshop was the young Albrecht Dürer. The *Nuremberg Chronicle* contains 13 comet illustrations, all printed from four woodblocks.

10. **Anonymous** (Mid–19th-century English designer)
FULL MOON AND COMET LETTER OPENER (Brass)
Lent by Roberta Olson and Alexander Johnson

11. **Rebeccah Emes and Edward Barnard I** (English, flourished from 1808 to c. 1825)
TANKARD, 1811 (Silver)
Clark Art Institute Inv. No. 114   **p. 430**

The decoration of this cylindrical tankard is quite extraordinary because instead of the more customary armorials, it is engraved with representations of a comet, three planets, and numerous stars. Its finial is formed as a 12-pointed star which rises out of a mass of clouds. The planets identified as Saturn, Uranus, and Jupiter are three of the seven known in 1811 when the tankard was made. The engraved comet is most likely the Great Comet of 1811, visible for a period of 17 months. Emes and Barnard (whose mark can be seen with the London hallmarks to the upper right of the handle) served a wide range of customers, including several prosperous retailers. Well-known as purveyors of uncommon wares, Emes and Barnard were a logical choice to execute an unusual commission. The only clue to the identity of the owner is a script initial F and a small crest, probably that of the Frend or Friend family of County Limerick in Ireland, engraved on the handle, just below the shell-shaped thumbpiece.

12. **Hendrik Goltzius** (Dutch, 1558-1617)
THE MUSE URANIA (Engraving)
Lent by the Williams College Museum of Art

Urania is the muse of Astronomy. The engraving on view is part of a series of the nine muses that was issued by Goltzius. The sphere has been traditionally associated as an attribute of Urania.

13. **Anonymous** (Italian)
Title Page of *Practica Musicae* **by Franchino Gaffurio**
MUSIC OF THE SPHERES, Milan: 1496 (Woodcut)
Lent by the Chapin Library of Rare Books, Williams College   **p. 72**

This wonderfully composed cut is an allegory, showing the derivation of Music from Apollo (seated at top), the Muses (left-hand figures), and the Celestial Bodies (right-hand figures). Each planet is drawn with its traditional astrological figure. Note, too, that Urania, the muse of astronomy, is the first of the muses.

14. **William Blake** (English, 1757-1827)
after a design by Henry Fuseli, Swiss, working in England, 1741-1825
Plate in *The Botanic Garden* **by Erasmus Darwin**
FERTILIZATION OF EGYPT, London: 1795 (Engraving)
Lent by the Chapin Library of Rare Books, Williams College   **p. 38**

For the third edition of this famous poem with scientific notes, all composed by the grandfather of Charles Darwin, Blake executed several new plates after designs by his friend and frequent collaborator, Henry Fuseli.

Here Sirius, or the Dog Star, is depicted as the portent of the summer rise and flooding of the Nile, which annually deposits rich soil in the river's flood plain.

15. **Jacob Andreas Friderich** (German, 1684-1751)
    after a design by Melchior Füssli, Swiss-German, 1677-1736
    Plate in *Kupfer-Bibel* **by Johann Jacob Scheuchzer TAB. I. GENESIS I: 1.** *CRE-ATIO UNIVERSI*
    Augsburg & Ulm: 1731, Vol. I. (Engraving)
    Lent by the Chapin Library of Rare Books, Williams College

    This artistic composite of astronomical knowledge includes a pre-Copernican cosmological diagram in the upper left, and a diagram of relative planet sizes in the lower right. Scheuchzer was one of the great scientific writers of his age, and in his six-volume *Physica Sacra* or *Kupfer-Bibel* he attempts to give the 18th-century scientific view — in word and picture — or every material thing of physical event described in the Bible.

16. **Johann Georg Pintz** (German, 1697-1767)
    TAB. DXIV. JOB IX: 9. *PLEJADES, ORION, URSA MINOR*   Opposite p. 1
    TAB. DXXIX. JOB XXXVIII: 4-6. *TERRA DEI ARCHITEKTONAEMA*
    TAB. DLXIII. PSALM CIV: 5-9. *TERRA SUIS FUNDATA BASIBUS* (Engravings)
    Lent by the Chapin Library of Rare Books, Williams College

    Extracted from the *Kupfer-Bibel*, these plates are part of the suite of 750 executed in the great engraving workshop of Johann Andreas Pfeffel, by appointment, engraver to the Emperor.

    Plate DXIV interprets the passage in Job's first speech of resignation: "it is God ... who made Aldebaran and Orion, the Pleiades and the Circle of the southern stars."

    Plate DXXIX is a baroque picturing of a passage drawn from God's final answer to Job: "Where were you when I laid the earth's foundations, who settled its dimensions, who stretched his measuring line over it?"

    Plate DLXIII continues the praise of God for creating the heavens with praise for his creation of the earth: "Thou didst fix the earth on its foundation so that it never can be shaken; the deep overspread it like a cloak...."

17. **Stefano della Bella** (Italian, 1610-1664)
    Frontispiece of *Dialogo*, **by Galileo Galilei**
    ARISTOTLE, PTOLEMY, AND COPERNICUS, Florence: 1632 (Etching)
    Lent by Jay M. Pasachoff   **Cover**

    Perhaps the most famous image in all of astronomy after the heliocentric diagram in the 1543 Copernicus, this etching captures the feeling of walking on eggshells that was embodied in *Dialogue concerning the Two Chief World Systems — Ptolemaic and Copernican*. Aristotle, with his back to the viewer, was an acceptable philosopher to the Catholic Church, and is seen here as the moderator. Though the book caused its author to stand trial for heresy for again defending Copernicanism, the young artist of the frontispiece had no such problems and enjoyed patronage and commissions in great number. It is estimated that della Bella executed nearly 1400 engravings during his lifetime.

18. **William Blake** (English, 1757-1827)
    Plate 14 in *Illustrations of the Book of Job*
    WHEN THE MORNING STARS SANG TOGETHER AND ALL THE SONS OF GOD SHOUTED FOR JOY
    London: March 1826. Proof copy (Engraving with etched border)
    Lent by the Chapin Library of Rare Books, Williams College

    Apprenticed as an engraver of other artists' designs, Blake chose engraving for his own numerous publications. Job, his final work, is considered his finest both technically and artistically.

Job 38:7 is illustrated here in the central panel. In the borders we see the artist's outline etchings of the days of creation. God, who is being quoted in the caption, forms the central figure, while Job, restored to his wife and friends, kneels below, and the sons of God rejoice among the stars above.

19. **Alexander Mair** (German, 1559-1625)
    Plates in *Uranometria* **by Johann Bayer, 1572-1625**
    ARGO NAVIS (ARGONAUT'S SHIP) and CYGNUS, Augsburg: 1603 (Engraving)
    Lent by Jay M. Pasachoff   **p. 210**

Some eight reissues of Mair's plates for Bayer's atlas appeared in the 17th century, but the plates are most brilliantly reproduced in this first edition of the first consciously artistic multi-sheet star atlas (48 maps). Bayer recorded at least 1700 stars from the Ptolemaic catalogue, his own observations, and the most recent observations of Tycho Brahe. Interestingly, Bayer was only an amateur astronomer; by training and occupation he was a lawyer.

20. **Friedrich Gottlieb Berger** (German, 1713-1800)
    After charts by Johann Elert Bode, 1747-1826
    Plate in *Uranographia, sive Astrorum descriptio* **by Johann Elert Bode**
    Tab. XIV. VIRGO, LIBRA, TURDUS SOLITARIUS, Berlin: 1801 (Engraving)
    Lent by Jay M. Pasachoff

Bode (1747-1826) was director of the observatory of the Berlin Academy of Sciences for forty years. Bode, in his final celestial atlas (he had published earlier atlases in 1768 and 1782), records 17,240 stars on his two hemisphere maps and eighteen maps of ninety-nine constellations, centered on the vernal and autumnal equinoxes. This is the last great star atlas — later popular ones used only stars visible to the naked eye, while ones prepared for astronomers reduced or eliminated the traditional, contrived constellation figures.

21. **Anonymous** (Italian)
    Page in *Poeticon Astronomicon* **by Hyginus**
    ARA AND HYDRA, Venice: 1482 (Woodcuts)
    Lent by the Chapin Library of Rare Books, Williams College

The thirty-nine constellation figures in this first illustrated edition of Hyginus undoubtedly were based on the simple figures found in manuscripts circulated since the 13th century. The medieval tradition of these pictures, rather than the Ptolemaic, is manifested by the demons drawn around the basic figure of the constellation Ara.

22. **Anonymous** (Italian)
    Frontispiece of *Epytoma in Almagestum Ptolomei* **by Johannes Regiomantanus**
    PTOLEMY AND REGIOMONTANUS WITH ARMILLARY SPHERE, Venice:
    1496 (Woodcut with wood engraving as border)
    Lent by the Chapin Library of Rare Books, Williams College   **p. 6**

Creator of the first observatory in northern Europe (at Nuremberg in 1471), Regiomontanus (Hans Mueller) was called to Rome as papal astronomer in 1475, but died in 1476.

Considered one of the very finest Venetian cuts of the late 15th century, the image accompanies Regiomontanus' condensation or digest of the *Almagest* or system of cosmology of Ptolemy. Since 1400, the rediscovered scientific writings of Ptolemy had been favorites of Renaissance Italian scholars.

23. **Anonymous** (16th-century Swiss artist)
    Comet on a Page from *Prodigiorum et Ostentorum Chronicon* (Woodcut)
    Lent by Roberta Olson and Alexander Johnson

This page comes from a book published in Basel in 1557. The author was the German philologist Conrad Wolffhart, a professor at the University of Basel, born about 1518 in Alsace, who died in 1561 after a distinguished career. As a humanist, Wolffhart had translated his surname into its Greek equivalent of Lycosthenes. In 1559, this work on portents and prodigies as well as the whole literary output of Wolffhart were condemned by Pope Paul IV and included in the Index of forbidden books.

AUTHOR INDEX

References in **bold face** are to papers; others are to comments.

438

SUBJECT INDEX

Specific astronomical topics and techniques are listed under "astronomy, subject matter" and "astronomy, techniques." Specific astronomical associations and observatories are listed under those headings. Major references are shown in **bold face**.

444

Keio Senior High School, 280–284

law, astronomy and, 55
literature, astronomy and, 53–54, 56.
    See also
        poetry, science fiction

misconceptions about astronomy, 2, 203–
        204, **229–248**, 259–260, 267
multiple-choice questions, **83–87**, 94
music, astronomy and, 48, 54–55, 367
mythology, astronomy and, 47–50, 344–
        345

newsletters on astronomy education
    American Journal of Physics, 36,
        320–321
    Il Giornale di Astronomia, 318
    IAU Commission 46, 4–5, 227, 330
    Les Cahiers Clairaut, 288–290, 293–
        294
    need for, 36
    STARnews, 265, 272–273
    The Universe in the Classroom: 306–
        309

observatories, public; role in education,
        1, 5, 29, 345–359
    Arcetri, 318–320, 352–353
    Auckland, 355
    Bochum, 350–351
    Carter, 355–356
    Maria Mitchell, 119–124
    McDonald, 313–315
    Mills, 345–347
    Sydney, 357–359
    University of London, 11–22, 109–
        119, 138
    Vatican, 408–413
    Ward, 354–357
Observer's Handbook, 225
observing sessions, 140–142, 154–158,
        184

Perkin-Elmer Corporation, iv

philosophy of astronomy, 7, 42–43, 68–
        71, 101–103, 154–158, 222, 229–
        238, 333–334
planetariums, 1, 5, 104, 222, 275, 336–
        339, 341–342, 343, 346, 350–
        351, **365–388**, 401
    Adler, 336–339
    as sources of information, 382
    Bochum, 350–351
    Dow, 373–374
    for the blind, 373
    in developing countries, 378–380
    in Germany, F.R., 374–376
    portable, 369–372
    in schools, 372
    role in teacher training, 373
    staff, 366, 380, 383
planetariums, "indoor," 160–162, 171–
        174, 177–178, 179–181
poetry, astronomy and, 53, 56, **57–61**
practical/laboratory work, 13–17, 36,
        37, 79, **109–158**, 202. *See
        also*
    astronomy, techniques; projects,
        research
    in distance education, 91–92
    in planetariums, 387–388
    in schools, 269–271, 404
projects, role of in astronomy educa-
        tion, 13, **109–119**. *See also*
    practical work, research
psychology, astronomy and, 55, 68–71,
        230–238, 243–246, 247–248
public education; popularization, 1, 5,
        323–364, 398–399. *See also*
    amateur astronomers; observa-
        tories, public; planetariums

religion, astronomy and, 243–246
research, undergraduate student par-
        ticipation in, 109–119, 119–124

school, astronomy in
    elementary, **1–5**, 23, 29, 104, 107,
        148–152, 200, 203–204, 213–